An Introduction to Statistical Analysis in Research

An Introduction to Statistical Analysis in Research

With Applications in the Biological and Life Sciences

Kathleen F. Weaver
Vanessa C. Morales
Sarah L. Dunn
Kanya Godde
Pablo F. Weaver

This edition first published 2018
© 2018 John Wiley & Sons, Inc

All rights reserved. No part of this publication may be reproduced, stored in a retrieval system, or transmitted, in any form or by any means, electronic, mechanical, photocopying, recording or otherwise, except as permitted by law. Advice on how to obtain permission to reuse material from this title is available at http://www.wiley.com/go/permissions.

The right of Kathleen F. Weaver, Vanessa C. Morales, Sarah L. Dunn, Kanya Godde, and Pablo F. Weaver to be identified as the authors of this work has been asserted in accordance with law.

Registered Offices
John Wiley & Sons, Inc., 111 River Street, Hoboken, NJ 07030, USA

Editorial Office
111 River Street, Hoboken, NJ 07030, USA

For details of our global editorial offices, customer services, and more information about Wiley products visit us at www.wiley.com.

Wiley also publishes its books in a variety of electronic formats and by print-on-demand. Some content that appears in standard print versions of this book may not be available in other formats.

Limit of Liability/Disclaimer of Warranty
The publisher and the authors make no representations or warranties with respect to the accuracy or completeness of the contents of this work and specifically disclaim all warranties, including without limitation any implied warranties of fitness for a particular purpose. This work is sold with the understanding that the publisher is not engaged in rendering professional services. The advice and strategies contained herein may not be suitable for every situation. In view of ongoing research, equipment modifications, changes in governmental regulations, and the constant flow of information relating to the use of experimental reagents, equipment, and devices, the reader is urged to review and evaluate the information provided in the package insert or instructions for each chemical, piece of equipment, reagent, or device for, among other things, any changes in the instructions or indication of usage and for added warnings and precautions. The fact that an organization or website is referred to in this work as a citation and/or potential source of further information does not mean that the author or the publisher endorses the information the organization or website may provide or recommendations it may make. Further, readers should be aware that websites listed in this work may have changed or disappeared between when this works was written and when it is read. No warranty may be created or extended by any promotional statements for this work. Neither the publisher nor the author shall be liable for any damages arising herefrom.

Library of Congress Cataloging-in-Publication Data

Names: Weaver, Kathleen F.
Title: An introduction to statistical analysis in research: with
   applications in the biological and life sciences / Kathleen F. Weaver [and four others].
Description: Hoboken, NJ: John Wiley & Sons, Inc., 2017. | Includes index.
Identifiers: LCCN 2016042830 | ISBN 9781119299684 (cloth) | ISBN 9781119301103 (epub)
Subjects: LCSH: Mathematical statistics–Data processing. | Multivariate
   analysis–Data processing. | Life sciences–Statistical methods.
Classification: LCC QA276.4 .I65 2017 | DDC 519.5–dc23 LC record available
   at https://lccn.loc.gov/2016042830

Cover image: Courtesy of the author
Cover design by Wiley

Set in 11/14pt WarnockPro by Aptara Inc., New Delhi, India

10 9 8 7 6 5 4 3 2 1

Contents

Preface *ix*
Acknowledgments *xi*
About the Companion Website *xiii*

1 Experimental Design *1*
1.1 Experimental Design Background *1*
1.2 Sampling Design *2*
1.3 Sample Analysis *7*
1.4 Hypotheses *9*
1.5 Variables *10*

2 Central Tendency and Distribution *13*
2.1 Central Tendency and Other Descriptive Statistics *13*
2.2 Distribution *18*
2.3 Descriptive Statistics in Excel *34*
2.4 Descriptive Statistics in SPSS *48*   ⎤
2.5 Descriptive Statistics in Numbers *52* ⎟ programs
2.6 Descriptive Statistics in R *57*   ⎦

3 Showing Your Data *61*
3.1 Background on Tables and Graphs *61*
3.2 Tables *62*
3.3 Bar Graphs, Histograms, and Box Plots *63*
3.4 Line Graphs and Scatter Plots *136*
3.5 Pie Charts *165*

4 Parametric versus Nonparametric Tests *191*
4.1 Overview *192*
4.2 Two-Sample and Three-Sample Tests *194*

5 *t*-Test *195*
5.1 Student's *t*-Test Background *195*

5.2	Examples *t*-Tests	*196*
5.3	Case Study	*201*
5.4	Excel Tutorial	*205*
5.5	Paired *t*-Test SPSS Tutorial	*209*
5.6	Independent *t*-Test SPSS Tutorial	*213*
5.7	Numbers Tutorial	*218*
5.8	R Independent/Paired-Samples *t*-Test Tutorial	*223*

6	**ANOVA**	*227*
6.1	ANOVA Background	*227*
6.2	Case Study	*236*
6.3	One-Way ANOVA Excel Tutorial	*241*
6.4	One-Way ANOVA SPSS Tutorial	*247*
6.5	One-Way Repeated Measures ANOVA SPSS Tutorial	*252*
6.6	Two-Way Repeated Measures ANOVA SPSS Tutorial	*261*
6.7	One-Way ANOVA Numbers Tutorial	*272*
6.8	One-Way R Tutorial	*288*
6.9	Two-Way ANOVA R Tutorial	*291*

7	**Mann–Whitney *U* and Wilcoxon Signed-Rank**	*297*
7.1	Mann–Whitney *U* and Wilcoxon Signed-Rank Background	*297*
7.2	Assumptions	*298*
7.3	Case Study – Mann—Whitney *U* Test	*299*
7.4	Case Study – Wilcoxon Signed-Rank	*302*
7.5	Mann–Whitney *U* Excel Tutorial	*305*
7.6	Wilcoxon Signed-Rank Excel Tutorial	*313*
7.7	Mann–Whitney *U* SPSS Tutorial	*319*
7.8	Wilcoxon Signed-Rank SPSS Tutorial	*324*
7.9	Mann–Whitney *U* Numbers Tutorial	*328*
7.10	Wilcoxon Signed-Rank Numbers Tutorial	*337*
7.11	Mann–Whitney *U*/Wilcoxon Signed-Rank R Tutorial	*350*

8	**Kruskal–Wallis**	*353*
8.1	Kruskal–Wallis Background	*353*
8.2	Case Study 1	*354*
8.3	Case Study 2	*358*
8.4	Kruskal–Wallis Excel Tutorial	*362*
8.5	Kruskal–Wallis SPSS Tutorial	*368*
8.6	Kruskal–Wallis Numbers Tutorial	*375*
8.7	Kruskal–Wallis R Tutorial	*386*

9	**Chi-Square Test**	*393*
9.1	Chi-Square Background	*393*
9.2	Case Study 1	*394*

9.3	Case Study 2 *401*	
9.4	Chi-Square Excel Tutorial *405*	
9.5	Chi-Square SPSS Tutorial *418*	
9.6	Chi-Square Numbers Tutorial *426*	
9.7	Chi-Square R Tutorial *429*	
10	**Pearson's and Spearman's Correlation** *435*	
10.1	Correlation Background *435*	
10.2	Example *435*	
10.3	Case Study – Pearson's Correlation *442*	
10.4	Case Study – Spearman's Correlation *445*	
10.5	Pearson's Correlation Excel and Numbers Tutorial *448*	
10.6	Spearman's Correlation Excel Tutorial *455*	
10.7	Pearson/Spearman's Correlation SPSS Tutorial *462*	
10.8	Pearson/Spearman's Correlation R Tutorial *467*	
11	**Linear Regression** *473*	
11.1	Linear Regression Background *473*	
11.2	Case Study *480*	
11.3	Linear Regression Excel Tutorial *484*	
11.4	Linear Regression SPSS Tutorial *497*	
11.5	Linear Regression Numbers Tutorial *508*	
11.6	Linear Regression R Tutorial *517*	
12	**Basics in Excel** *523*	
12.1	Opening Excel *524*	
12.2	Installing the Data Analysis ToolPak *525*	
12.3	Cells and Referencing *529*	
12.4	Common Commands and Formulas *532*	
12.5	Applying Commands to Entire Columns *534*	
12.6	Inserting a Function *536*	
12.7	Formatting Cells *537*	
13	**Basics in SPSS** *539*	
13.1	Opening SPSS *539*	
13.2	Labeling Variables *541*	
13.3	Setting Decimal Placement *543*	
13.4	Determining the Measure of a Variable *544*	
13.5	Saving SPSS Data Files *545*	
13.6	Saving SPSS Output *547*	
14	**Basics in Numbers** *551*	
14.1	Opening Numbers *551*	
14.2	Common Commands *553*	

14.3	Applying Commands	*555*
14.4	Adding Functions	*557*

15	**Basics in R**	*561*
15.1	Opening R	*561*
15.2	Getting Acquainted with the Console	*562*
15.3	Loading Data	*566*
15.4	Installing and Loading Packages	*570*
15.5	Troubleshooting	*576*

16	**Appendix**	*579*
	Flow Chart	*579*

Literature Cited *581*
Glossary *585*
Index *591*

Preface

This book is designed to be a practical guide to the basics of statistical analysis. The structure of the book was born from a desire to meet the needs of our own science students, who often felt disconnected from the mathematical basis of statistics and who struggled with the practical application of statistical analysis software in their research. Thus, the specific emphasis of this text is on the conceptual framework of statistics and the practical application of statistics in the biological and life sciences, with examples and case studies from biology, kinesiology, and physical anthropology.

In the first few chapters, the book focuses on experimental design, showing data, and the basics of sampling and populations. Understanding biases and knowing how to categorize data, process data, and show data in a systematic way are important skills for any researcher. By solidifying the conceptual framework of hypothesis testing and research methods, as well as the practical instructions for showing data through graphs and figures, the student will be better equipped for the statistical tests to come.

Subsequent chapters delve into detail to describe many of the parametric and nonparametric statistical tests commonly used in research. Each section includes a description of the test, as well as when and how to use the test appropriately in the context of examples from biology and the life sciences. The chapters include in-depth tutorials for statistical analyses using Microsoft Excel, SPSS, Apple Numbers, and R, which are the programs used most often on college campuses, or in the case of R, is free to access on the web. Each tutorial includes sample datasets that allow for practicing and applying the statistical tests, as well as instructions on how to interpret the statistical outputs in the context of hypothesis testing. By building confidence through practice and application, the student should gain the proficiency needed to apply the concepts and statistical tests to their own situations.

The material presented within is appropriate for anyone looking to apply statistical tests to data, whether it is for the novice student, for the student looking to refresh their knowledge of statistics, or for those looking for a practical step-by-step guide for analyzing data across multiple platforms. This book is designed for undergraduate-level research methods and biostatistics courses and would also be useful as an accompanying text to any statistics course or course that requires statistical testing in its curriculum.

Examples from the Book

The tutorials in this book are built to show a variety of approaches to using Microsoft Excel, SPSS, Apple Numbers, and R, so the student can find their own unique style in working with statistical software, as well as to enrich the student learning experience through exposure to more and varied examples. Most of the data used in this book were obtained directly from published articles or were drawn from unpublished datasets with permission from the faculty at the University of La Verne. In some tutorials, data were generated strictly for teaching purposes; however, data were based on actual trends observed in the literature.

Acknowledgments

This book was made possible by the help and support of many close colleagues, students, friends, and family; because of you, the ideas for this book became a reality. Thank you to Jerome Garcia and Anil Kapoor for incorporating early drafts of this book into your courses and for your constructive feedback that allowed it to grow and develop. Thank you to Priscilla Escalante for your help in researching tutorial design, Alicia Guadarrama and Jeremy Wagoner for being our tutorial testers, and Margaret Gough and Joseph Cabrera for your helpful comments and suggestions; we greatly appreciate it. Finally, thank you to the University of La Verne faculty that kindly provided their original data to be used as examples and to the students who inspired this work from the beginning.

About the Companion Website

This book is accompanied by a companion website:

www.wiley.com/go/weaver/statistical_analysis_in_research

The website features:

- R, SPSS, Excel, and Numbers data sets from throughout the book
- Sample PowerPoint lecture slides
- End of the chapter review questions
- Software video tutorials that highlight basic statistical concepts
- Student workbook including material not found in the textbook, such as probability, along with an instructor manual

1

Experimental Design

> **Learning Outcomes**
>
> By the end of this chapter, you should be able to:
>
> 1. Define key terms related to sampling and variables.
> 2. Describe the relationship between a population and a sample in making a statistical estimate.
> 3. Determine the independent and dependent variables within a given scenario.
> 4. Formulate a study with an appropriate sampling design that limits bias and error.

1.1 Experimental Design Background

As scientists, our knowledge of the natural world comes from direct observations and experiments. A good experimental design is essential for making inferences and drawing appropriate conclusions from our observations. Experimental design starts by formulating an appropriate question and then knowing how data can be collected and analyzed to help answer your question. Let us take the following example.

Case Study

Observation: A healthy body weight is correlated with good diet and regular physical activity. One component of a good diet is consuming enough fiber; therefore, one **question** we might ask is: do Americans who eat more fiber on a daily basis have a healthier body weight or body mass index (BMI) score?

How would we go about answering this question?

In order to get the most accurate data possible, we would need to design an experiment that would allow us to survey the entire **population** (all possible test subjects – all people living in the United States) regarding their eating habits and then match those to their BMI scores. However, it would take a lot of time and money to survey every person in the country. In addition, if too much time elapses from the beginning to the end of collection, then the accuracy of the data would be compromised.

More practically, we would choose a representative sample with which to make our inferences. For example, we might survey 5000 men and 5000 women to serve as a representative **sample**. We could then use that smaller sample as an **estimate** of our population to evaluate our question. In order to get a proper (and unbiased) sample and estimate of the population, the researcher must decide on the best (and most effective) sampling design for a given question.

1.2 Sampling Design

Below are some examples of sampling strategies that a researcher could use in setting up a research study. The strategy you choose will be dependent on your research question. Also keep in mind that the sample size (N) needed for a given study varies by discipline. Check with your mentor and look at the literature to verify appropriate sampling in your field.

Some of the sampling strategies introduce bias. **Bias** occurs when certain individuals are more likely to be selected than others in a sample. A biased sample can change the predictive accuracy of your sample; however, sometimes bias is acceptable and expected as long as

Figure 1.1 A representation of a random sample of individuals within a population.

it is identified and justifiable. Make sure that your question matches and acknowledges the inherent bias of your design.

Random Sample

In a random sample all individuals within a population have an equal chance of being selected, and the choice of one individual does not influence the choice of any other individual (as illustrated in Figure 1.1). A random sample is assumed to be the best technique for obtaining an accurate representation of a population. This technique is often associated with a random number generator, where each individual is assigned a number and then selected randomly until a preselected sample size is reached. A random sample is preferred in most situations, unless there are limitations to data collection or there is a preference by the researcher to look specifically at subpopulations within the larger population.

In our BMI example, a person in Chicago and a person in Seattle would have an equal chance of being selected for the study.

Likewise, selecting someone in Seattle does not eliminate the possibility of selecting other participants from Seattle. As easy as this seems in theory, it can be challenging to put into practice.

Systematic Sample

A systematic sample is similar to a random sample. In this case, potential participants are ordered (e.g., alphabetically), a random first individual is selected, and every kth individual afterward is picked for inclusion in the sample. It is best practice to randomly choose the first participant and not to simply choose the first person on the list. A random number generator is an effective tool for this. To determine k, divide the number of individuals within a population by the desired sample size.

This technique is often used within institutions or companies where there are a larger number of potential participants and a subset is desired. In Figure 1.2, the third person (going down the first column) is the first individual selected and every sixth person afterward is selected for a total of 7 out of 40 possible.

Figure 1.2 A systematic sample of individuals within a population, starting at the third individual and then selecting every sixth subsequent individual in the group.

Figure 1.3 A stratified sample of individuals within a population. A minimum of 20% of the individuals within each subpopulation were selected.

Stratified Sample

A stratified sample is necessary if your population includes a number of different categories and you want to make sure your sample includes all categories (e.g., gender, ethnicity, other categorical variables). In Figure 1.3, the population is organized first by category (i.e., strata) and then random individuals are selected from each category.

In our BMI example, we might want to make sure all regions of the country are represented in the sample. For example, you might want to randomly choose at least one person from each city represented in your population (e.g., Seattle, Chicago, New York, etc.).

Volunteer Sample

A volunteer sample is used when participants volunteer for a particular study. Bias would be assumed for a volunteer sample because people who are likely to volunteer typically have certain characteristics in common. Like all other sample types, collecting demographic data would be important for a volunteer study, so that you can determine most of the potential biases in your data.

Sample of Convenience

A sample of convenience is not representative of a target population because it gives preference to individuals within close proximity. The reality is that samples are often chosen based on the availability of a sample to the researcher.

Here are some examples:

- A university researcher interested in studying BMI versus fiber intake might choose to sample from the students or faculty she has direct access to on her campus.
- A skeletal biologist might observe skeletons buried in a particular cemetery, although there are other cemeteries in the same ancient city.
- A malacologist with a limited time frame may only choose to collect snails from populations in close proximity to roads and highways.

In any of these cases, the researcher assumes that the sample is biased and may not be representative of the population as a whole.

> **Replication** is important in all experiments. Replication involves repeating the same experiment in order to improve the chances of obtaining an accurate result. Living systems are highly variable. In any scientific investigation, there is a chance of having a sample that does not represent the norm. An experiment performed on a small sample may not be representative of the whole population. The experiment should be replicated many times, with the experimental results averaged and/or the median values calculated (see Chapter 2).

For all studies involving living human participants, you need to ensure that you have submitted your research proposal to your campus' Institutional Review Board (IRB) or Ethics Committee prior to initiating the research protocol. For studies involving animals, submit your research proposal to the Institutional Animal Care and Use Committee (IACUC).

Counterbalancing

When designing an experiment with paired data (e.g., testing multiple treatments on the same individuals), you may need to consider counterbalancing to control for bias. Bias in these cases may take the form of the subjects learning and changing their behavior between trials, slight differences in the environment during different trials, or some other variable whose effects are difficult to control between trials. By counterbalancing we try to offset the slight differences that may be present in our data due to these circumstances. For example, if you were investigating the effects of caffeine consumption on strength, compared to a placebo, you would want to counterbalance the strength session with placebo and caffeine. By dividing the entire test population into two groups (A and B), and testing them on two separate days, under alternating conditions, you would counterbalance the laboratory sessions. One group (A) would present to the laboratory and undergo testing following caffeine consumption and then the other group (B) would present to the laboratory and consume the placebo on the same day. To ensure washout of the caffeine, each group would come back one week later on the same day at the same time and undergo the strength tests under the opposite conditions from day 1. Thus, group B would consume the caffeine and group A would consume the placebo on testing day 2. By counterbalancing the sessions you reduce the risk of one group having an advantage or a different experience over the other, which can ultimately impact your data.

1.3 Sample Analysis

Once we take a sample of the population, we can use **descriptive statistics** to characterize the population. Our estimate may include the **mean** and **variance** of the sample group. For example, we may want to compare the mean BMI score of men who intake greater than 38 g of dietary fiber per day with those who intake less than 38 g of dietary fiber per day (as indicated in Figure 1.4). We cannot sample all men; therefore, we might randomly sample 100 men from the larger population for each category (<38 g and >38 g). In this study, our sample group, or subset, of 200 men ($N = 200$) is assumed to be representative of the whole.

Although this estimate would not yield the exact same results as a larger study with more participants, we are likely to get a good estimate that approximates the population mean. We can then use

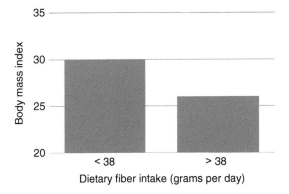

Figure 1.4 Bar graph comparing the body mass index (BMI) of men who eat less than 38 g of fiber per day to men who eat more than 38 g of fiber per day.

inferential statistics to determine the quality of our estimate in describing the sample and determine our ability to make predictions about the larger population.

If we wanted to compare dietary fiber intake between men and women, we could go beyond descriptive statistics to evaluate whether the two groups (populations) are different, as in Figure 1.5. Inferential statistics allows us to place a **confidence interval** on whether the two samples are from the same population, or whether they are really two different populations. To compare men and women, we could use an independent t-test for statistical analysis. In this case, we would receive both the means for the groups, as well as a ***p*-value**,

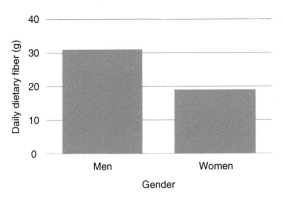

Figure 1.5 Bar graph comparing the daily dietary fiber (g) intake of men and women.

which would give us an estimated degree of confidence in whether the groups are different from each other.

1.4 Hypotheses

In essence, statistics is hypothesis testing. A **hypothesis** is a testable statement that provides a possible explanation to an observable event or phenomenon. A good, testable hypothesis implies that the independent variable (established by the researcher) and dependent variable (also called a response variable) can be measured. Often, hypotheses in science laboratories (general biology, cell biology, chemistry, etc.) are written as "If…then…" statements; however, in scientific publications, hypotheses are rarely spelled out in this way. Instead, you will see them formulated in terms of possible explanations to a problem. In this book, we will introduce formalized hypotheses used specifically for statistical analysis. Hypotheses are formulated as either the null hypothesis or alternative hypotheses. Within certain chapters of this book, we indicate the opportunity to formulate hypotheses using this symbol Ⓗ.

In the simplest scenario, the **null hypothesis** (H_0) assumes that there is no difference between groups. Therefore, the null hypothesis assumes that any observed difference between groups is based merely on variation in the population. In the dietary fiber example, our null hypothesis would be that there is no difference in fiber consumption between the sexes.

The **alternative hypotheses** (H_1, H_2, etc.) are possible explanations for the significant differences observed between study populations. In the example above, we could have several alternative hypotheses. An example for the first alternative hypothesis, H_1, is that there will be a difference in the dietary fiber intake between men and women.

Good hypothesis statements will include a rationale or reason for the difference. This rationale will correspond with the background research you have gathered on the system.

It is important to keep in mind that difference between groups could be due to other variables that were not accounted for in our experimental design. For example, if when you were surveying men and women over the telephone, you did not ask about other dietary choices (e.g., Atkins, South Beach, vegan diets), you may have introduced bias unexpectedly. If by chance, all the men were on a high protein diet and the women were vegan, this could bring bias into your sample. It is important to plan out your experiments and consider all variables that may influence the outcome.

1.5 Variables

An important component of experimental design is to define and identify the variables inherent in your sample. To explain these variables, let us look at another example.

Case Study

In 1995, wolves were reintroduced to Yellowstone National Park after an almost 70-year absence. Without the wolf, many predator–prey dynamics had changed, with one prominent consequence being an explosion of the elk population. As a result, much of the vegetation in the park was consumed, resulting in obvious changes, such as to nesting bird habitat, but also more obscure effects like stream health. With the reintroduction of the wolf, park rangers and scientists began noticing dramatic and far reaching changes to food webs and ecosystems within the park. One question we could ask is how trout populations were impacted by the reintroduction of the wolf. To design this experiment, we will need to define our variables.

The **independent variable**, also known as the treatment, is the part of the experiment established by or directly manipulated by the research that causes a potential change in another variable (the dependent variable). In the wolf example, the independent variable is the presence/absence of wolves in the park.

The **dependent variable**, also known as the response variable, changes because it "depends" on the influence of the independent variable. There is often only one independent variable (depending on the level of research); however, there can potentially be several dependent variables. In the question above, there is only one dependent variable – trout abundance. However, in a separate question, we could examine how wolf introduction impacted populations of beavers, coyotes, bears, or a variety of plant species.

Controlled variables are other variables or factors that cause direct changes to the dependent variable(s) unrelated to the changes caused by the independent variable. Controlled variables must be carefully monitored to avoid error or bias in an experiment. Examples of controlled variables in our example would be abiotic factors (such as sunlight) and biotic factors (such as bear abundance). In the Yellowstone wolf/trout example, researchers would need to survey the same streams at the same time of year over multiple seasons to minimize error.

Here is another example: In a general biology laboratory, the students in the class are asked to determine which fertilizer is best for promoting plant growth. Each student in the class is given three plants; the plants are of the same species and size. For the experiment, each plant is given a different fertilizer (A, B, and C). What are the other variables that might influence a plant's growth?

Let us say that the three plants are not receiving equal sunlight, the one on the right (C) is receiving the most sunlight and the one on the left (A) is receiving the least sunlight. In this experiment, the results would likely show that the plant on the right became more mature with larger and fuller flowers. This might lead the experimenter to determine that company C produces the best fertilizer for flowering plants. However, the results are biased because the variables were not controlled. We cannot determine if the larger flowers were the result of a better fertilizer or just more sunlight.

Types of Variables

Categorical variables are those that fall into two or more categories. Examples of categorical variables are nominal variables and ordinal variables.

Nominal variables are counted not measured, and they have no numerical value or rank. Instead, nominal variables classify information into two or more categories. Here are some examples:

- Sex (male, female)
- College major (Biology, Kinesiology, English, History, etc.)
- Mode of transportation (walk, cycle, drive alone, carpool)
- Blood type (A, B, AB, O)

Ordinal variables, like nominal variables, have two or more categories; however, the order of the categories is significant. Here are some examples:

- Satisfaction survey (1 = "poor," 2 = "acceptable," 3 = "good," 4 = "excellent")
- Levels of pain (mild, moderate, severe)
- Stage of cancer (I, II, III, IV)
- Level of education (high school, undergraduate, graduate)

Ordinal variables are ranked; however, no arithmetic-like operations are possible (i.e., rankings of poor (1) and acceptable (2) cannot be added together to get a good (3) rating).

Quantitative variables are variables that are counted or measured on a numerical scale. Examples of quantitative variables include height, body weight, time, and temperature. Quantitative variables fall into two categories: discrete and continuous.

Discrete variables are variables that are counted:

- Number of wing veins
- Number of people surveyed
- Number of colonies counted

Continuous variables are numerical variables that are measured on a continuous scale and can be either ratio or interval.

Ratio variables have a true zero point and comparisons of magnitude can be made. For instance, a snake that measures 4 feet in length can be said to be twice the length of a 2 foot snake. Examples of ratio variables include: height, body weight, and income.

Interval variables have an arbitrarily assigned zero point. Unlike ratio data, comparisons of magnitude among different values on an interval scale are not possible. An example of an interval variable is temperature (Celsius or Fahrenheit scale).

2

Central Tendency and Distribution

> **Learning Outcomes**
>
> By the end of this chapter, you should be able to:
>
> 1. Define and calculate measures of central tendency.
> 2. Describe the variance within a normal population.
> 3. Interpret frequency distribution curves and compare normal and non-normal populations.

2.1 Central Tendency and Other Descriptive Statistics

Sampling Data and Distribution

Before beginning a project, we need an understanding of how populations and data are distributed. How do we describe a population? What is a bell curve, and why do biological data typically fall into a normal, bell-shaped distribution? When do data not follow a normal distribution? How are each of these populations treated statistically?

Measures of Central Tendencies: Describing a Population

The **central tendency** of a population characterizes a typical value of a population. Let us take the following example to help illustrate this concept. The company Gallup has a partnership with Healthways to collect information about Americans' perception of their health

Table 2.1 Americans' perceptions of health and happiness collected from the company Gallup.

US Well-Being Index	Score	Change versus Last Measure
Well-being index	67.0	+0.2
Emotional health	79.7	+0.3
Physical health	76.9	+0.6
Life evaluation	50.5	+0.5
Work environment	48.5	+0.6

and happiness, including work environment, emotional and physical health, and basic access to health care services. This information is compiled to calculate an overall well-being index that can be used to gain insight into people at the community, state, and national level. Gallup pollers call 1000 Americans per day, and their researchers summarize the results using measures of central tendency to illustrate the typical response for the population. Table 2.1 is an example of data collected by Gallup.

The central tendency of a population can be measured using the arithmetic **mean**, **median**, or **mode**. These three components are utilized to calculate or specify a numerical value that is reflective of the distribution of the population. The measures of central tendency are described in detail below.

Mean

The mean, also known as average, is the most common measure of central tendency. All values are used to calculate the mean; therefore, it is the measure that is most sensitive to any change in value. (This is not the case with median and mode.) As you will see in a later section, normally distributed data have an equal mean, median, and mode.

The mean is calculated by adding all the reported numerical values together and dividing that total by the number of observations. Let us examine the following scenario: suppose a professor is implementing a new teaching style in the hope of improving students' retention of class material. Because there are two courses being offered, she decides to incorporate the new style in one course and use the other course without changes as a control. The new teaching style involves a "flip-the-class" application where students present a course topic to their peers, and the instructor behaves more as a mentor than a lecturer. At the end of the semester, the professor compared exam scores and determined which class had the higher mean score.

Table 2.2 summarizes the exam scores for both classes (control and treatment). Calculating the mean requires that all data points are taken into account and used to determine the average value. Any change in a single value directly changes the calculated average.

Table 2.2 Exam scores for the control and the "flip-the-class" students.

	Exam Scores (Control)	Exam Scores (Treatment)
	76	89
	72	87
	70	87
	71	90
	70	91
	82	88
	70	90
	78	84
	78	88
	79	83
	83	87
	80	88
	81	90
	79	83
	75	90
Sum	1144	1315
Observations	15	

Calculate the mean for the control group: $(1144)/15 = \mathbf{76.3}$.
Calculate the mean for the treatment group: $(1315)/15 = \mathbf{87.7}$.

Table 2.3 Reported ribonucleic acid (RNA) concentrations for eight samples.

Sample	RNA Concentration (ng/μL)
1	125
2	146
3	133
4	152
5	12
6	129
7	117
8	155
Sum	**969**

Mean = (969)/8 = **121.1 ng/μL**.

Although the mean is the most commonly used measure of central tendency, outliers can easily influence the value. As a result, the mean is not always the most appropriate measure of central tendency. Instead, the median may be used.

Let us look at Table 2.3 and calculate the average ribonucleic acid (RNA) concentrations for all eight samples. Although 121.1 ng/μL is an observed mean RNA concentration, it is considered to be on the lower end of the range and does not clearly identify the most representative value. In this case, the mean is thrown off by an outlier (the fifth sample = 12 ng/μL). In cases with extreme values and/or skewed data, the median may be a more appropriate measure of central tendency.

Median

The **median** value may also be referred to as the "middle" value. Medians are most applicable when dealing with skewed datasets and/or outliers; unlike the mean, the median is not as easily influenced. To find the median, data points are first arranged in numerical order. If there is an odd number of observations, then the middle data point will serve as the median. Let us determine the median value for an earlier example looking at student exam scores. There were 15 observations:

76, 72, 70, 71, 70, 82, 70, 78, 78, 79, 83, 80, 81, 79, and 75

To calculate the median, first arrange the exam scores in order. In this case, we placed them in ascending order:

70, 70, 70, 71, 72, 75, 76, **78**, 78, 79, 79, 80, 81, 82, and 83

The middle number of the distribution is 78; this is the median value for the exam score example. It is easy to find the median value for an odd number of data points; however, if there is an even number of data points, then the average of the two middle values must be calculated after scores are arranged in order. In the previous RNA example, there were eight readings in ng/μL.

125, 146, 133, 152, 12, 129, 117, and 155

To calculate the median, arrange the RNA values (ng/μL) in order. Here, we put them in ascending order:

12, 117, 125, 129, 133, 146, 152, 155

The fourth (129 ng/μL) and the fifth (133 ng/μL) fall in the middle. Therefore, in order to determine the median, calculate the mean of these two readings:

(129 + 133)/2 = 131 ng/μL

Where the mean (121.1 ng/μL) may be considered misleading, the properties of the median (131 ng/μL) allow for datasets that have more values in a particular direction (high or low) to be compared and analyzed. Nonparametric statistics such as the Mann–Whitney U test and the Kruskal–Wallis test, which are introduced in later chapters, utilize the medians to make comparisons between groups.

Mode

Another measure of central tendency is the mode. The **mode** is the most frequently observed value in the dataset. The mode is not as

Table 2.4 Most common zip code reported by 10 university students.

Student	Zip Code
1	91750
2	90815
3	91709
4	91750
5	90807
6	96750
7	80030
8	91750
9	45620
10	91709

Mode = **91750**.

commonly used as the mean and median, but it is still useful in some situations. For example, we would use the mode to determine the most common zip code for all students at a university, as depicted in Table 2.4. In this case, zip code is a categorical variable, and we cannot take a mean. Likewise, the central tendency of nominal variables, such as male or female, can also be estimated using the mode.

2.2 Distribution

A **frequency distribution curve** represents the frequency of each measured variable in a population. For example, the average height of a woman in the United States is 5'5". In a normally distributed population, there would be a large number of women at or close to 5'5" and relatively fewer with a height around 4' or 7' tall. Frequency distributions are typically graphed as a **histogram** (similar to a bar graph). Histograms display the frequencies of each measured variable, allowing us to visualize the central tendency and spread of the data.

In living things, most characteristics vary as a result of genetic and/or environmental factors and can be graphed as histograms. Let us review the following example:

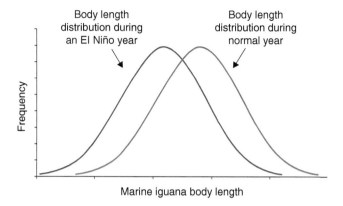

Figure 2.1 Frequency distribution of the body length of the marine iguana during a normal year and an El Niño year.

In the Galapagos Islands, one of the more charismatic residents is the marine iguana (*Amblyrhynchus cristatus*). Marine iguanas are the world's only marine lizards, and they dive under water to feed on marine algae. The abundance of algae varies year to year and can be especially sensitive to El Niño cycles, which dramatically reduce the amount of available food for marine iguanas. In response to food shortages, marine iguanas are one of the few animals with the ability to shrink in body size (bones and all!). If we were studying the effects of El Niño cycling on marine iguana body size, we could create frequency distributions of iguana size year by year (as illustrated in Figure 2.1).

Summary

In a normal year, we would expect to see a distribution with the highest frequency of medium-sized iguanas and lower frequencies as we approached the extremes in body size. However, as the food resources deplete in an El Niño year, we would expect the distribution of iguana body size to shift toward smaller body sizes (Figure 2.1). Notice in the figure the bell shape of the curve is the same; therefore, a concept known as the variance (which will be described below) has not changed. The only difference between El Niño years is the shifted distribution in body size.

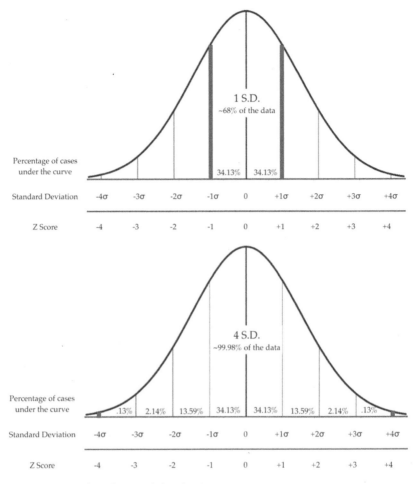

Figure 2.2 Display of normal distribution.

Nuts and Bolts

The distribution curve focuses on two parameters of a dataset: the mean (average) and the variance. The mean (μ) describes the location of the distribution heap or average of the samples. In a normally distributed dataset (see Figure 2.2), the values of central tendencies (e.g., mean, median, and mode) are equal. The variance (σ^2) takes into account the spread of the distribution curve; a smaller variance will reflect a narrower distribution curve, while a larger variance coincides with a wider distribution curve. Variance can most easily be described as the spread around the mean. Standard deviation (σ) is defined as the square root of the variance. When satisfying a normal distribution, both the mean and standard deviation can be utilized to make specific conclusions about the spread of the data:

- 68% of the data fall within one standard deviation of the mean
- 95% of the data fall within two standard deviations of the mean
- 99.7% of the data fall within three standard deviations of the mean

Describing the Shapes of Histograms

Histograms can take on many different physical characteristics or shapes. The shape of the histogram, a direct representation of the sampling population, can be an indication of a normal distribution or a non-normal distribution. The shapes are described in detail below.

Normal distributions are bell-shaped curves with a symmetric, convex shape and no lingering tail region on either side. The data clearly are derived from a homogenous group with the local maximum in the middle, representing the average or mean (see Figure 2.3). An example of a normal distribution would be the head size of newborn babies or finger length in the human population.

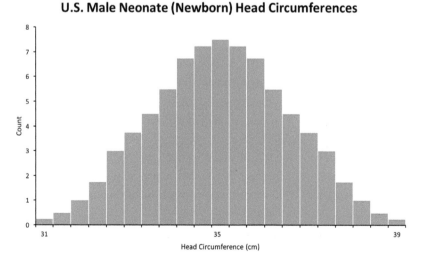

Figure 2.3 Histogram illustrating a normal distribution.

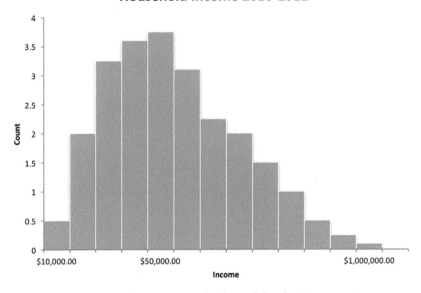

Figure 2.4 Histogram illustrating a right skewed distribution.

Skewed distributions are asymmetrical distribution curves, with the distribution heap either toward the left or right with a lingering tail region. These asymmetrical shapes are generally referred to as having "skewness" and are not normally distributed. Outside factors influencing the distribution curve cause the shift in either direction, indicating that values from the dataset weigh more heavily on one side than they do on the other. Distribution curves skewed to the right are considered "positively skewed" and imply that the values of the mean and median are greater than the mode (see Figure 2.4). An example of a positively skewed distribution is household income from 2010 to 2011. The American Community Survey reported the US median household income at approximately $50,000 for 2011; however, there are households that earned well over $1,000,000 for that same year. Therefore, you would expect a distribution curve with a **right skew**. To determine the direction of skewness (i.e., right or left skewed), pay close attention to the positioning of the lingering tail region, whether it is to the right or left side of the distribution heap will determine the type of skewness (e.g., a long right tail means the data are right skewed).

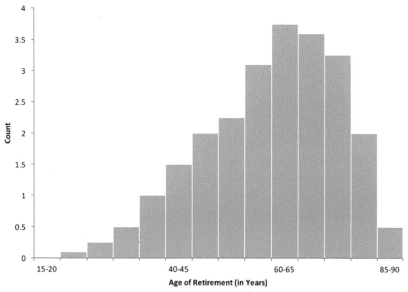

Figure 2.5 Histogram illustrating a left skewed distribution.

Distribution curves skewed to the left are considered "negatively skewed" and imply that the values of the mean and median are less than the mode (see Figure 2.5). An example of this would be the median age of retirement, which the Gallup poll reported was 61 years of age in the year 2013. Here, we would expect a **left skew** because people typically retire later in life, but many also retire at younger ages, with some even retiring before 40 years of age.

Skewness is easily evaluated using statistical software that calculates a p-value. This assists the practitioner in deciding whether the data are normal or not. A significant p-value rejects the null hypothesis that the calculated skew value is equal to zero (or some other value as defined by the selected algorithm), or that there is no skew. In other words, a significant p-value indicates the data are skewed. The sign of the skew value will denote the direction of the skew (see Table 2.5).

Kurtosis refers to symmetrical curves that are not shaped like the normal bell-shaped curve (when normal, the data are mesokurtic), and the tails are either heavier or lighter than expected. If the curve has kurtosis, the data do not follow a normal distribution. Generally,

Table 2.5 Interpretation of skewness based on a positive or negative skew value.

Direction of Skew	Sign of Skew Value
Right	Positive
Left	Negative

there are two forms of data that deviate from a normal distribution with kurtosis: platykurtic, where **tails are lighter** (as illustrated in Figure 2.6), or leptokurtic, where the **tails are heavier** (as illustrated in Figure 2.7). Table 2.6 provides an interpretation of the kurtosis value.

With lighter tails, there are fewer outliers from the mean. An example would be a survey of age for traditional undergraduate students (Figure 2.6). Typically, the age range is between 18 and 21 years with very few students far outside this range. Heavier tails indicate more outliers than a normal distribution. An example is birth weight in U.S. infants (Figure 2.7). While most babies are of average birth weight, many are born below or above the average.

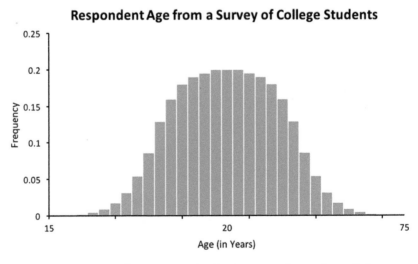

Figure 2.6 Histogram illustrating a platykurtic curve where tails are lighter.

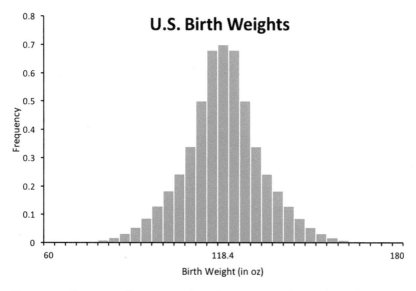

Figure 2.7 Histogram illustrating a leptokurtic curve where tails are heavier.

Bimodal (double-Peaked) distributions are split between two or more distributions (see Figure 2.8). Having more than one distribution, or nonhomogeneous groups within one dataset, can cause the split. Bimodal distributions can be asymmetrical or symmetrical depending on the dataset and can have two or more local maxima, or peaks. Distribution curves depicted with more than two local maxima are considered to be multimodal. An example of a bimodal distribution can be found in a population with two groups or categories. For example, if we look at the height of Americans, we would see two distinct populations of height measurements – men and women.

Plateau distributions are extreme versions of multimodal distributions because each bar is essentially its own node (or peak);

Table 2.6 Interpretation of the shape of the curve based on a positive or negative kurtosis value.

Shape of Curve	Sign of Kurtosis Value
Leptokurtic	Positive
Platykurtic	Negative

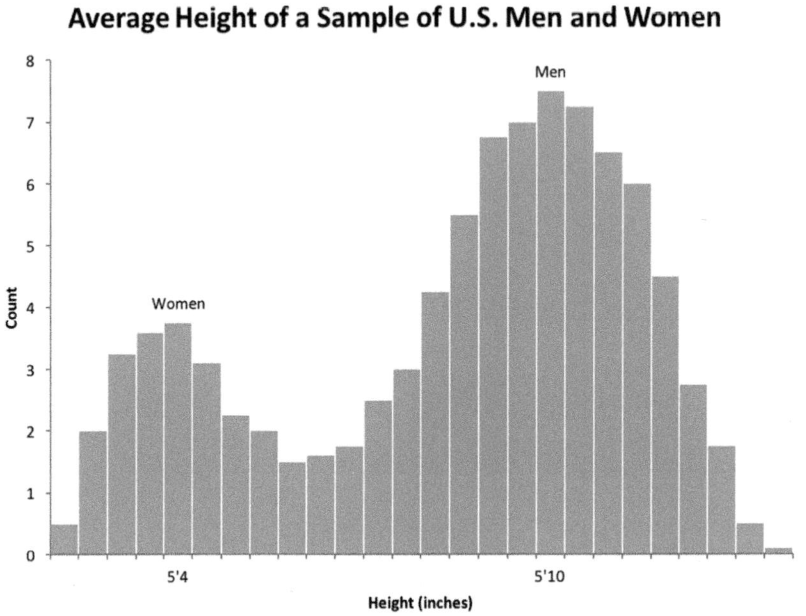
Average Height of a Sample of U.S. Men and Women
Men
Women
Count
0
1
2
3
4
5
6
7
8
5'4
5'10
Height (inches)

Figure 2.8 Histogram illustrating a bimodal, or double-peaked, distribution.

therefore, there is no single clear pattern (as illustrated in Figure 2.9). The curve lacks a convex shape, which makes the distribution heap, or local maximum, difficult to identify. This type of distribution implies a wide variation around the mean and lacks any useful insight about the sampling data.

Outliers

Histograms provide a way to graphically show the distribution of a dataset. They answer questions such as: "Are my sampling data part of a normal distribution?" or "Are my data skewed?" "If so, are they skewed to the left or to the right?" These are questions that must be addressed a priori in order to determine which statistical test is appropriate for the dataset. In addition, by graphing the distribution curve, a researcher has the opportunity to identify any possible outliers that were not obvious when looking at the numerical dataset.

Outliers can be defined as the numerical values extremely distant from the norm or rest of the data. In other words, outliers are the extreme cases that do not "fit" with the rest of the data. By now, we understand that there will be variation of numerical values around the mean, and this variation can either be an indication of a large or

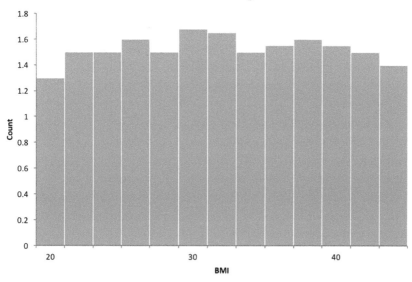

Figure 2.9 Histogram illustrating a plateau distribution.

small variance (spread around the mean). However, extreme values, such as outliers, fall well beyond the levels of variance observed for a particular dataset and are then classified as special observations.

Outliers can be handled in several ways, including deeming the observation as an error and subsequent removal of the outlying data point. However, before you remove a data point, assuming it is an error, talk with your mentor or an expert in the field to consider what might be the best option with the given data. If possible, re-collect that data point to verify accuracy. If not possible, consider using the median and running a nonparametric analysis.

Quantifying the *p*-Value and Addressing Error in Hypothesis Testing

One of the most vital, yet often confusing aspects of statistics is the concept of the *p*-value and how it relates to hypothesis testing. The following explanation uses simplified language and examples to illustrate the key concepts of inferential statistics, hypothesis testing, and statistical significance.

In previous sections, we have discussed descriptive statistics, such as the mean, median, and mode, which describe key features of our sample population. What if, however, we wanted to use these statistics to make some inferences about how well our sample represents the overall population, or whether two samples were different from each other, at an acceptable level of statistical probability? In these cases, we would be using inferential statistics, which allow us to extend our interpretations from the sample and tell us whether there is some sort of interesting phenomenon occurring in the larger population.

Let us take the following example to illustrate some of these points: An anthropologist has discovered a human skeleton near the coast of Peru. She recently developed a method to estimate lung volume from the skeleton. Based on anatomical reconstructions, the lung volume (vital capacity) of this individual is estimated to be approximately 590 mL. The researcher thinks this is odd, considering that the lung volume of modern sea level populations in the area is typically smaller, with a mean of 420 mL. How abnormal would this measurement of 590 mL be, in relation to human variation in lung volume for the sea level population? Figure 2.10 shows the frequency distribution curve for lung volume in the sea level population.

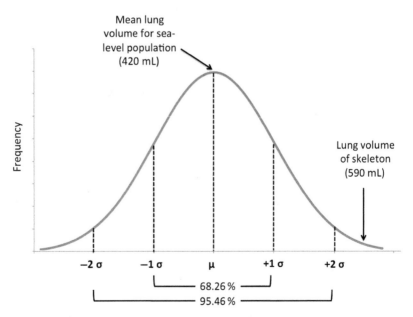

Figure 2.10 Estimated lung volume of the human skeleton (590 mL), compared with the distribution of lung volumes in the nearby sea level population.

According to the distribution, we can see that the probability of a person from the area having a lung volume of 590 mL is very low (less than 1%). In other words, over 99% of the population in the area has a smaller lung volume than the sample. Because the chances of this sample coming from the population are low, we could then make an inference that the skeleton may be from another population that has a different lung volume capacity.

In her research, the anthropologist learns of the indigenous Aymara people, from the mountains of Peru and Chile. The Aymara people have adapted to low oxygen in many ways, including enlarged lung capacity (the mean lung volume for the Aymara is approximately 580 mL). The researcher wants to determine if the populations are significantly different from each other (see Figure 2.11). At this point, we can develop some simple hypotheses that we can continue to apply throughout the remainder of the book.

The first step is to develop the **null hypothesis (H_0)**. In this case, the null hypothesis would be that there is no statistical difference between our two groups (the lung volume of the sea level and Aymara populations are not different). Our **alternative hypothesis (H_1)**, would be that there is a real, statistically significant difference between the groups. We must choose whether to reject or not reject our null hypothesis (H_0), based on some objective criterion, which

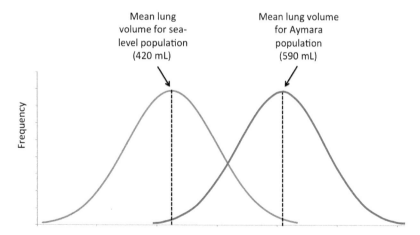

Figure 2.11 Distributions of lung volumes for the sea level population (mean = 420 mL), compared with the lung volumes of the Aymara population (mean = 590 mL).

Table 2.7 Demonstration of the relationship between the percent confidence and error in determining the alpha (α).

Confidence (%)	Error or Chance of Occurrence (%)	Alpha (α)
99	1	0.01
95	5	0.05
90	10	0.10

we define ahead of time as alpha (α). In most cases, this α value will be set to 0.05. It may be helpful to think of your α value as a degree of confidence that we are making the right call, with regard to rejecting our null hypothesis. An α of 0.05 represents a 95% confidence that you will not mistakenly reject the null hypothesis. Also, think of using the p-value as evidence against the null hypothesis. Table 2.7 demonstrates the relationship between confidence (%) and error or chance of occurrence (%).

In science, we can almost never be 100% certain of an outcome being correct, and some error is acceptable and expected. An alpha of 0.05 is most common in life science research; however, you may occasionally see an alpha of 0.10 (e.g., cancer research when you are willing to accept more error or a decrease in confidence in exchange for an outcome or cure).

> *Note*: Depending on the field or level of research, the significance level may vary. For example, in areas of cancer research where people are willing to take more risk on a potential treatment, a much larger p-value (0.09 or 0.10) is supported.

To test our hypotheses, we would run a statistical test that compares the lung volumes of the two populations (sea level and Aymara). In this case, the test gives us a probability value (p-value) of 0.01. The p-value reflects the probability (0.01 or 1%) of obtaining the test result if the null hypothesis was true, based on our data. Stated differently, we are 99% confident that we can reject the null hypothesis. Because we established our α value to be 0.05, any calculated p-value ≤ 0.05 means that we can reject the null hypothesis. However, if the p-value is large ($p > 0.05$), we would fail to reject the null hypothesis, meaning that there would be no statistically significant difference between the populations. A p-value greater than the expected α (e.g., $p > 0.05$), represents less than 95% confidence in the ability to reject the null hypothesis.

Summary

Based on the example, we would conclude that there is a statistically significant difference between the lung volumes of the sea level population and the Aymara population ($p = 0.01$).

With any conventional hypothesis testing, the possibility of error must be kept in mind. In hypothesis testing, error falls into two main categories, Type I and Type II, which refer to the probability of yielding a false positive or a false negative conclusion, respectively:

Type I error occurs when the null hypothesis is rejected incorrectly (false positive).

Type II error occurs when the null hypothesis fails to be rejected when it should have been (false negative).

The level of significance ($\alpha = 0.05$) implies that if the null hypothesis is true, then we would have a 5% or 1 in 20 chance of mistakenly rejecting it, and thus committing a Type I error.

The probability of committing a Type II error is represented by the symbol β and is related to statistical power. Statistical power is the probability of correctly rejecting a false null hypothesis. Increasing the statistical power essentially lowers the probability of committing a Type II error and increases the possibility of detecting changes or relationships within a study. There are three factors that influence the statistical power of a study and can be strategically addressed in order to help increase the power:

1. **Sample size (N)** – a large total sample size (represented with a capital N) increases the statistical power because you are increasing your finite representation of a potentially infinite population. If you have a smaller subset of the total number of participants, you would represent the subset with a lower case (n). If 50 people were surveyed for fitness center usage the group would be represented

as $N = 50$. If you were to divide that group into males ($n = 22$) and females ($n = 28$) the groups would be represented with a lower case (n).
2. **Significance criterion (α)** – increasing α (e.g., from 0.01 to 0.05 or from 0.05 to 0.1) decreases the chance of making a Type II error. However, it also increases the chance of a Type I error.
3. **Low variability** – low variation from sample to sample within the dataset (samples are tightly distributed around the dataset mean).

Error and power are also important to consider when deciding between a parametric and nonparametric statistical method. If the assumptions have been satisfied, then the probability of committing a Type I error is equally likely for both parametric and nonparametric methods. Due to the fact that parametric statistics rely on stricter assumptions, there is less forgiveness when these assumptions are violated.

There are two possible alternatives for addressing violated assumptions. First, in the case where the dataset does not follow a normal distribution (this happens often in human research), one could carry out data transformations on the collected measurements. In doing so, the original data can be transformed to satisfy the test assumptions, and a parametric statistical method can be applied on the transformed data. Data transformations do not always prove to be successful, and in such cases, you may need to use a nonparametric method to analyze the data.

Nonparametric methods make fewer assumptions about the data and are commonly known for ranking the measured variables rather than using the actual measurements themselves to carry out the statistical analysis. In doing so, the scale of the data is partially lost causing the nonparametric method to be less powerful; however, ranking the data makes nonparametric methods useful for dealing with outliers.

Data Transformation

Data transformation can be applied on a dataset that strongly violates statistical assumptions. As a result, this helps to "normalize" the data. This is a familiar situation for human biology researchers dealing with the assessment of hormones as hormones do not always follow a normal distribution. The logarithm of the data may better suit the situation, allowing for a more normal distribution and satisfying the original assumptions. If the transformations do in fact fit a normally distributed population, then proceed with applying the parametric statistical method on the transformed data. Keep in mind that whichever mathematical formula was applied to one data point must be applied in the same form to all the remaining data points within a variable. In other words, what is done to one must be done to all. The three commonly used data transformations are the log transformation, the square-root transformation, and the arcsine transformation.

Log Transformation

Log transformations take the log of each observation, whether in the form of the natural log or base-10 log. It is the most commonly used data transformation in the sciences, specifically in biology, physical anthropology, and health fields and tends to work exceptionally well when dealing with ratios or products of variables. In addition, log transformations work well with datasets that have a frequency distribution heavily skewed to the right. The following formula serves as a mathematical representative for taking the natural log (ln):

$$Y' = \ln[Y]$$

Square-Root Transformation

Square-root transformations take the square root of each observation and are commonly used in counts (e.g., cell biology and immunology). A square-root transformation would be suitable for counts of bacterial colonies on a plated petri dish or the number of viable cells after a 24-hour incubation period. The following formula applies a square-root transformation:

$$Y' = \sqrt{Y}$$

Arcsine Transformation

Arcsine transformations take the arcsine of the square root of each observation and are commonly used on datasets involving proportions. Use the following formula to calculate the arcsine of certain proportional observations:

$$Y' = \arcsine\left[\sqrt{Y}\right]$$

Building Blocks

Understanding the fundamentals of distribution and error are not only important when analyzing the raw data, but also when deciding which test best fits the dataset. The following chapters will review the concepts learned in this chapter, such as outliers and variance, and will use these fundamentals to guide you through different statistical approaches.

Tutorials

2.3   Descriptive Statistics in Excel

To run descriptive statistics in Microsoft Excel, utilize the following tutorial. More information on tips and tools for Excel can be found in Chapter 12. For each of the calculations, we will use the following LDL levels from 20 male patients.

A	B
Patient	Male LDL level (mg/dL)
1	110
2	129
3	121
4	116
5	123
6	111
7	114
8	126
9	119
10	127
11	101
12	112
13	127
14	122
15	113
16	102
17	106
18	116
19	129
20	132

*This example was taken from the research conducted by Dr. Sarah Dunn.

Mean

1. To calculate the mean, use the average function by typing in **=average** into an empty cell. Double click on the average function from the drop down menu.

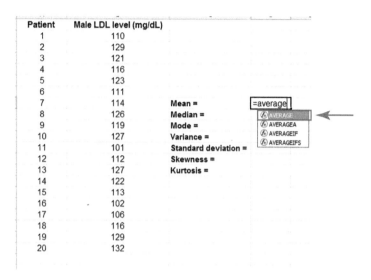

2. Highlight the cells that will be used to calculate the mean. Press enter/return.

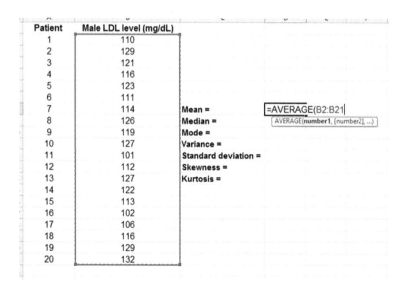

3. The output is the **mean**.

Patient	Male LDL level (mg/dL)		
1	110		
2	129		
3	121		
4	116		
5	123		
6	111		
7	114	Mean =	117.8
8	126	Median =	
9	119	Mode =	
10	127	Variance =	
11	101	Standard deviation =	
12	112	Skewness =	
13	127	Kurtosis =	
14	122		
15	113		
16	102		
17	106		
18	116		
19	129		
20	132		

Median

1. To calculate the median, use the median function by typing in **=median** into an empty cell. Double click on the median function from the drop down menu.

Patient	Male LDL level (mg/dL)		
1	110		
2	129		
3	121		
4	116		
5	123		
6	111		
7	114	Mean =	117.8
8	126	Median =	=median
9	119	Mode =	ⓕMEDIAN
10	127	Variance =	
11	101	Standard deviation =	
12	112	Skewness =	
13	127	Kurtosis =	
14	122		
15	113		
16	102		
17	106		
18	116		
19	129		
20	132		

2. Highlight the cells that will be used to calculate the median. Press enter/return.

Patient	Male LDL level (mg/dL)		
1	110		
2	129		
3	121		
4	116		
5	123		
6	111		
7	114	Mean =	117.8
8	126	Median =	=MEDIAN(B2:B21
9	119	Mode =	MEDIAN(number1, [number2], ...)
10	127	Variance =	
11	101	Standard deviation =	
12	112	Skewness =	
13	127	Kurtosis =	
14	122		
15	113		
16	102		
17	106		
18	116		
19	129		
20	132		

3. The output is the **median**.

Patient	Male LDL level (mg/dL)		
1	110		
2	129		
3	121		
4	116		
5	123		
6	111		
7	114	Mean =	117.8
8	126	Median =	117.5
9	119	Mode =	
10	127	Variance =	
11	101	Standard deviation =	
12	112	Skewness =	
13	127	Kurtosis =	
14	122		
15	113		
16	102		
17	106		
18	116		
19	129		
20	132		

Mode

1. To calculate the mode, use the mode function by typing in **=mode** into an empty cell. Double click on the mode function from the drop down menu.

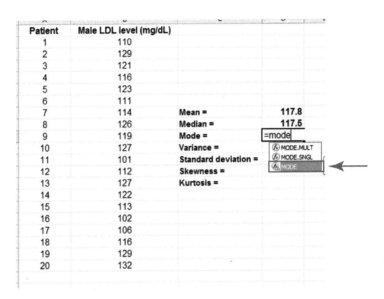

2. Highlight the cells that will be used to calculate the mode. Press enter/return.

3. The output is the **mode**.

Patient	Male LDL level (mg/dL)				
1	110				
2	129				
3	121				
4	116				
5	123				
6	111				
7	114	Mean =		117.8	
8	126	Median =		117.5	
9	119	Mode =		129	
10	127	Variance =			
11	101	Standard deviation =			
12	112	Skewness =			
13	127	Kurtosis =			
14	122				
15	113				
16	102				
17	106				
18	116				
19	129				
20	132				

Variance

1. To calculate the variance, use the variance function by typing in **=var** into an empty cell. Double click on the variance function from the drop down menu.

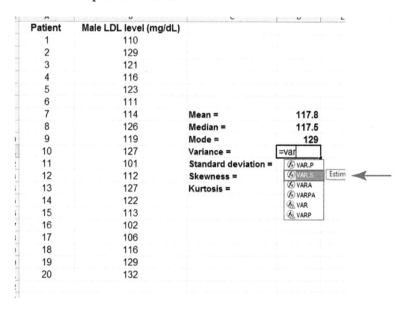

2. Highlight the cells that will be used to calculate the variance. Press enter/return.

Patient	Male LDL level (mg/dL)		
1	110		
2	129		
3	121		
4	116		
5	123		
6	111		
7	114	Mean =	117.8
8	126	Median =	117.5
9	119	Mode =	129
10	127	Variance =	=VAR.S(B2:B21
11	101	Standard deviation =	VAR.S(number1, [number2], ...)
12	112	Skewness =	
13	127	Kurtosis =	
14	122		
15	113		
16	102		
17	106		
18	116		
19	129		
20	132		

3. The output is the **variance**.

Patient	Male LDL level (mg/dL)		
1	110		
2	129		
3	121		
4	116		
5	123		
6	111		
7	114	Mean =	117.8
8	126	Median =	117.5
9	119	Mode =	129
10	127	Variance =	84.27368
11	101	Standard deviation =	
12	112	Skewness =	
13	127	Kurtosis =	
14	122		
15	113		
16	102		
17	106		
18	116		
19	129		
20	132		

Standard Deviation

1. To calculate the standard deviation, use the standard deviation function by typing in **=stdev** into an empty cell. Double click on the standard deviation function from the drop down menu.

A	B	C	D	E
Patient	Male LDL level (mg/dL)			
1	110			
2	129			
3	121			
4	116			
5	123			
6	111			
7	114	Mean =	117.8	
8	126	Median =	117.5	
9	119	Mode =	129	
10	127	Variance =	84.27368	
11	101	Standard deviation =	=stdev	
12	112	Skewness =	STDEV.P	
13	127	Kurtosis =	STDEV.S	Estin
14	122		STDEVA	
15	113		STDEVPA	
16	102		STDEV	
17	106		STDEVP	
18	116			
19	129			
20	132			

2. Highlight the cells that will be used to calculate the standard deviation. Press enter/return.

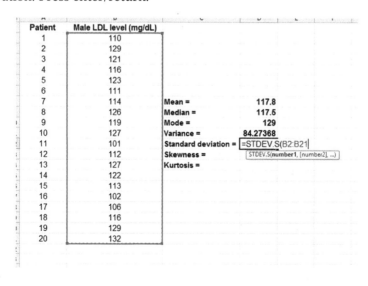

3. The output is the **standard deviation**.

Patient	Male LDL level (mg/dL)		
1	110		
2	129		
3	121		
4	116		
5	123		
6	111		
7	114	Mean =	117.8
8	126	Median =	117.5
9	119	Mode =	129
10	127	Variance =	84.27368
11	101	Standard deviation =	9.180070
12	112	Skewness =	
13	127	Kurtosis =	
14	122		
15	113		
16	102		
17	106		
18	116		
19	129		
20	132		

Skewness

1. To calculate skewness, use the skewness function by typing in =**skew** into an empty cell. Double click on the skewness function from the drop down menu.

Patient	Male LDL level (mg/dL)			
1	110			
2	129			
3	121			
4	116			
5	123			
6	111			
7	114	Mean =	117.8	
8	126	Median =	117.5	
9	119	Mode =	129	
10	127	Variance =	84.27368	
11	101	Standard deviation =	9.180070	
12	112	Skewness =	=skew	
13	127	Kurtosis =	SKEW	Returns
14	122			
15	113			
16	102			
17	106			
18	116			
19	129			
20	132			

2. Highlight the cells that will be used to calculate the skewness. Press enter/return.

Patient	Male LDL level (mg/dL)		
1	110		
2	129		
3	121		
4	116		
5	123		
6	111		
7	114	Mean =	117.8
8	126	Median =	117.5
9	119	Mode =	129
10	127	Variance =	84.27368
11	101	Standard deviation =	9.180070
12	112	Skewness =	=SKEW(B2:B21
13	127	Kurtosis =	SKEW(number1, [number2], ...)
14	122		
15	113		
16	102		
17	106		
18	116		
19	129		
20	132		

3. The output is the **skewness** value.

Patient	Male LDL level (mg/dL)		
1	110		
2	129		
3	121		
4	116		
5	123		
6	111		
7	114	Mean =	117.8
8	126	Median =	117.5
9	119	Mode =	129
10	127	Variance =	84.27368
11	101	Standard deviation =	9.180070
12	112	Skewness =	-0.261476
13	127	Kurtosis =	
14	122		
15	113		
16	102		
17	106		
18	116		
19	129		
20	132		

Because the skewness value is close to zero, the data are not significantly skewed.

Kurtosis

1. To calculate kurtosis, use the kurtosis function by typing in **=kurt** into an empty cell. Double click on the kurtosis function from the drop down menu.

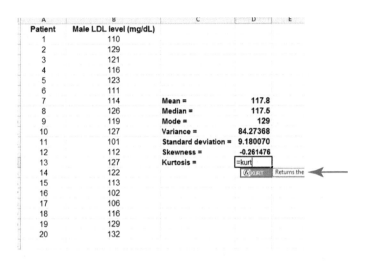

2. Highlight the cells that will be used to calculate the kurtosis. Press enter/return.

3. The output is the kurtosis value.

Patient	Male LDL level (mg/dL)		
1	110		
2	129		
3	121		
4	116		
5	123		
6	111		
7	114	Mean =	117.8
8	126	Median =	117.5
9	119	Mode =	129
10	127	Variance =	84.27368
11	101	Standard deviation =	9.180070
12	112	Skewness =	-0.261476
13	127	Kurtosis =	-0.882119
14	122		
15	113		
16	102		
17	106		
18	116		
19	129		
20	132		

Because the kurtosis value is not close to zero, the data are kurtic. The negative sign indicates they are platykurtic (light tails).

Descriptive Statistics Command

1. If you need additional information about the data (e.g., standard error, minimum, maximum) or you would like Excel to automatically compute the descriptive statistics for you, use the **Data Analysis ToolPak**. Click on an empty cell, then select **Data Analysis**, which should be located along the toolbar under the **Data** tab.

Note: If it does not appear, then you must manually install it (see Chapter 12).

2. The following window will appear. Select the **Descriptive Statistics** function, then click OK.

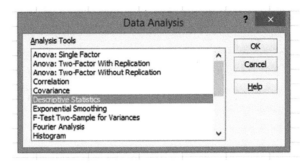

3. Set the **Input Range** by selecting the corresponding icon to the right.

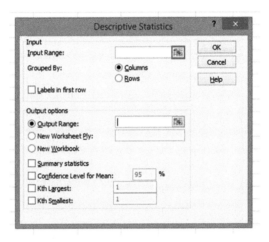

4. Input the range by highlighting the cells that will be used to calculate the descriptive statistics. Select the small icon to the right when finished.

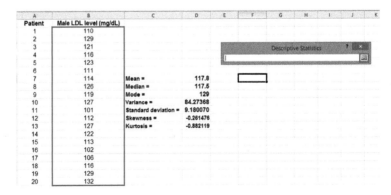

5. Once you have returned to the following window, select your output options. In the example below, **Output Range** will generate the data on the same Excel sheet. Click on the icon to the right to select an empty cell to which your data will be generated in.

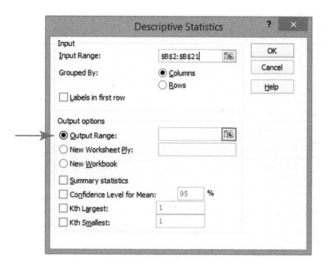

6. Now that the **Output options** has been set, select **Summary statistics**. Click **OK**.

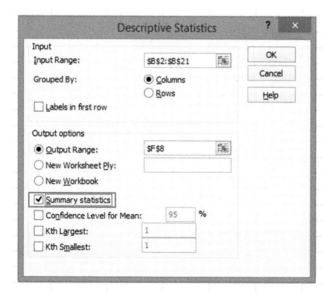

7. The following output will appear.

Patient	Male LDL level (mg/dL)						
1	110						
2	129						
3	121						
4	116						
5	123						
6	111						
7	114	Mean =	117.8		Column1		
8	126	Median =	117.5				
9	119	Mode =	129		Mean	117.8	
10	127	Variance =	84.27368		Standard Error	2.05272604	
11	101	Standard deviation =	9.180070		Median	117.5	
12	112	Skewness =	-0.261476		Mode	129	
13	127	Kurtosis =	-0.882119		Standard Deviation	9.18006995	
14	122				Sample Variance	84.2736842	
15	113				Kurtosis	-0.8821191	
16	102				Skewness	-0.2614758	
17	106				Range	31	
18	116				Minimum	101	
19	129				Maximum	132	
20	132				Sum	2356	
					Count	20	

8. As you will notice, the **Descriptive Statistics** function provides additional information about the data.

Column1	
Mean	117.8
Standard Error	2.05272604
Median	117.5
Mode	129
Standard Deviation	9.18006995
Sample Variance	84.2736842
Kurtosis	-0.8821191
Skewness	-0.2614758
Range	31
Minimum	101
Maximum	132
Sum	2356
Count	20

2.4 Descriptive Statistics in SPSS

To calculate the descriptive information (mean, standard deviation, range, variance, skewness, and kurtosis) of a dataset in SPSS, follow the example provided using the dietary intake information pertaining to the ingestion of omega-3 fatty acids in grams per day from a group of 25 young males and females. Refer to Chapter 13 for getting started and understanding SPSS.

1. The data should look similar to the following image. The label includes the variable name, and in this case the unit of measure for the variable **Omega3gperday** (omega-3 intake in grams per day).

	Omega3gperday
1	.78
2	.54
3	.76
4	.39
5	.25
6	.24
7	1.17
8	1.45
9	1.20
10	.75
11	.24
12	.92
13	1.07
14	.50
15	.70
16	2.99
17	1.93
18	1.24
19	.78
20	2.35
21	.85
22	.78
23	1.93
24	.60
25	.88
26	

*This example was taken from the research conducted by Dr. Sarah Dunn.

2. Click on the **Analyze** tab on the toolbar located along the top of the page. Scroll down to **Descriptive Statistics** and select **Descriptives**.

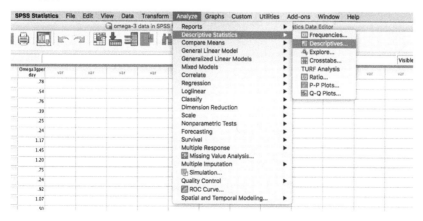

3. The following screen will appear.

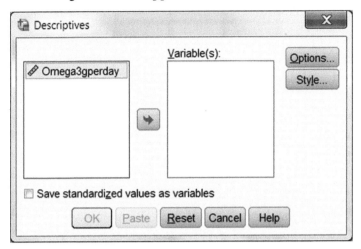

4. Make sure to move the variable over to the Variable(s) box. To do so, click on the variable (**Omega3gperday**). Then click on the arrow to the right corresponding to the **Variable(s)** box.

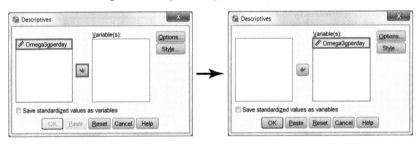

5. Select **Options** on the right side of the window.

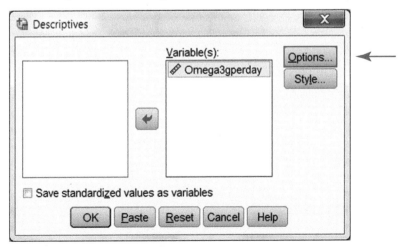

Central Tendency and Distribution

6. Use the checkboxes to choose the options you want to view in the output. At minimum, you will want to select **Mean**, **Std. deviation**, **Variance**, **Kurtosis**, and **Skewness**. Once all the option boxes are checked, click **Continue**.

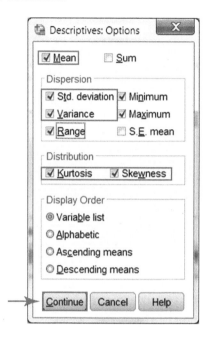

7. Click **OK**.

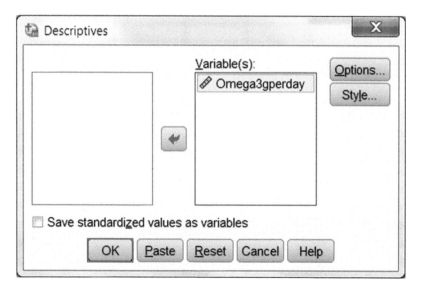

8. A separate output will appear.

Descriptives

		Descriptive Statistics										
	N	Range	Minimum	Maximum	Mean	Std. Deviation	Variance	Skewness		Kurtosis	
	Statistic	Statistic	Statistic	Statistic	Statistic	Statistic	Statistic	Statistic	Std. Error	Statistic	Std. Error
Omega3gperday	25	2.75	.24	2.99	1.0167	.67129	.451	1.431	.464	2.105	.902
Valid N (listwise)	25										

The skewness and kurtosis values are not close to zero, indicating the data are skewed to the right and leptokurtic (heavy tails).

2.5 Descriptive Statistics in Numbers

For each of the calculations below, we will use the following height (cm) data from 25 females. Make sure you are comfortable with the programming introduced in Chapter 14 prior to starting this tutorial.

Female Height (cm)
176
169
176
174
167
166
167
168
169
163
163
170
171
164
179
178
171
171
178
168
169
163
163
170
169

*This example was taken from the research conducted by Dr. Sarah Dunn.

Central Tendency and Distribution | 53

Mean

1. Use the average function **=average**. Highlight the cells that will be used to calculate the mean.

Ind #	Female Height (cm)					
1	176					
2	169					
3	176					
4	174					
5	167					
6	166	**Measures of Central Tendency**				
7	167	Mean =	· fx ˅	AVERAGE ▼	B2:B26 ▼	✕ ✓
8	168	Median =				
9	169	Mode =				
10	163					
11	163					
12	170					

2. Click on the check button or press enter/return.

Median

1. Use the median function **=median**. Highlight the cells that will be used to calculate the median.

Ind #	Female Height (cm)					
1	176					
2	169					
3	176					
4	174					
5	167					
6	166	**Measures of Central Tendency**				
7	167	Mean =		169.68		
8	168	Median =	· fx ˅	MEDIAN ▼	B2:B26 ▼	✕ ✓
9	169	Mode =				
10	163					
11	163					
12	170					
13	171					
14	164					

2. Click on the check button or press enter/return.

Mode

1. Use the mode function **=mode**. Highlight the cells that will be used to calculate the mode.

Ind #	Female Height (cm)
1	176
2	169
3	176
4	174
5	167
6	166
7	167
8	168
9	169
10	163
11	163
12	170
13	171

Measures of Central Tendency	
Mean =	169.68
Median =	169
Mode =	• fx ⌄ MODE ▼ B2:B26 ▼

2. Click on the check button or press enter/return.
3. Summary of the outputs.

Measures of Central Tendency	
Mean =	169.68
Median =	169
Mode =	169

Central Tendency and Distribution | 55

Variance

1. Use the variance function **=var**. Highlight the cells that will be used to calculate the variance.

Ind #	Female Height (cm)					
1	176					
2	169					
3	176					
4	174					
5	167					
6	166	**Measuring the spread of data**				
7	167	Variance =	• fx ⌄	VAR ▼	B2:B26 ▼	⊗ ✓
8	168	Standard Deviation =				
9	169					
10	163					

2. Click on the check button or press enter/return.

Standard Deviation

1. Use the standard deviation function **=stdev**. Highlight the cells that will be used to calculate the standard deviation.

Ind #	Female Height (cm)					
1	176					
2	169					
3	176					
4	174					
5	167					
6	166	**Measuring the spread of data**				
7	167	Variance =	23.9766666666667			
8	168	Standard Deviation =	• fx ⌄	STDEV ▼	B2:B26 ▼	⊗ ✓
9	169					
10	163					

2. Click on the check button or press enter/return.
3. Summary of the outputs.

Measuring the spread of data	
Variance =	23.9766666666667
Standard Deviation =	4.89659745809952

4. Finally, you should change the significant figures or number of digits behind the decimal point for consistency. Click **Cell** in the pane on the right hand side of your screen. Then, click the drop down menu under **Data Format** and select **Number**. Use the arrows to the right of the box next to **Decimals** to select the appropriate number of decimal points. The default is auto; click the up arrow to change the number.

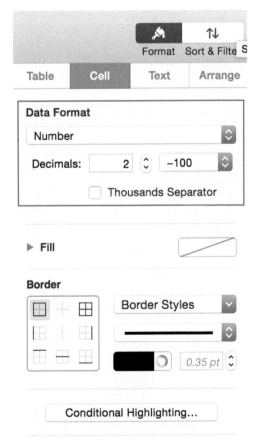

5. Here is a final summary output:

Measures of Central Tendency	
Mean =	169.68
Median =	169.00
Mode =	169.00
Measuring the spread of data	
Variance =	23.98
Standard Deviation =	4.90

Note: To run skewness or kurtosis, you will need to use another program. See the tutorials within Excel, SPSS, or R.

2.6 Descriptive Statistics in R

For each of the calculations below, we will use the following protein data (grams per day) from 25 females. Refer to Chapter 15 for R specific terminology and instructions on how to invoke and construct code.

Mean

1. Load the data into R by storing the numbers below to a vector name, for example "**protein**."

```
> protein<-c(166.1237688, 121.2233691, 67.35898296, 65.81328015, 74.55393022,
42.74269337, 66.84138156, 70.96573431, 70.96573431, 80.62756931, 42.74269337,
58.8848863, 90.86836089, 41.16542229, 62.85778643, 151.5806621, 145.6547939,
100.6836615, 156.1237688, 117.6942273, 363.0614423, 166.1237688, 145.6547939,
67.35898296)
```

2. Apply the **mean()** function to the vector **protein**. Press enter/return.

```
> mean(protein)
[1] 105.7363
```

The output for the mean value is placed below the R code you typed in.

Median

1. If you have not already loaded the data as in step 1 of the instructions for calculating the mean, do so now. Then, apply the **median()** function to the **protein** vector. Press enter/return.

```
> median(protein)
[1] 77.59075
```

The output for the median value is placed below the R code you typed in.

Mode

Not all statistical programs have automated features for statistical calculations. In the case of mode, R does not have an automated feature. Instead, code can be developed to measure the mode. The development of the code is beyond the scope of this chapter.

Variance

1. Load the data as in step 1 of the instructions for calculating the mean. Then, apply the **var()** function to the **protein** vector. Press enter/return.

```
> var(protein)
[1] 4703.435
```

The output for the variance value is placed below the R code you typed in.

Standard Deviation

1. Load the data as in step 1 of the instructions for calculating the mean. Then, apply the `sd()` function to the **protein** vector. Press enter/return.

```
> sd(protein)
[1] 68.58159
```

The output for the standard deviation value is placed below the R code you typed in.

Skewness

1. Load the **moments** package in R using the `library()` function. If you do not have the package installed, you will get a message that reads "Error in library (moments): there is no package called 'moments.'" Refer to Chapter 15 for instructions on how to install packages. Apply the `library()` function to load the moments package. Press enter/return.

```
> library(moments)
```

2. Load the data as in step 1 of the instructions for calculating the mean. Then, apply the `agostino.test()` function to the **protein** vector to run the D'Agostino skewness test. We will add the `alternative =` argument to specify using a two-sided test for generating a p-value, which will allow us to measure skewness in either direction. Press enter/return.

```
> agostino.test(protein, alternative=c("two.sided"))

        D'Agostino skewness test

data:  protein
skew = 2.2741, z = 4.0264, p-value = 5.663e-05
alternative hypothesis: data have a skewness
```

Note: The coding for a two sided test is a string, so we need to use the `c()` function and quotation marks to select the numbers of tails.

The output for the skewness value is placed below the R code you typed in. The *p*-value is less than a *p*-value$_{cutoff}$ of 0.05. Because the *p*-value is significant, we can conclude the skewness is greater than zero and the data are skewed. The skew value is positive, indicating the data are skewed to the right.

3. This result can be verified by plotting a histogram using the **hist()** argument applied to the **protein** vector following the tutorial in Chapter 3.

Kurtosis

1. Follow the instructions in step 1 of calculating skewness to load the **moments** package. Press enter/return.
2. Load the data as in step 1 of the instructions for calculating the mean. Then, apply the **anscombe.test()** function to the **protein** vector to run the Anscombe–Glynn kurtosis test. We will add the **alternative =** argument to specify using a two-sided test for generating a *p*-value, which will allow us to measure kurtosis in either direction. Press enter/return.

```
> anscombe.test(protein, alternative=c("two.sided"))

       Anscombe-Glynn kurtosis test

data:  protein
kurt = 9.2241, z = 3.5810, p-value = 0.0003423
alternative hypothesis: kurtosis is not equal to 3
```

Note: The coding for a two sided test is a string, so we need to use the **c()** function and quotation marks to select the numbers of tails.

The output for the kurtosis value is placed below the R code you typed in. The *p*-value is less than a *p*-value$_{cutoff}$ of 0.05. Because the *p*-value is significant, we can conclude the kurtosis is greater than three and the data are kurtotic. The kurtosis value is positive, indicating the data are leptokurtic (heavy tails).

3. This result can be verified by plotting a histogram using the **hist()** argument applied to the **protein** vector following the tutorial in Chapter 3.

3

Showing Your Data

> **Learning Outcomes**
>
> By the end of this chapter, you should be able to:
>
> 1. Create readable tables and figures for a variety of examples.
> 2. Determine which table or graph (bar, line, scatter, or pie) is appropriate for a given question, example, or study.

3.1 Background on Tables and Graphs

In order to communicate your data in a clear manner, you must create tables and graphs that can be easily interpreted by the reader. A common mistake that beginning researchers make is to show information that is not essential for conveying their message, resulting in confusing tables or graphs. There are instances where a full table is needed. For example, a list of collection localities or a copy of source data is required for a thesis, and may be appropriate as part of the supplementary material of many publications.

When presenting your data in other media, such as when giving a presentation, a large table can be hard to read. Learning to summarize your results in a way that others can easily interpret is essential. When possible, utilize figures, graphs, or charts. In this chapter, you will learn how to present your data and decide which table or figure is best suited for your results. For more tips on displaying data in a meaningful way, please read "*The Science of Scientific Writing*"

An Introduction to Statistical Analysis in Research: With Applications in the Biological and Life Sciences,
First Edition. Kathleen F. Weaver, Vanessa C. Morales, Sarah L. Dunn, Kanya Godde and Pablo F. Weaver.
© 2018 John Wiley & Sons, Inc. Published 2018 by John Wiley & Sons, Inc.
Companion Website: www.wiley/com/go/weaver/statistical_analysis_in_research

by Gopen and Swan (1990). The authors emphasize meaningful and logical presentation of data that is appealing and expected by a reader.

3.2 Tables

A table is a good way of presenting raw data, statistical output, or other numerical trends. The following three points are important to keep in mind when constructing a table:

- The reader expects to see the independent variable on the left. This allows the reader to follow the progression of logic from what is being manipulated to the effects on the dependent variable.
- Keep the amount of information presented to a minimum. Too much information may cause the reader to miss the important patterns within the data.
- Highlight important patterns in the table. Use symbols (*), color, or bolding to draw the reader's eye.

Data presented in the following table are from a study that focused on adult shell morphology in two groups of *Oreohelix* mountain snails in the state of Idaho, USA. Table 3.1 is a good example of how to formulate your data in a simple and clear manner.

Table 3.1 Shell shape, height, width, and depth measurements used in a morphometric study of snails in the genus *Oreohelix* from two locations in Idaho.

Location	Shell Shape (1 = Globular, 2 = Intermediate, 3 = Flat)	Height of Shell (mm)	Width of Shell (mm)	Depth of Shell (mm)
Pocatello	2	15.79	17.90	10.57
Pocatello	2	14.94	16.36	10.13
Pocatello	2	12.86	14.57	9.16
Pocatello	2	14.99	16.68	10.35
Pocatello	1	13.62	15.32	10.11
Rathdrum	2	11.32	12.59	7.31
Rathdrum	2	11.07	12.80	7.52
Rathdrum	2	10.72	11.86	7.65
Rathdrum	2	12.50	14.48	8.94
Rathdrum	1	13.35	15.08	9.67

3.3 Bar Graphs, Histograms, and Box Plots

Background

A bar chart (also called a bar graph) is a simple, yet powerful way to visually depict your data. Bar charts consist of bars, either vertical or horizontal, that represent the values of some type of categorical variables. These values could represent the means or medians for the groups, as well as the actual value for a single observation. The visual representation of a bar chart allows the viewer to easily interpret patterns in the data and typically consists of four types of charts: clustered bar charts, clumped bar charts, stacked bar charts, and histograms.

Types of Bar Graphs

Clustered Bar Charts

An alpine researcher is interested in the snowfall patterns in selected mountain regions of the western United States. She records the snowfall for mountain forests in Mammoth, CA; Mount Baker, WA; and Alyeska, AK. In the graph below, Figure 3.1, the x-axis consists of the independent, categorical variable (locality) and the y-axis represents the dependent variable (mean snowfall in meters).

Notice that Figure 3.1 gives a clear depiction of the differences in the mean snowfall at the three localities. By adding error bars

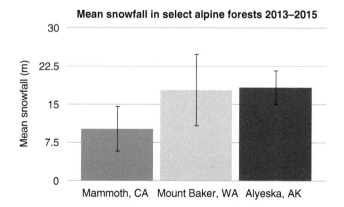

Figure 3.1 Clustered bar chart comparing the mean snowfall of alpine forests between 2013 and 2015 in Mammoth, CA; Mount Baker, WA; and Alyeska, AK.

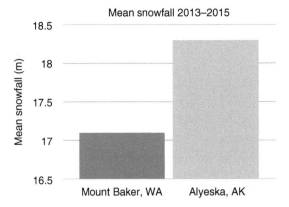

Figure 3.2 Clustered bar chart comparing the mean snowfall of alpine forests between 2013 and 2015 in Mount Baker, WA and Alyeska, AK. An improperly scaled axis exaggerates the differences between groups.

(standard deviations), the researcher is also able to illustrate the variance in each one of the groups of data. For instance, the snowfall in Alyeska, AK is less variable than the snowfall in Mount Baker, WA.

One of the most important considerations when displaying data with a bar graph is the scaling of the axes. Unfortunately, graphs built in the programs Excel and Numbers are often created with an improperly scaled y-axis. If the y-axis does not begin with zero, then the differences between groups appear exaggerated. By "zooming in" on this smaller set of y-axis values, the graph can be misleading. Take the previous example of snowfall measurements. It is clear that the average snowfall in Mount Baker, WA and Alyeska, AK are very similar. However, if the graph of these two localities is built with a modified y-axis as in Figure 3.2, with a minimum value set to 16.5, the differences appear dramatic, when in reality they are not.

Clumped Bar Charts

If this same researcher was interested in illustrating the trends in snowfall patterns over a 3-year period, a clumped bar chart would be useful. Figure 3.3 shows snowfall patterns within each locality over a 3-year period. By using a clumped bar chart, the researcher can demonstrate trends within each category of data. For example, we

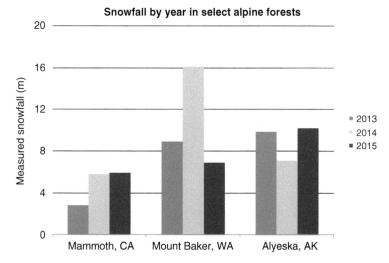

Figure 3.3 Clumped bar chart comparing the mean snowfall of alpine forests by year (2013, 2014, and 2015) in Mammoth, CA; Mount Baker, WA; and Alyeska, AK.

can see that the snowfall was exceptionally high in 2014 at the Mount Baker, WA location; however, the snowfall at the Mammoth, CA location was fairly stable over time.

Stacked Bar Charts

Next the researcher wants to illustrate differences in the timing of snowfall by month within each site. For this example, a stacked bar chart is helpful in illustrating the relative contributions of parts to the whole. Figure 3.4 shows the amount of snow that fell within the months of January, February, and March, 2015. Notice that in Mammoth, CA there was zero snowfall in the month of January.

Histograms

Histograms are another form of bar charts used to display continuous categories, like a consecutive range of values for age. If your data are made up of quantitative variables, then consider constructing a histogram. The format is similar to that of a bar chart; however, the categories along the bottom are represented with a set range of values. Hence, both axes will be represented on a numerical scale.

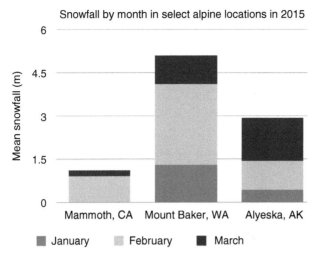

Figure 3.4 Stacked bar chart comparing the mean snowfall of alpine forests by month (January, February, and March) for 2015 in Mammoth, CA; Mount Baker, WA; and Alyeska, AK.

Also, the aesthetics are slightly different because there are no spaces between the bars. In a histogram, there will never be space between bars because the horizontal axis is representing continuous values (Figure 3.5). If a space does exist between bars, then it means that there are no values for that range.

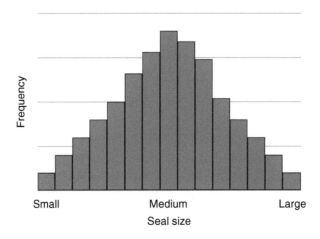

Figure 3.5 Histogram of seal size.

Box Plots

The box plot (also called a box and whisker plot) is a convenient way to illustrate several key descriptive statistics from a dataset. Box plots show the median, as well as the distribution of the data through the use of quartiles, which divide ranked data into four equal groups, each consisting of a quarter of the data.

Consider the following dataset:

19, 18, 34, 17, 38, 22, 28, 26, 31, 33, 18, 21

The first step in developing a box plot for these data is to define the quartiles. Several methods are currently debated regarding how to define quartiles; the following example uses the simplest and most intuitive method. In the sample dataset above, the numbers must first be rearranged so that they are in order:

17, 18, 18, 19, 21, 22, 26, 28, 31, 33, 34, 38

Second, find the median, which is also defined as the second quartile (Q2). In the current example, there is an even number of data points, so the median is calculated as the average of the middle two numbers (Q2 = 24). If there were an odd number of points, the median would be excluded for the next step. Third, calculate the median of each half of the data (on either side of the median); these medians are the first and third quartiles (Q1 and Q3):

Q1 = 18.5   Q2 = 24   Q3 = 32

17, 18, 18, | 19, 21, 22, | 26, 28, 31, | 33, 34, 38

The box component of a box plot spans the first quartile to the third quartile, and is known as the interquartile range (IQR); the median is shown inside the box at the position of the second quartile, as illustrated in Figures 3.6 and 3.7.

By showing the median, as well as the position of the first and third quartiles, box plots give information about the degree of dispersion, as well as the skewness of the data. Box plots often also have lines (the whiskers) extending from the box to represent the variability of the data outside of the upper and lower quartiles. The whiskers

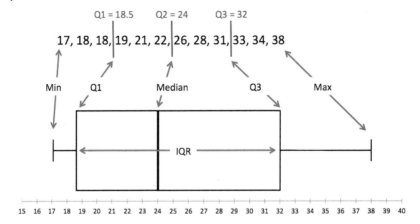

Figure 3.6 Example box plot showing the median, first and third quartiles, as well as the whiskers.

usually mark the minimum and maximum values for the dataset. However, if the dataset contains outliers, the whiskers will extend only up to a certain point, defined as Q1 − 1.5 × IQR or Q3 + 1.5 × IQR (Figure 3.7). Outliers will be depicted as points outside of the whiskers (Figure 3.8).

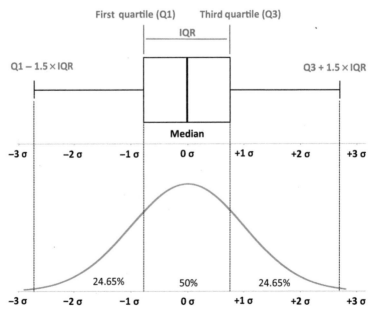

Figure 3.7 Comparison of the box plot to the normal distribution of a sample population.

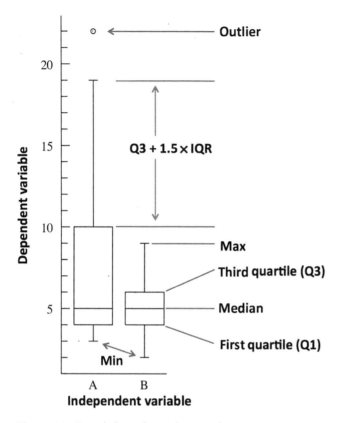

Figure 3.8 Sample box plot with an outlier.

The box plots on previous pages, Figures 3.7 and 3.8, have been drawn for illustrative purposes in a horizontal orientation, but are most often shown vertically, as in Figure 3.8. In Figure 3.8, descriptive information from two groups of data is depicted. Although the medians for the two groups are the same, the differences in the dispersion and skew of the data are apparent. While group B shows a normal distribution, group A shows a "positive skew," with a tail that extends in the positive direction. The box plot for group A also shows the position of an outlier, whose value is beyond the range of the whiskers.

Generating box plots is straightforward in both SPSS and R and is included in this book's tutorials. However, generating box plots in Excel and Numbers is both lengthy and complex, and involves manipulating stacked bar charts. If you do not have access to SPSS or R, we recommend looking for a free, online box plot generator, which is an easy and quick solution for creating box plots of your data.

Tutorials

How to Make a Bar Chart in Excel

The following tutorial will walk you through the construction of a bar chart (also known as column graph or bar plot) using Excel. The data involve the number of rows of snail radula.

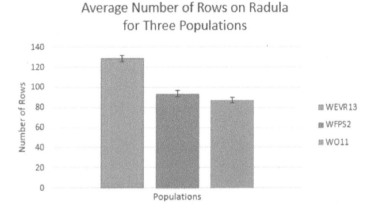

*Data taken from the research of Vanessa C. Morales, Robert Candelaria, and Dr. Kathleen Weaver.

Refer to Chapter 12 for tips and tools when using Excel.

Excel offers two methods to construct a simplified bar chart with error bars. While the first method may be more challenging at first, the lessons learned will give you greater mastery and flexibility. Calculate the average and standard deviation of the radula from each population prior to beginning the tutorial.

Method 1

1. Arrange data in columns on the spreadsheet.

Population	# of Rows	Average	Standard Deviation
WEVR13	128	128.6667	3.055050463
	132		
	126		
WFPS2	97	93.66667	2.886751346
	92		
	92		
WO11	89	87.33333	2.886751346
	89		
	84		

Showing Your Data | 71

2. Click on an empty cell. Select **Insert**, **Column**, and select the first **2-D Column** option. There are several types of bar graphs available. Use the one appropriate for the data you want to display.

3. A blank canvas will appear.

4. Right click on the blank canvas and choose the **Select Data** option.

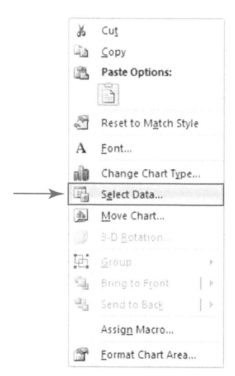

5. Under **Legend Entries**, select **Add**.

Note: Add each data point as separate series so that the standard deviation bars can be entered separately.

6. Select the icon corresponding to the **Series name** subheading.

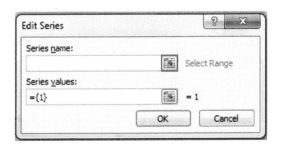

7. Select the first series title then click on the icon to the right.

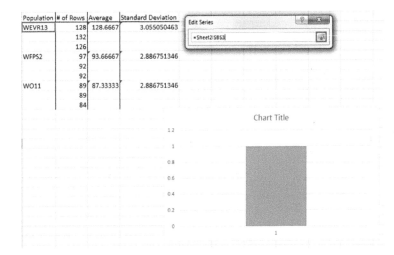

Showing Your Data | 73

8. Select the icon corresponding to the **Series values**.

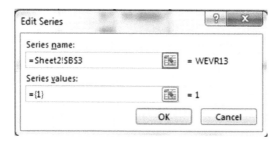

9. Select the first value then click the icon on the right.

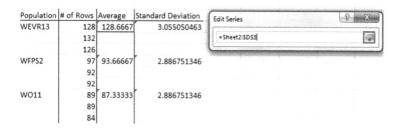

10. Click **OK**.

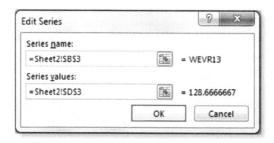

11. You will be directed to the original popup. Repeat steps 5–10 to input the remaining values.

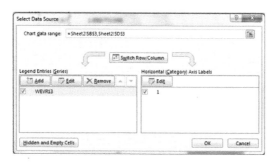

12. After the second variable is added, you should be left with a graph that looks like the following.

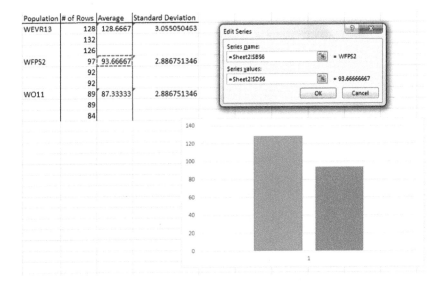

After the third variable:

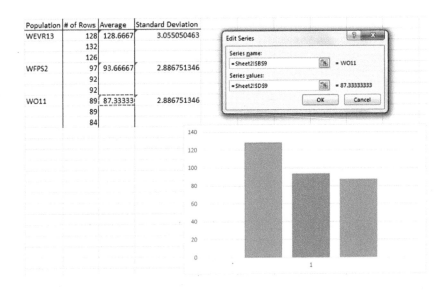

13. Once all the variables have been added to the graph, click **OK**.

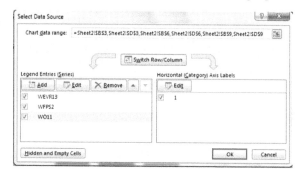

14. A very basic column graph will appear, similar to the one below.

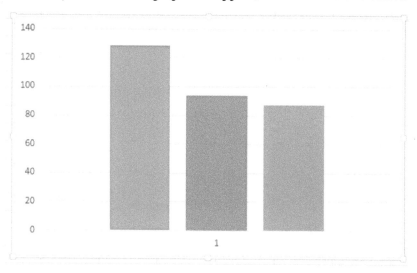

15. As a default, Excel labels the *x*-axis as "1." To delete this label, select label "1." A box will appear. Then, press delete on your keyboard.
16. To add proper labels and change any other aesthetic detail, go to **Chart Tools** on the toolbar and select **Design**. The graph must be selected in order for this option to appear.

17. Predesigned layouts to add a title, legend, and axes can be found in the **Quick Layout** menu under **Chart Tools, Design**. Some options under the **Quick Layout** menu will automatically add Axes Titles, Chart Title, and Legend. The following steps will instruct how to add each element individually.

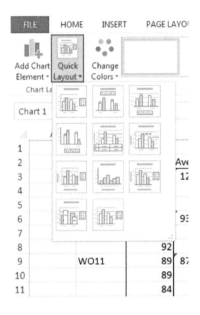

18. Insert a relevant title that describes your data by selecting **Chart Tools, Design, Add Chart Element, Chart Title**, and the option **Above Chart**.

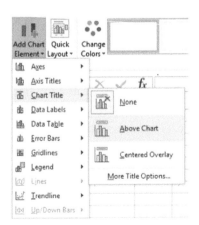

19. Input an appropriate title. Your graph should appear similar to the one below.

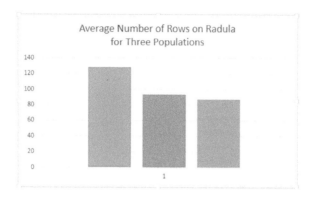

20. Insert axis titles by selecting **Chart Tools, Design, Add Chart Element, Axis Titles**, and the **Primary Vertical** option.

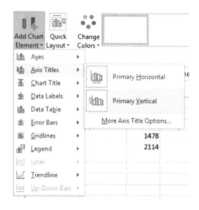

21. Label the *y*-axis with an appropriate title that corresponds to the data. The graph should look similar to the following.

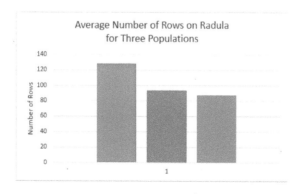

22. To insert the x-axis, select **Chart Tools**, **Design**, **Add Chart Element**, **Axis Titles**, and the **Primary Horizontal** option.

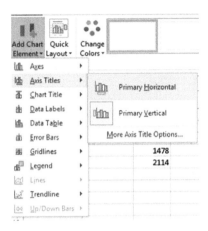

23. Label the x-axis with an appropriate title. The finished labeled graph should look similar to the following.

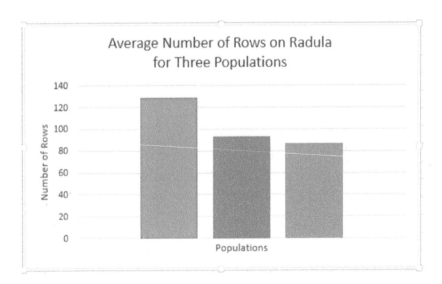

24. To insert a legend, select **Add Chart Element** and then select **Legend**. There are several options for the position of the legend. For this example, we will use the option labeled **Right**.

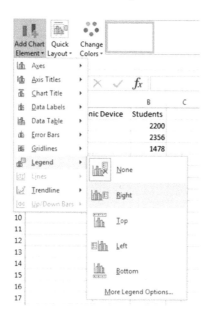

25. The legend will automatically label each column of the graph based on the series titles inputted during the creation of the graph. The graph should look similar to the following.

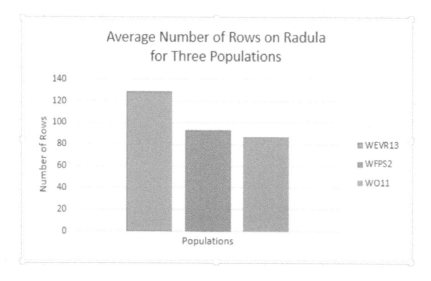

Adding Error Bars

Error bars must be added to each bar of your graph individually, as the standard deviation differs for each variable. The following steps will guide you through the process of adding error bars to your graph.

1. To add an error bar, click on a single bar from the graph.

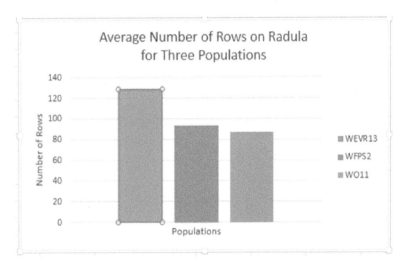

2. Go to the toolbar, and under **Chart Tools**, select **Design**. Then, select the **Add Chart Element** option. Select the **Error Bars** option, then select More **Error Bars Options**.

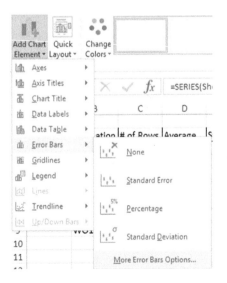

Showing Your Data | 81

3. The following menu will appear to the right of the spreadsheet.

4. Under the **Error Amount** subheading, select **Custom** and press the **Specify Value** button.

5. Click on the icon corresponding to the **Positive Error Value**.

6. Select the standard deviation for the bar that is selected. Click on the icon to the right when finished.

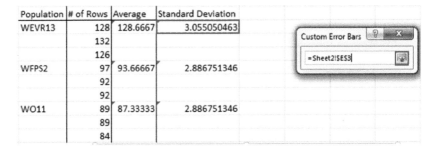

7. Set the **Negative Error Value** by clicking on the corresponding icon.

8. Select the same standard deviation value. Click on the icon to the right when finished.

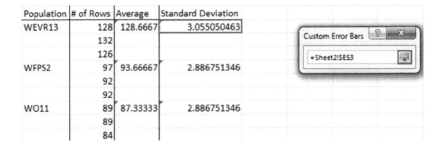

9. The positive and negative error values should now be set. Click **OK**.

10. The single error bar will appear in the selected bar on the graph.

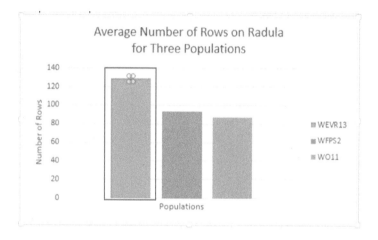

11. To insert custom error bars for each variable, repeat steps 1–10.
12. After all error bars have been added, the finished graph should look similar to the following graph.

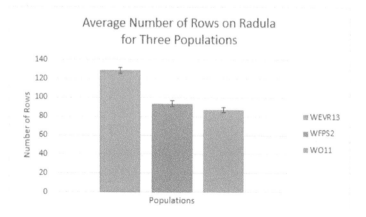

Method 2

1. We will use the same data presented in Method 1. Rearrange the data so that it appears similar to the following.

Population	# of Rows	Average	Standard Deviation
WEVR13	128	128.6667	3.055050463
	132		
	126		
WFPS2	97	93.66667	2.886751346
	92		
	92		
WO11	89	87.33333	2.886751346
	89		
	84		

↓

	WEVR	WFPS	W01
Average	128.6666667	93.66666667	87.33333333
Standard Deviation	3.055050463	2.886751346	2.886751346

2. Highlight the values (computed average) and labels to be included in the graph.

	WEVR	WFPS	W01
Average	128.6666667	93.66666667	87.33333333
Standard Deviation	3.055050463	2.886751346	2.886751346

3. Click on an empty cell. Select **Insert, Column,** and select the first **2-D Column** option. There are several types of bar graphs available. Use the one appropriate for the data you want to display.

Showing Your Data | 85

4. A very basic column graph will appear, similar to the one below. Notice that the labels are presented in the graph. Select the following icon to the right of the graph.

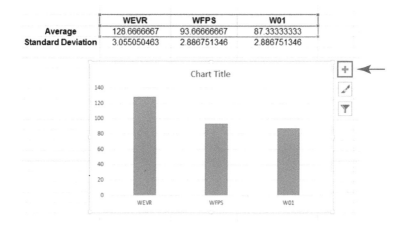

5. The following menu will appear to the right of the graph.

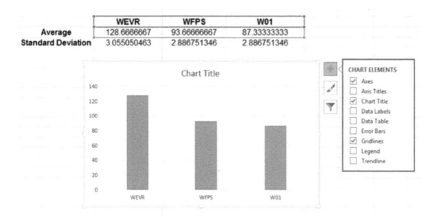

6. Select the boxes for **Axis Titles** and **Error Bars**.

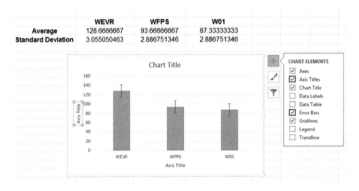

7. Select the arrow to the right of **Error Bars** and scroll down to **More Options**.

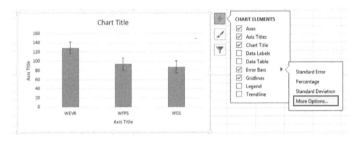

8. The following menu will appear to the right of the spreadsheet.

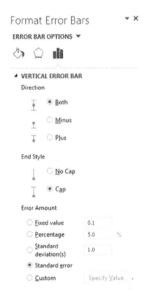

Showing Your Data | 87

9. Under the **Error Amount** subheading, select **Custom** and press the **Specify Value** button.

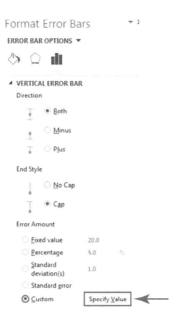

10. Click on the icon corresponding to the **Positive Error Value**.

11. Highlight all the values for standard deviation, then click on the icon corresponding to the **Positive Error Value**.

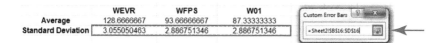

12. Set the **Negative Error Value** by clicking on the corresponding icon.

13. Select the same standard deviation values. Click on the icon to the right when finished.

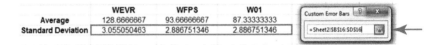

14. The positive and negative error values should now be set. Click **OK**.

15. A similar graph will appear.

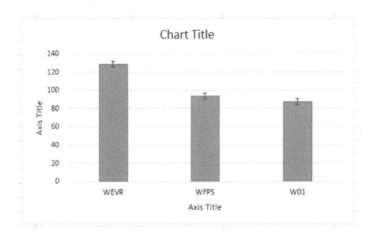

16. Double click the **Axis Title** corresponding to the *x*-axis and label with an appropriate title.

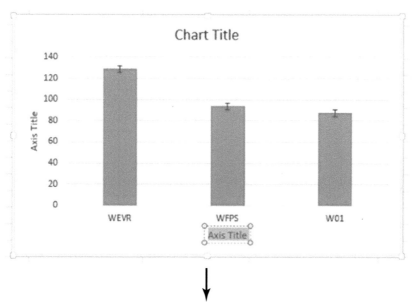

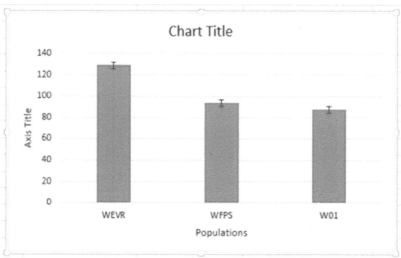

17. Label the *y*-axis and Chart Title. Your graph should look similar to the following.

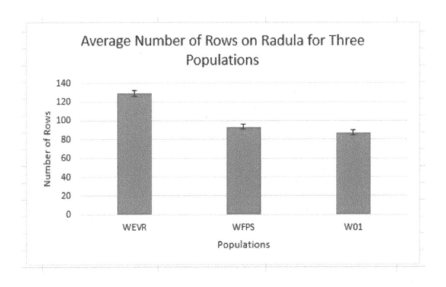

18. To change the colors of the bars, double click one column so that it is the only one highlighted. Then, right click to have the following menu appear. Select **Fill** and choose a desired color.

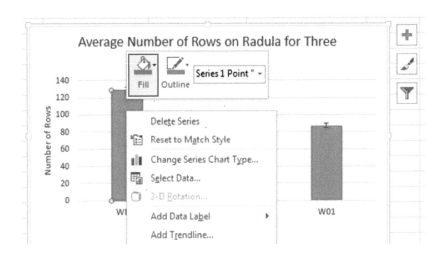

19. Do the same for the remaining columns to produce a final graph similar to the one below.

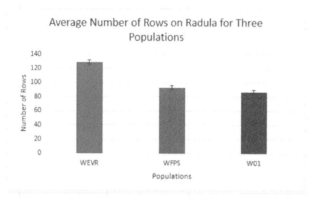

How to Make a Bar Chart in SPSS

The following tutorial will walk you through the construction of a proper bar chart (also known as column graph or bar plot) using SPSS.

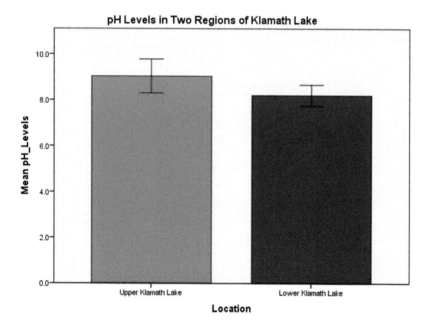

Refer to Chapter 13 for getting started and understanding SPSS.

1. Arrange the data so that it appears as the following.

	Location	pH_Levels
1	1.0	8.0
2	1.0	8.2
3	1.0	9.7
4	1.0	9.5
5	1.0	9.5
6	1.0	9.3
7	2.0	7.6
8	2.0	7.7
9	2.0	8.7
10	2.0	8.6
11	2.0	8.4
12	2.0	8.2

2. Along the toolbar at the top, select **Graphs** followed by **Chart Builder**.

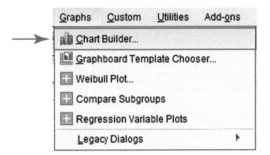

3. The following dialog box will appear. Click **OK**.

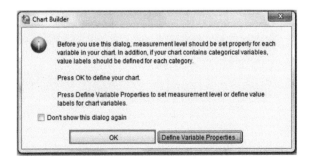

4. The following window will appear. Under the **Gallery** tab, locate **Bar**.

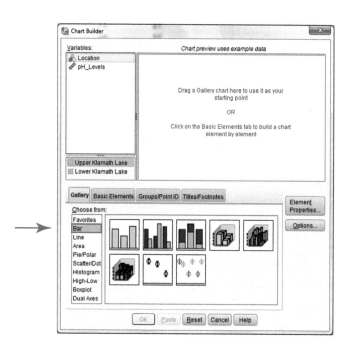

5. Select the **Simple Bar** icon and drag it to the **Chart Preview** window.

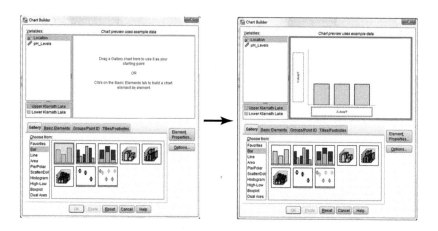

6. Select **Location** (independent variable) and drag it to the box labeled **X-Axis?** in the chart preview window.

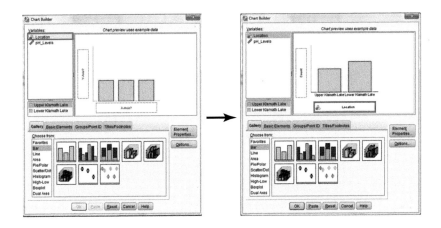

7. Select **pH Levels** (dependent variable) and drag it to the box labeled **Count** in the chart preview window.

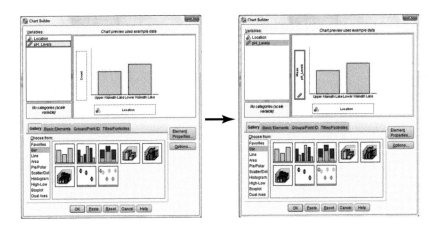

Note: The preview window does not use your data in the preview image. Your graph may look different to the graph in the preview.

8. To give the graph a title, click on **Titles/Footnotes** in the Chart Builder window. Select **Title 1** followed by **Element Properties**.

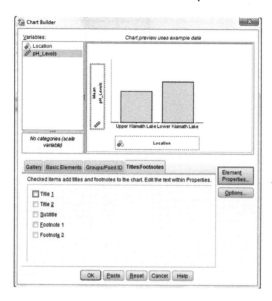

9. In the Element Properties window, type an appropriate title for the graph in the **Content** box. Then, click **Apply** at the bottom of the window.

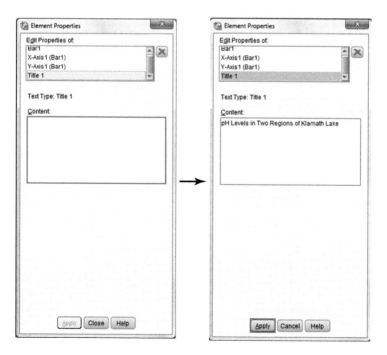

Adding Error Bars

1. To add error bars, select **Bar1** under the Edit Properties of: menu. Check the box **Display Error Bars**, select **Standard Deviation** and ensure that the multiplier is set to 1. Then select **Apply**.

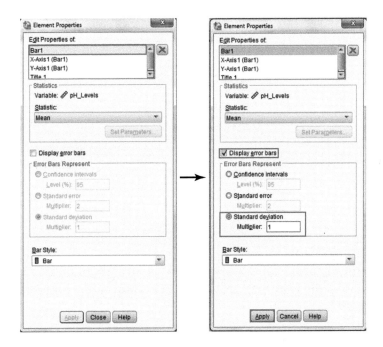

2. Click **OK** at the bottom of the Chart Builder window.

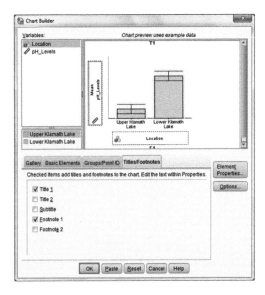

3. The graph will appear in a separate output and should look similar to the following.

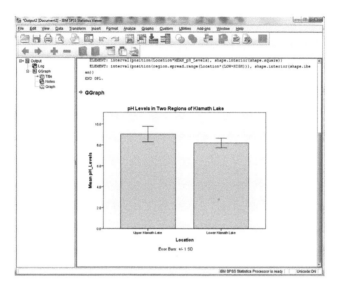

4. To change the colors of the bars on the graph, double click the graph in the output window. A **Chart Editor** window will appear.

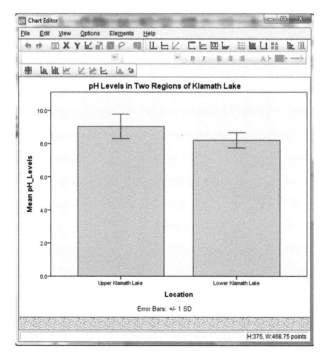

5. Double click on a bar. The **Properties** window will appear to the right of the Chart Editor window.

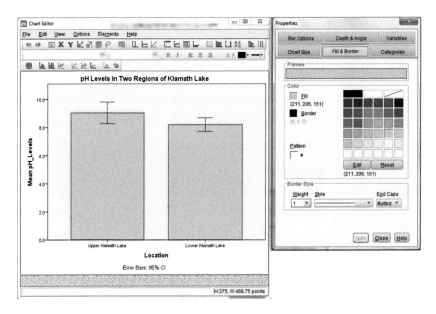

6. Choose a color for the bar, and click **Apply**.

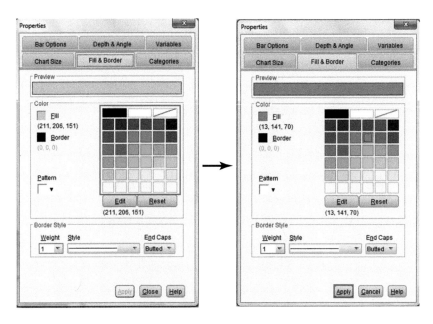

7. The bar will now appear in the chart editor window with the new color.

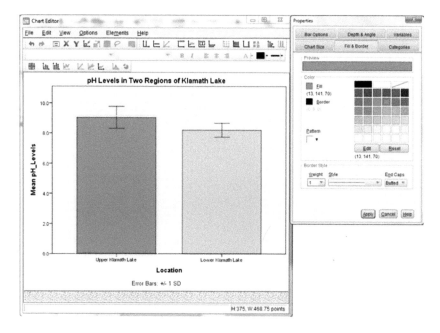

8. Double click on another bar and repeat the process to change that bar's color.
9. The final graph will look similar to the following.

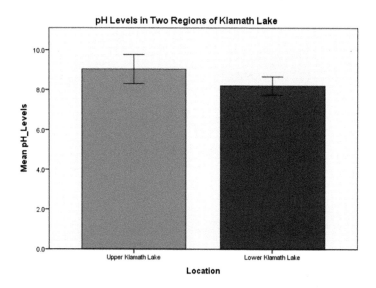

How to Make a Bar Chart in Numbers

The following tutorial will walk you through the construction of a bar chart (also known as column graph or bar plot) using Numbers.

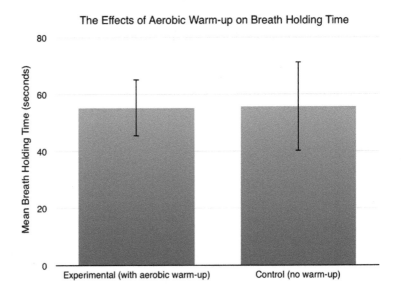

More information on programming in Numbers is found in Chapter 14.

A study was conducted to determine whether the acute effects of an aerobic warm-up (jump rope) had a significant influence on a participant's average breath-hold time.

1. Arrange the data so that it appears as the following.

	Mean Breathe Holding Time (seconds)	Standard Deviation	n (men & women)
Experimental (with aerobic warm-up)	55.25	9.97	16
Control (no warm-up)	55.75	15.64	16

2. Highlight the **Mean Breath Holding Time**.

	Mean Breathe Holding Time (seconds)	Standard Deviation	n (men & women)
Experimental (with aerobic warm-up)	55.25	9.97	16
Control (no warm-up)	55.75	15.64	16

3. Select the **Chart** function.

4. Select the 2D Bar Chart.

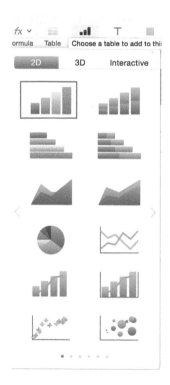

5. The following graph will appear.

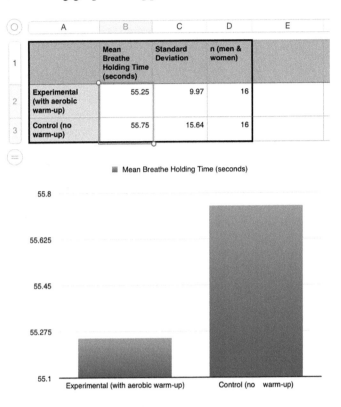

6. Under **Chart**, select **Title** and insert the title on the graph. You can change the coloring here too.

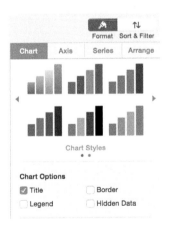

7. Under **Axis**, add the **Axis Name** for the Value (Y).

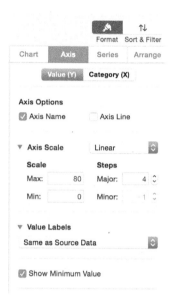

8. Under **Axis**, add the category labels, choose **Auto-Fit Category Labels**. **Label References** can be changed here.

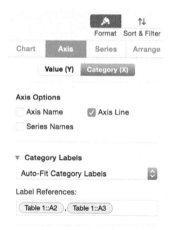

9. The graph will look like the following.

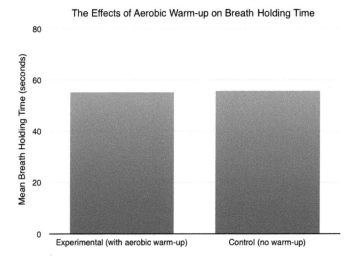

10. To add Error Bars, go to **Series** and select the pull down for **Error Bars**. Chose **Positive and Negative**. Under the **Positive and Negative** box, select the standard deviation values from the table.

	A	B	C	D
1		Mean Breathe Holding Time (seconds)	Standard Deviation	n (men & women)
2	Experimental (with aerobic warm-up)	55.25	9.97	16
3	Control (no warm-up)	55.75	15.64	16

11. The final graph with error bars will look like the following.

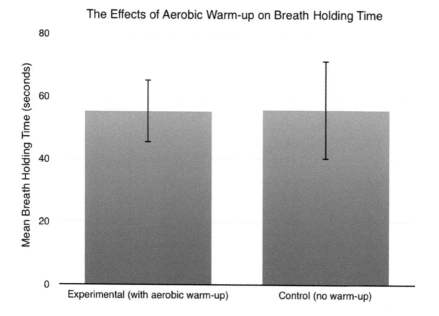

How to Make a Bar Chart in R

The following tutorial will walk you through the construction of a bar chart (also known as column graph or bar plot) using R.

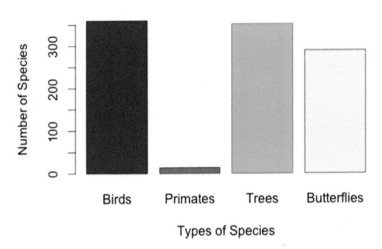

Refer to Chapter 15 for R specific terminology and instructions on how to invoke and construct code.

The Kibale Forest in Uganda is famous for its biodiversity, and more specifically, it boasts the highest number of primate species with overlapping home ranges (the total distance a species occupies and travels over a period of time).

1. Create a vector of the number of species by type. In this case, let us name it **number**. Press enter/return.

    > number<-c(359, 13, 351, 289)

2. Construct the script using the `barplot()` function. You will use the arguments `names.arg=`, `ylab=`, `xlab=`, `main=`, and `col=`, which instruct R what each number represents: the titles of each bar, the *y*-axis label, the *x*-axis label, chart title, and colors for each bar.

Note: For **names.arg=** and **col=**, you will need to apply the symbols that tell R to store multiple values, which are the letter "c" and parentheses: **c()**. For the remaining arguments, we will be applying a string, so the text should be set inside quotation marks.

We will first apply all the arguments except **col=**. Press enter/return.

Names of species that correspond to the number we saved in step 1. These must be in the same order as the values to which they correspond in the **number** vector.

```
> barplot(number, names.arg=c("Birds", "Primates", "Trees", "Butterflies"),
  ylab="Number of Species", xlab="Types of Species", main ="Biodiversity at
  Kibale Forest")
```

Y-axis label

X-axis label

Chart Tile

The output consists of a gray scale chart such as the one below.

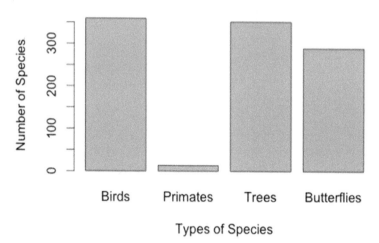

3. You can add color to any chart in R using the **col=** function. To examine the variety of colors R has available, type the following code into R to display the color names. Press enter/return.

```
> colors(distinct=FALSE)
```

4. After selecting four colors for the four bars, add them to the line of programming from step 2. Place the colors in the order you wish to have them appear on the graph. Press enter/return.

```
> barplot(number, names.arg=c("Birds", "Primates", "Trees", "Butterflies"),
ylab="Number of Species", xlab="Types of Species", main ="Biodiversity at
Kibale Forest", col=c("darkblue", "red", "green", "yellow"))
```

The output consists of a color chart such as the one below.

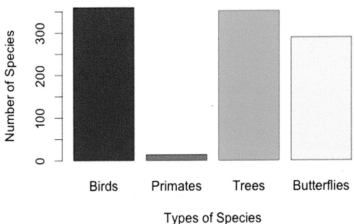

Note: There is no automatic way to add error bars to bar charts in R. If error bars are necessary for your research, consider graphing your data in another program, such as Excel.

R plots are highly customizable. The help facility and Internet search engines are fantastic resources for learning additional customizations that can be applied.

How to Make a Box Plot in SPSS

The following tutorial will walk you through the construction of a proper box plot using SPSS.

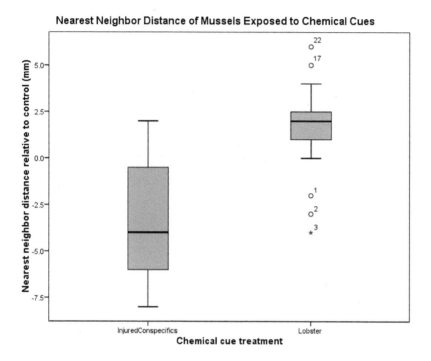

Data on nearest neighbor distance are based on Castorani and Hovel (2016). The data have been estimated from published graphs for use in this tutorial.

Refer to Chapter 13 for getting started and understanding SPSS.

1. Arrange the data so that they appear as the following.

	InjuredConspecifics	Lobster
1	-8	-2
2	-7	-3
3	-6	-4
4	1	0
5	0	0
6	-1	1
7	-7	1
8	-1	1
9	-4	2
10	-4	2
11	-4	2
12	-5	2
13	-5	2
14	-5	2
15	-5	3
16	-6	4
17	-2	5
18	2	4
19	-8	2
20	-8	2
21	0	2
22	0	6
23	0	3
24		

2. In the toolbar, select **Graphs** → **Legacy Dialogs** → **Boxplot**.

3. The box plot box will appear. Select **Simple** and **Summaries of Separate Variables**, and then click **Define**.

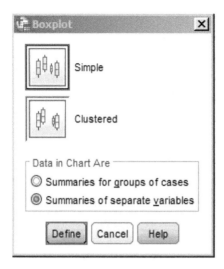

4. Use the arrow to move the two categorical variables into **Boxes Represent** and click on **Options**.

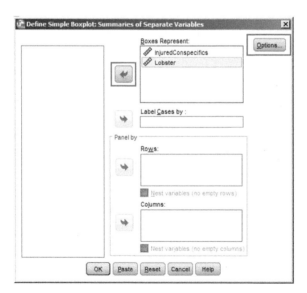

5. Within Options, choose **Exclude cases variable by variable** and click **Continue**, and then click **OK**.

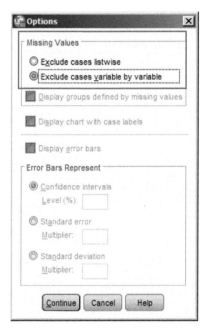

6. The following box plot will appear in the SPSS Statistics Viewer output.

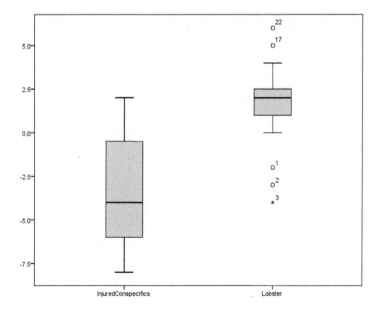

7. To format the graph to add axes and a title, double click on the graph and the **Chart Editor** will appear. Within the editor, select the **X**.

8. Select **Display axis title**. Click **Apply** and then enter the title.

9. Within the editor, select the **Y** and select **Display axis title**. Click **Apply** and then enter the title.

10. Within the editor, select the button to insert a chart title. The title will appear on the chart, then enter the chart title.

11. To change the color of the box plot bars, double click on either box plot. Within **Fill & Border**, choose a new color and click **Apply**.

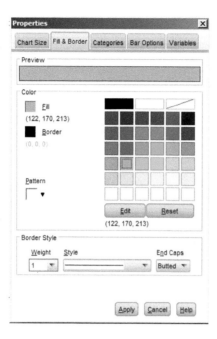

12. Finally, exit out of the Chart Editor, and the following graph will appear.

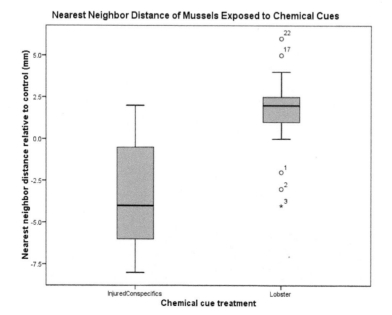

13. Finally, to save the figure, either screen shot it or export the output. See Chapter 13 for more details.

How to Make a Box Plot in R

The following tutorial will walk you through the construction of a proper box plot using R.

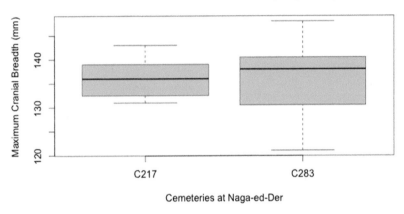

Box Plot of Maximum Cranial Breadth (mm) at Naga-ed-Der

Cemeteries at Naga-ed-Der

*This example was taken from the research conducted by Dr. Kanya Godde.

Refer to Chapter 15 for R specific terminology and instructions on how to invoke and construct code.

Human cranial measurements describe the underlying genetics and epigenetics of an individual. Crania excavated from two cemeteries (217 and 218) at the site of Naga-ed-Dêr in Egypt were measured for maximum cranial breadth (mm).

1. Enter the data into two vectors named for the two cemeteries. Remember that a "." represents missing data in continuous variables, but "NA" will also work and will allow us to type missing data directly into R (whereas the "." can be read by R in uploaded CSV files). In this case, let us use NA. Press enter/return.

```
> C217<-c(131, 133, 136, 136, 142, 132, 143, 136)
> C283<-c(131, 141, 140, 148, 130, 121, 138, NA)
```

2. Create a data frame in R to store both vectors in a single dataset. Use the **data.frame()** argument to complete this step. Press enter/return.

```
> egypt<-data.frame(C217, C283)
```

To illustrate how your data frame works, type in the name of your data frame and press enter/return. The output should be similar to the one below.

```
> egypt
  C217 C283
1  131  131
2  133  141
3  136  140
4  136  148
5  142  130
6  132  121
7  143  138
8  136   NA
```

3. Plot the box plot with the **boxplot()** function applied to the **egypt** data frame. Use the **xlab=** and **ylab=** arguments to label the x- and y-axes. The argument **main=** applies a title to the whole chart. Press enter/return.

> boxplot(egypt, xlab="Cemeteries at Naga-ed-Der", ylab="Maximum Cranial Breadth (mm)", main="Box Plot of Maximum Cranial Breadth (mm) at Naga-ed-Der")

↑
Chart title

The output consists of a gray scale chart such as the one below.

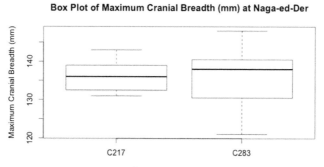

4. You can add color to any chart in R using the `col=` function. To examine the variety of colors R has available, type the following code into R to display the color names. Press enter/return.

> `colors(distinct=FALSE)`

5. After selecting a color for the boxes, add them to the line of programming from step 3. Press enter/return.

> `boxplot(egypt, xlab="Cemeteries at Naga-ed-Der", ylab="Maximum Cranial Breadth (mm)", main="Box Plot of Maximum Cranial Breadth (mm) at Naga-ed-Der", col="turquoise2")`

The output consists of a color chart such as the one below.

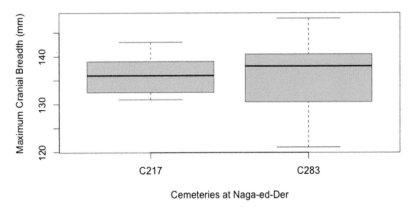

Note: R plots are highly customizable. The help facility and Internet search engines are fantastic resources for learning additional customizations that can be applied.

How to Make a Histogram in Excel

The following tutorial will walk you through the construction of a histogram in Excel.

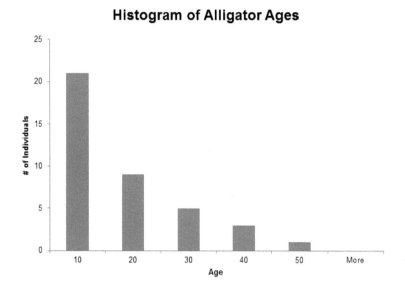

Refer to Chapter 12 for tips and tools when using Excel.

1. Enter the following data on alligator ages into Excel as a single column (listed within three columns for readability).

	A	B	C	D
1	8	2	3	
2	4	7	6	
3	7	12	5	
4	45	38	3	
5	3	29	1	
6	9	13	11	
7	18	13	32	
8	16	22	17	
9	34	10	18	
10	14	4	21	
11	9	7	5	
12	2	8	28	
13	4	25	9	
14				

2. Add in the Bins, which correspond to each bar in the graph, for age ranges (10, 20, 30, 40, 50).

	A	B	C
1	8		
2	4	Bin	
3	7		10
4	45		20
5	3		30
6	9		40
7	18		50
8	16		
9	34		
10	14		
11	9		
12	2		
13	4		
14	2		
15	7		
16	12		
17	38		

3. Within **Data**, select the **Data Analysis** button.

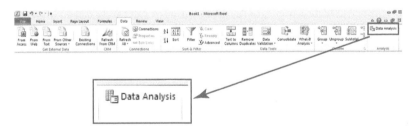

4. Select **Histogram** and click **OK**.

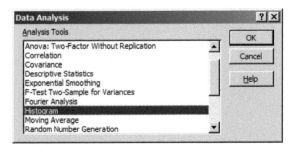

5. In the Histogram box, input the following: **Input Range** (source data), **Bin Range** (bin values 10, 20, 30, 40, 50), and **Output Range** (any other cell for the output). Select the box for **Chart Output**.

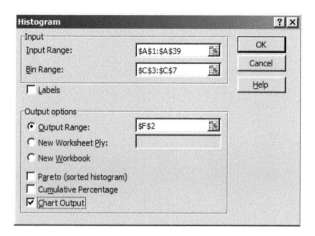

6. The following is the output. To customize the axis labels and title, double click on the word that is currently in place and change the text. To remove the legend, click on it and hit delete.

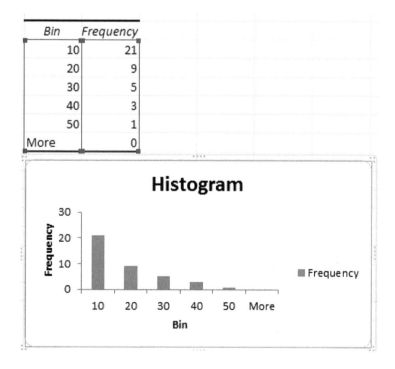

7. Here is the final graph.

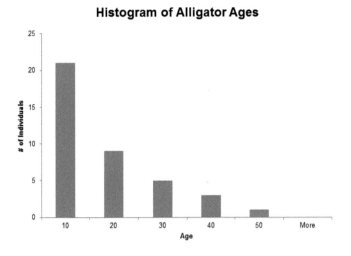

How to Make a Histogram in SPSS

The following tutorial will walk you through the construction of a histogram in SPSS.

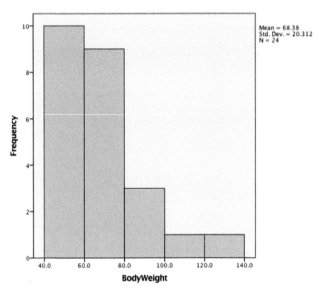

*This example was taken from the research conducted by Dr. Sarah L. Dunn.

Refer to Chapter 13 for getting started and understanding SPSS.

1. Arrange the data so that it appears similar to the following.

	BodyWeight
1	56.6
2	65.6
3	50.0
4	134.6
5	58.0
6	54.4
7	109.6
8	82.8
9	67.4
10	59.6
11	44.2
12	63.4
13	61.2
14	59.0
15	75.6
16	80.4
17	69.6
18	52.8
19	53.8
20	63.4
21	90.0
22	60.6
23	77.4
24	51.2

2. Click on the **Graphs** tab in the toolbar located along the top of the page. Scroll down to the **Legacy Dialogs** then select **Histogram**.

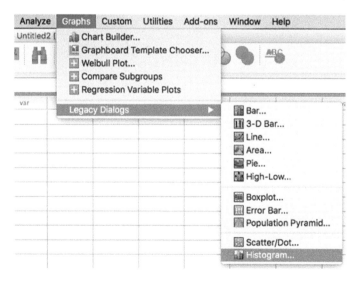

3. The following window will appear.

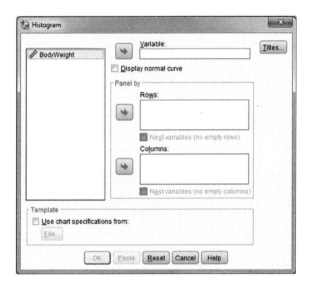

4. Click on the variable you want graphed (**BodyWeight**), then click the corresponding arrow to move the selection over to the **Variable** section.

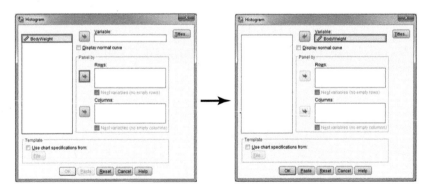

5. If you want the normal distribution curve superimposed over the data, then check the box labeled as **Display normal curve**. Otherwise, select **OK**.

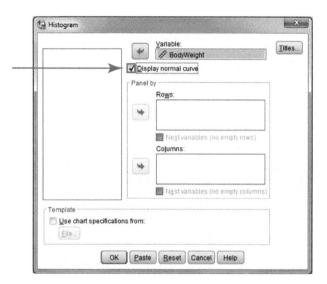

6. The graph will appear as a separate output.

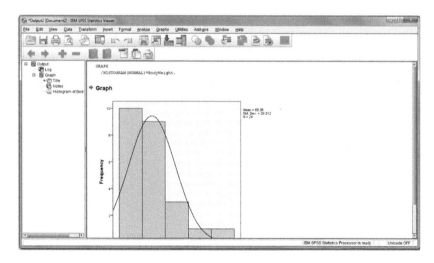

7. If you selected to display the normal curve, then your graph will look similar to the following graph.

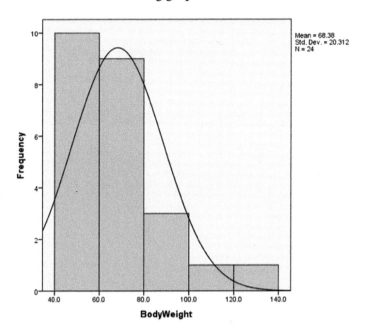

However, if you did not make that selection, then your graph will appear similar to the one below.

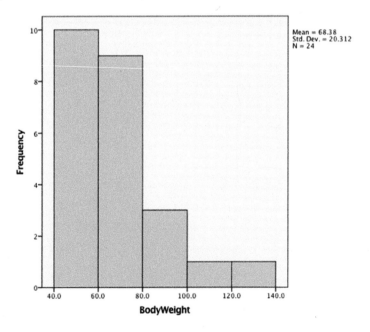

How to Make a Histogram in Numbers

The following tutorial will walk you through the construction of a histogram in Numbers.

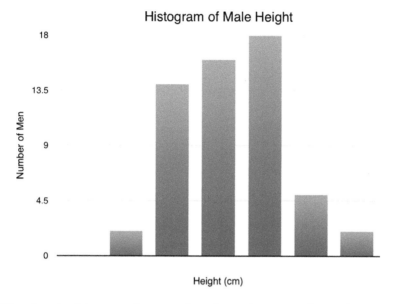

*The data in this example was taken from the research conducted by Dr. Sarah L. Dunn.

More information on programming in Numbers is found in Chapter 14.

1. Enter the following data into Numbers as a single column (listed within five columns for readability).

Males (cm)				
188	186	188	180	190
186	180	180	173	178
186	181	173	180	183
199	178	182	177	183
182	182	178	187	188
179	178	192	187	186
181	192	188	181	186
187	188	181	190	199
195	181	178	193	182
183	178	188	185	179
181	195	186	181	
187	183	180	178	

2. Sort the data. Select **Sort Entire Table** and **Add a Column**. Click on the drop down box under Sort by and select Males (cm) and ascending (1, 2, 3, …).

3. Next to your **Male** height column, add three columns for **Height Ranges**, **Interval #**, and **Histogram Values**. Based on the data, we can divide the data into the following seven intervals: <170, 171–175, 176–180, 181–185, 186–190, 191–195, and 196–200. Label the intervals 1–7.

Males (cm)	Height Ranges	Interval #	Histogram Values
173			
173			
177	170	1	
178	175	2	
178	180	3	
178	185	4	
178	190	5	
178	195	6	
178	200	7	
179			
179			
180			
180			
180			
180			
180			
181			
181			
181			
181			
181			

4. To determine how many males fall within each of the intervals of our histogram, use the following equation.

Males (cm)				
173				
173	Height Ranges	Interval #	Histogram Values	
177	170	1	• fx ∨ INDEX ▾ FREQUENCY ▾ data-values , interval-values ✕ ✓
178	175	2		
178	180	3		
178	185	4		
178	190	5		
178	195	6		
178	200	7		
179				
179				

5. For data values, select your source data. Then click the down arrow on the right side of the bubble to drop down the menu and click on all the boxes to preserve row and column within start and end (Preserve keeps these specific cells selected in the equation when you do a drag in step 8).

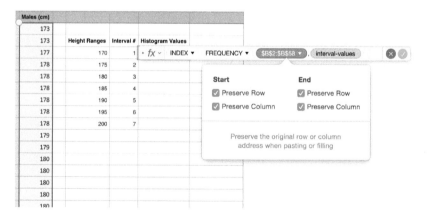

6. For the interval values, select the height ranges. Then click the down arrow on the right side of the bubble to drop down the menu and click on all the boxes to preserve row and column within start and end.

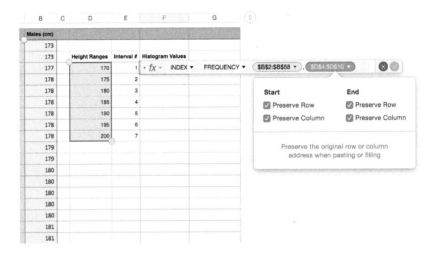

7. Add the interval to the index portion of the equation (notice that we did not preserve the row and column in this step).

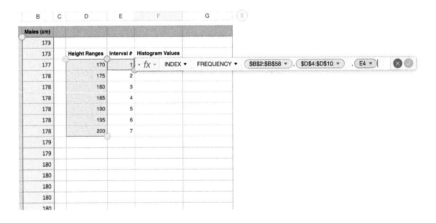

8. Drag the yellow circle down to fill in the rest of the histogram values.

Males (cm)		Height Ranges	Interval #	Histogram Values
173				
173		170	1	0
177		175	2	
178		180	3	
178		185	4	
178		190	5	
178		195	6	
178		200	7	
178				
179				
179				
180				

9. Here is a summary of all the counts for the histogram.

Height Ranges	Interval #	Histogram Values
170	1	0
175	2	2
180	3	14
185	4	16
190	5	18
195	6	5
200	7	2

10. Now let us build the actual histogram. Select the histogram values.

Height Ranges	Interval #	Histogram Values
170	1	0
175	2	2
180	3	14
185	4	16
190	5	18
195	6	5
200	7	2

11. From within **Charts**, select the **Bar Chart** button.

12. Within the format area, click **Chart** and add a title to the graph by clicking the **Title** box.

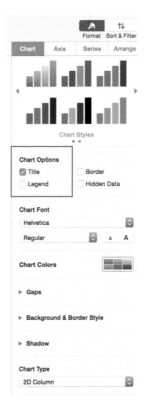

13. Within the format area, click **Axes** and select **Axis Name** and **Axis Line** for both the Y and X axes (which are in separate tabs just below **Axes**). Click on **Edit Data References** at the bottom of the chart. To customize the axis labels and title, double click on the word that is currently in place and change the text.

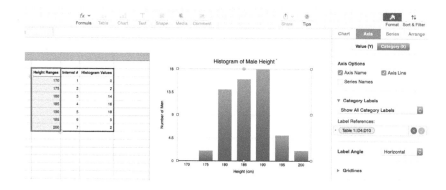

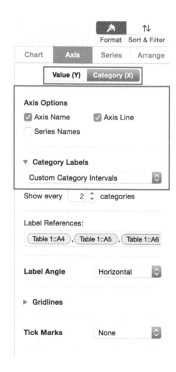

14. Here is the final graph.

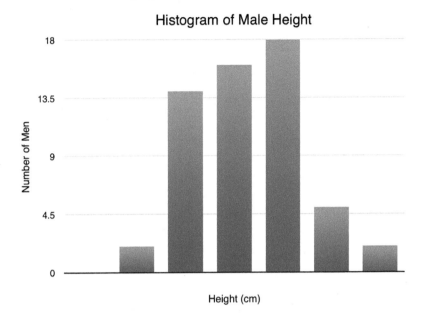

How to Make a Histogram in R

The following tutorial will walk you through the construction of a histogram in R.

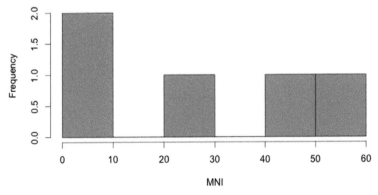

Refer to Chapter 15 for R specific terminology and instructions on how to invoke and construct code.

In a closely related area to biological anthropology, zooarchaeology, animal bones and their representation at ancient sites are studied to understand past lifeways. At several European archaeological sites, the minimum number of individuals (MNI) represented by bones were calculated from excavation of fishes.

1. Create a vector to store the **Minimum Number of Individuals (MNI)** of fishes found at various European archaeological sites. Press enter/return.

   > mni<-c(6, 30, 58, 2, 44)

2. Construct your script using the `hist()` function. List the vector (**MNI**) and then use the arguments `xlab=` and `main=`, which instruct R what titles to place on the *x*-axis and the whole chart. The arguments use strings, so the text should be set off with quotation marks. Press enter/return.

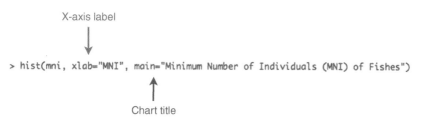

The output should be a graph such as the one below. This is a histogram.

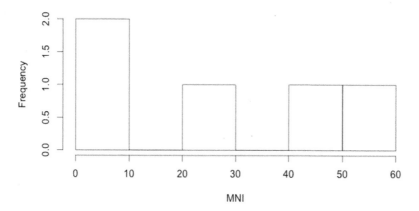

3. You can add color to any chart in R using the `col=` function. To examine the variety of colors R has available, type the following code into R to display the color names. Press enter/return.

> colors(distinct=FALSE)

4. After selecting a color for the bars, add them to the line of programming from step 2. Press enter/return.

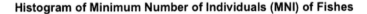
> hist(mni, xlab="MNI", main="Histogram of Minimum Number of Individuals (MNI) of Fishes ", col="violet")

The output consists of a color chart such as the one below.

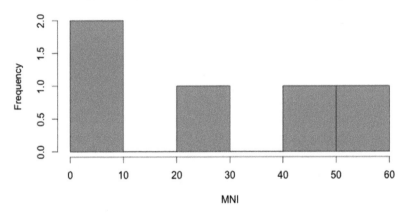

Histogram of Minimum Number of Individuals (MNI) of Fishes

Note: R plots are highly customizable. The help facility and Internet search engines are fantastic resources for learning additional customizations that can be applied.

3.4 Line Graphs and Scatter Plots

Background

Much like bar charts, line graphs and scatter plots allow us to visualize trends in the data and the relationships between the independent variable (established by the researcher) and the dependent variable (also known as the treatment or response variable). However, unlike bar charts, which compare categories of data, line graphs and scatter plots are used when both the independent and dependent variables are continuous.

Line Graphs

In line graphs and scatter plots, the independent variable is measured across a range of continuous values and is depicted on the *x*-axis. Examples of continuous independent variables include temperature, elevation, time, concentration, etc.; in each of these cases, the values are not constrained by categories, but can be measured on a continuous scale. The dependent variable is also measured on a continuous scale and is plotted on the *y*-axis (Figure 3.9).

As the name implies, a line graph utilizes a line to connect the data points and is most appropriate when the dependent variables are in some way connected, either through a time series or other type of progression. Observations are plotted using their Cartesian

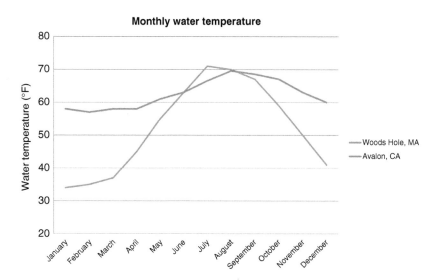

Figure 3.9 Line graph comparing the monthly water temperatures (°F) for Woods Hole, MA and Avalon, CA.

(x and y) coordinates, and then the points are connected by a single line through the data, which forms a kink at each data point.

In the example in Figure 3.9, a marine biologist was interested in seasonal patterns of water temperatures in Avalon, CA, compared with Woods Hole, MA. Lines connect the observations for the two sites, allowing for a visual comparison of the trends. It is apparent from the line graph shown above that temperatures fluctuate more dramatically with season at the Woods Hole site than they do at the Avalon site.

Scatter Plots

A scatter plot, also called a scatter chart or scatter diagram, is similar in concept to the line graph but is more useful when looking at overall trends in a scattering of unrelated data points across a continuous scale. In the case of a scatter plot, each data point is plotted using its Cartesian coordinates and is represented as a single dot on the graph. By using a best fit trend line through the data points, the researcher is able to illustrate both the spread of the data around the line, as well as a model for the relationship between the independent and dependent variables. Straight trend lines through the data take the form of the linear equation, **y=mx+b**, where the value "y" is equal to the slope "m" multiplied by the value for "x," plus the value of the y intercept "b." Trend lines may not always be linear, but that topic is outside the scope of this text. By creating a trend line, we are in essence creating a model for the relationship between the two variables, which allows for predictions to be made from the data. More details on the relationship between two variables will come in Chapters 10 and 11.

The following case study uses a scatter plot to illustrate the relationship between two continuous variables. A study by Yund et al. (2015) examined the distribution of blue mussels across the sea shelf in the Gulf of Maine. Two species of mussel, *Mytilus edulis* and *Mytilus trossulus* comprise the mussel community, and their ranges overlap considerably across the sea shelf. As part of their study, the

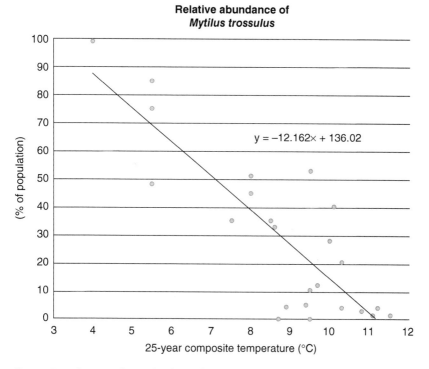

Figure 3.10 Scatter plot with a line of best fit showing the relationship between temperature (°C) and the relative abundance of *Mytilus trossulus* to *Mytilus edulis* (from 0 to 100%).

researchers evaluated the relationships between temperature and the relative abundance of each species.

In Figure 3.10, the relative abundance of *M. trossulus* (from 0 to 100%) was plotted against water temperature data from a 25-year period. A trend line has been added to visualize the nature of the relationship, and the equation for the line has also been added to the plot. The negative slope value is indicative of a negative relationship between water temperature and the relative abundance of *M. trossulus*. In other words, *M. trossulus* was less abundant compared with *M. edulis* as the water got warmer along the sea shelf.

Tutorials

How to Make a Line Graph and Scatter Plot in Excel

The following tutorial will walk you through the construction of a line chart, also known as a line graph (or scatter plot, if without the line), using Excel.

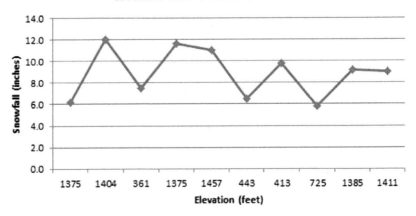

*The following data were gathered from the National Weather Service.

Refer to Chapter 12 for tips and tools when using Excel.

1. Arrange data into columns in the spreadsheet.

	A	B
1	Elevation (feet)	Snowfall (inches)
2	1375	6.2
3	1404	12.0
4	361	7.5
5	1375	11.6
6	1457	11.0
7	443	6.5
8	413	9.8
9	725	5.8
10	1385	9.2
11	1411	9.0

Note: You will need to click on an empty cell before continuing to the next step.

2. Select **Insert** Tab, move to the **Charts** heading, select the line graph, and choose either the line graph option without data points or the option with data points.

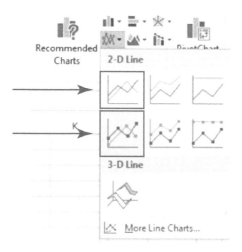

Note: If you want to build a scatter plot diagram, the steps are the same, except you would select the scatter plot option.

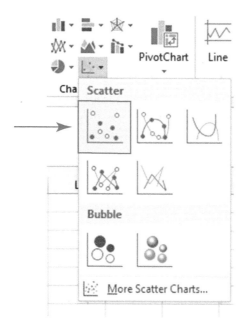

3. A blank canvas will appear. Right click on the blank canvas and choose the **Select Data** option.

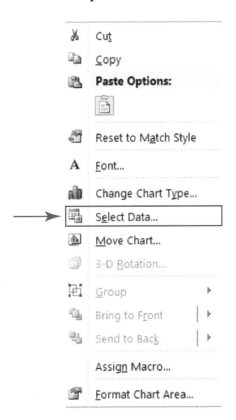

4. Under **Legend Entries**, select **Add**.

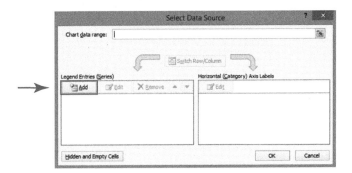

5. Click on the icon correlating to the subheading **Series Name**.

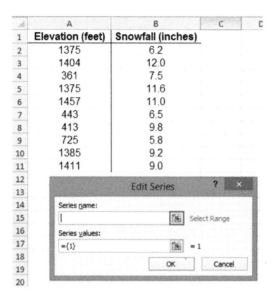

6. Select the series title (**Snowfall**), then click on the icon to the right.

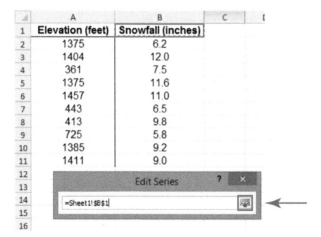

7. Set the **Series Y Values** by clicking on the corresponding icon like you did for selecting the **Series name**.

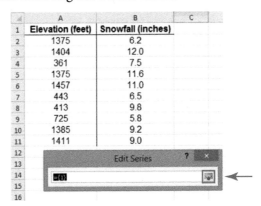

8. Select the y value range and click the icon to the right.

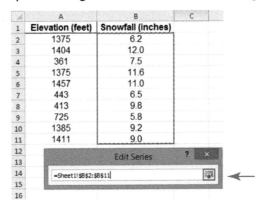

9. Click **OK**.

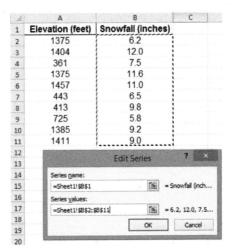

10. The original popup will appear. Click **Edit** under the **Horizontal (Category) Axis Label** subheading.

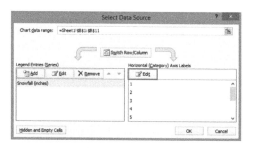

11. Click on the icon to the right to set the label range for the horizontal axis as in steps 7 and 8.

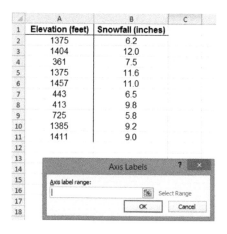

12. Select the range and click on the icon to the right when finished.

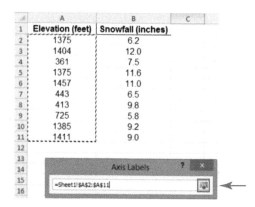

13. Click **OK**.

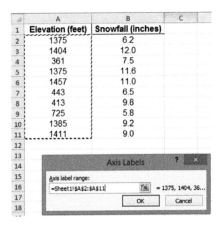

14. Click **OK**.

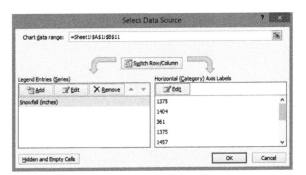

15. A very basic graph will appear, similar to the one below.

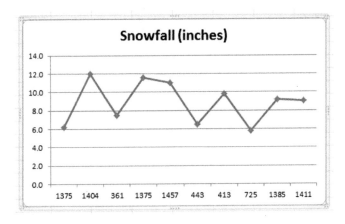

16. To add proper labels and change any other aesthetic detail, go to **Chart Tools** in the toolbar and select **Design**. If the **Chart Tools** option does not appear in the toolbar, then click on the graph. The graph must be selected in order for this option to appear.

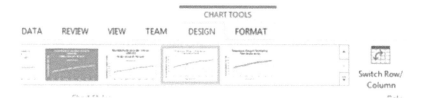

Note: Some chart options under the **Quick Layout** menu will automatically add Axes Titles, Chart Title, and Legend. To add each item individually, follow steps 17–23.

17. First input a title by selecting **Add Chart Element**, then **Chart Title**. Normally titles are placed above the chart; however, other options are available if desired.

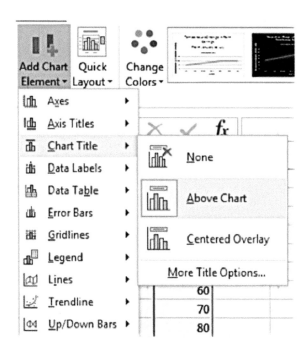

18. Input an appropriate title similar to the one below.

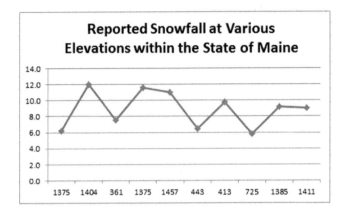

19. Next label the axis by selecting **Axis Titles**. Select **Primary Horizontal**.

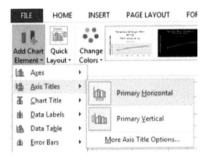

20. Label the *x*-axis with an appropriate title. The graph should look similar to the following.

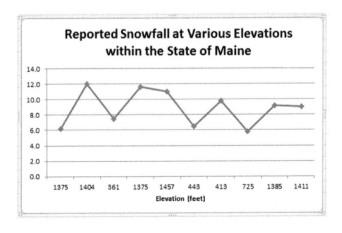

21. Next, label the *y*-axis by selecting the **Primary Vertical** option. Label the *y*-axis with an appropriate title.

22. The finished labeled graph should look similar to the one below.

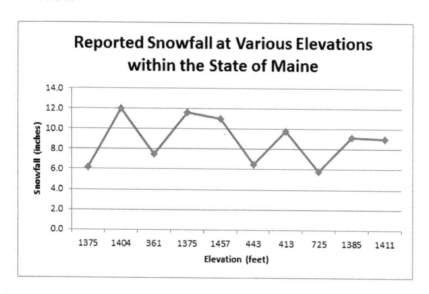

23. For any other aesthetic changes that you desire to make to the graph, such as creating a legend, inputting labels above each data point, or inputting gridlines, options are available under the **Add Chart Element** menu.

24. To change the color of the graph, select the **Design** tab and select **Change Colors**. Scroll over the colors with the mouse to view the possible themes. When you find the color you desire, click the option and it will permanently change the graph.

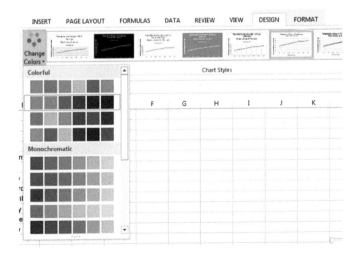

25. To give the graph a border or change the text type, select the **Format** tab.

How to Make a Line Graph and Scatter Plot in SPSS

The following tutorial will walk you through the construction of a line chart, also known as a line graph (or scatter plot, if without the line), and a scatter plot using SPSS.

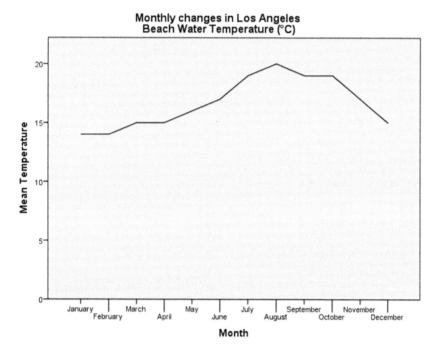

Refer to Chapter 13 for getting started and understanding SPSS.

1. Arrange the data in the columns on the spreadsheet.

	Month	Temperature
1	1	14
2	2	14
3	3	15
4	4	15
5	5	16
6	6	17
7	7	19
8	8	20
9	9	19
10	10	19
11	11	17
12	12	15

2. Along the toolbar at the top, click on **Graphs**. Then, select **Chart Builder**.

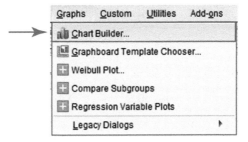

3. The following dialog box will appear. Click **OK**.

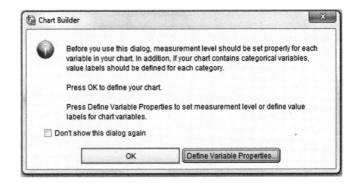

Showing Your Data | 153

4. A Chart Builder window will appear. Under the wording **Choose From**, click on **Line**.

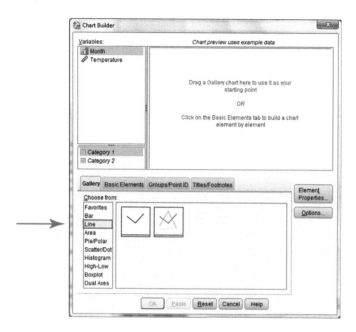

Note: To create a scatter plot, select **Scatter/Dot** from the "Choose From" menu and follow steps 5–11.

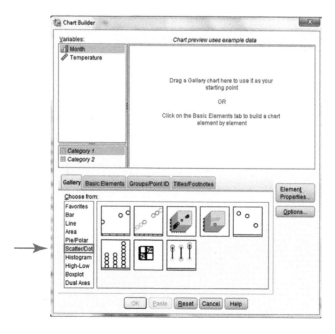

5. Click on Simple Line and drag it to the Chart Preview window.

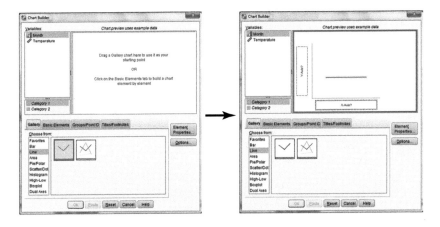

6. Click on **Month** (independent variable) and drag it to the box labeled **X-Axis?** in the chart preview window.

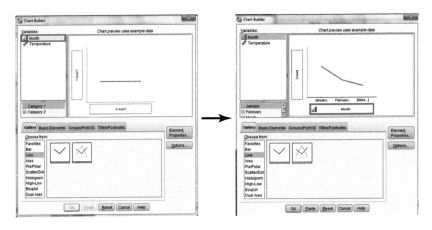

7. Click on **Temperature** (dependent variable) and drag it to the box labeled **Count** in the chart preview window.

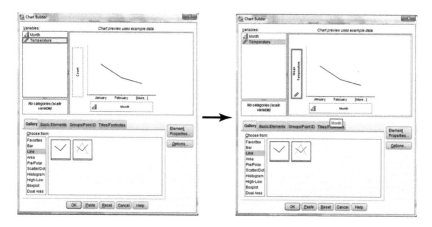

Note: The preview window does not use your data in the preview image. Your graph may look different to the graph in the preview.

8. To give the graph a title, click on **Titles/Footnotes** in the **Chart Builder** window. Then, click **Title 1** followed by **Element Properties**.

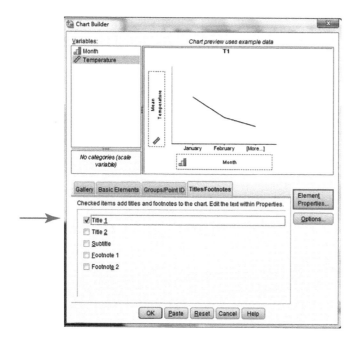

9. In the Element Properties window, type in a relevant title for the graph in the **Content** window. Click **Apply**.

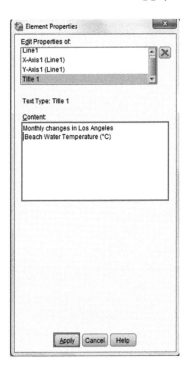

10. Click **OK** at the bottom of the Chart Builder window.

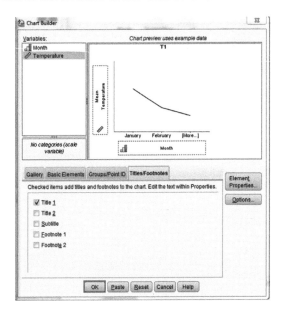

11. Your new graph will appear in a separate output window.

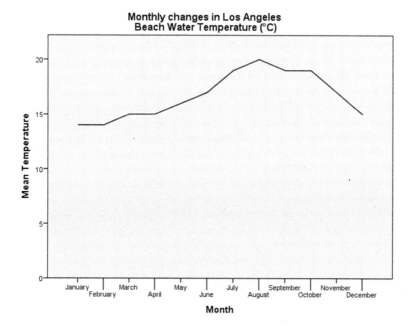

Note: After selecting Scatter/Dot, your output will appear similar to the following image.

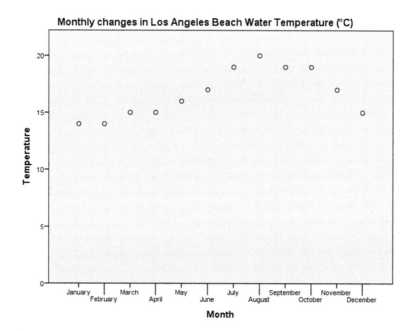

How to Make a Line Graph and Scatter Plot in Numbers

The following tutorial will walk you through the construction of a line chart, also known as a line graph and scatter plot, using Numbers.

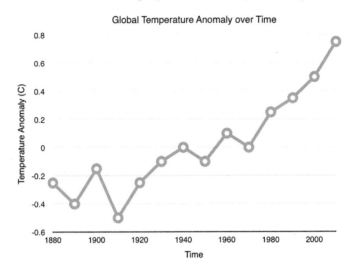

*Data adapted from NASA's Goddard Institute for Space Studies (GISS) from http://climate.nasa.gov/

More information on programming in Numbers is found in Chapter 14.

1. Arrange your data like the following.

Global Temperature

Year	Temperature Anomaly (C)
1880	-0.25
1890	-0.4
1900	-0.15
1910	-0.5
1920	-0.25
1930	-0.1
1940	0
1950	-0.1
1960	0.1
1970	0
1980	0.25
1990	0.35
2000	0.5
2010	0.75

2. Highlight the *y*-variables (**Temperature Anomaly**) in the table.

Year	Temperature Anomaly (C)
1880	-0.25
1890	-0.4
1900	-0.15
1910	-0.5
1920	-0.25
1930	-0.1
1940	0
1950	-0.1
1960	0.1
1970	0
1980	0.25
1990	0.35
2000	0.5
2010	0.75

3. Click on **Chart** in the top bar and select line graph within the drop down menu.

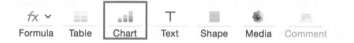

160 | *An Introduction to Statistical Analysis in Research*

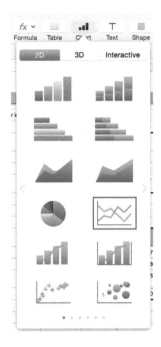

Note: If you wanted to run a scatter plot diagram, the steps are the same, except you would select the scatter plot option.

4. The following graph will appear.

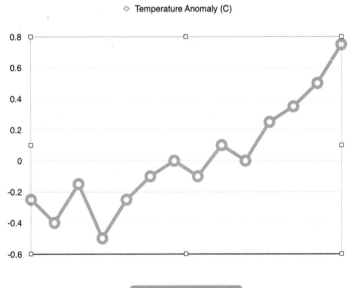

5. Within the **Format** bar on the right, click **Chart** and check the box to add a **Title**. You can enter the title directly on the graph.

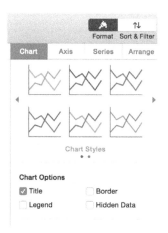

6. Click on **Axis** and select **Value (Y)** and then check the box for **Axis Name**. Enter the title **Temperature Anomaly (C)** in the text box on the *y*-axis.

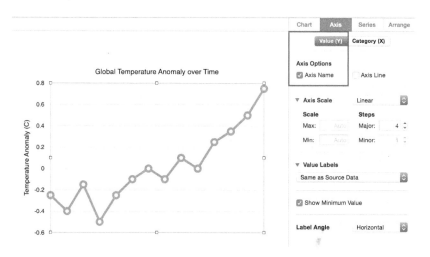

7. Then, select **Category (X)** and check the box for **Axis Name**. Enter the title **Time** by double clicking on the *x*-axis.

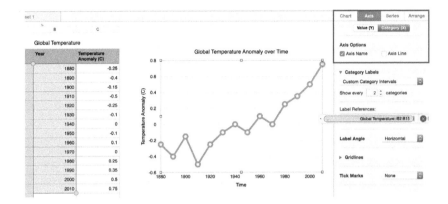

8. To label the *x*-axis with the years, click on **Label References**, delete the contents of the cell and then highlight the years within the data and click the green check or press enter/return. To make the axis look streamlined, under category labels, select **Custom Category Intervals** and then adjust **Show every** to 2 intervals.

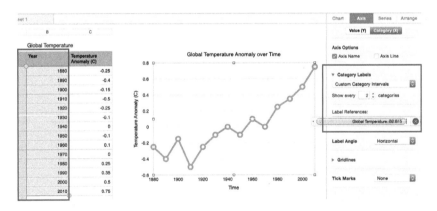

9. The final graph will look like the following.

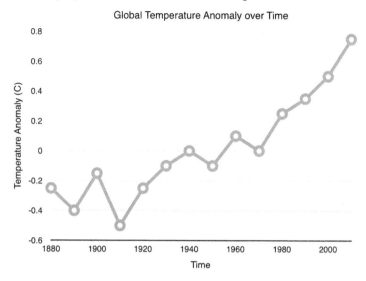

How To Make a Line Graph and Scatter Plot in R

The following tutorial will walk you through the construction of a proper line chart, also known as a line graph and scatter plot, using R.

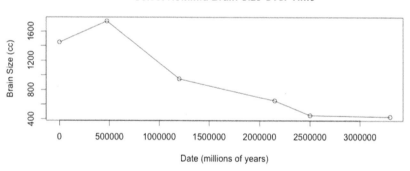

Refer to Chapter 15 for R specific terminology and instructions on how to invoke and construct code.

In paleoanthropology, anthropologists study fossil hominids/hominins, or our human ancestors. An anthropologist was interested in creating a line graph depicting brain size in several key hominids that lived at different points in history. She used the following brain sizes in cubic centimeters (cc) and point estimates for the dates they lived to construct a line graph.

1. Create two vectors, one to store **brain sizes** and the other to store **dates**. The order of the brain sizes and dates must match within the vectors so that a specific date corresponding to a particular brain size is found in the same positions in the two vectors. This ensures that the correct brain size is matched to the correct date. Press enter/return.

   ```
   > brainsize<-c(430, 450, 650, 950, 1740, 1450)
   > dates<-c(3300000, 2500000, 2150000, 1200000, 475000, 0)
   ```

2. Construct your script using the `plot()` function. The vectors are listed in the function in the order of the data to appear on the *x*-axis followed by the *y*-axis. We want dates to be on the *x*-axis. You will use the arguments **ylab=**, **xlab=**, and **main=**, which instruct R what titles to place on the *y*-axis, the *x*-axis, and whole chart. The arguments use strings, so the text should be set off with quotation marks. Press enter/return.

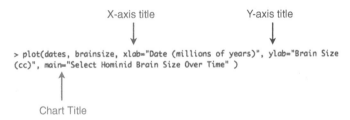

The output should be a graph such as the one below. This is a scatter plot.

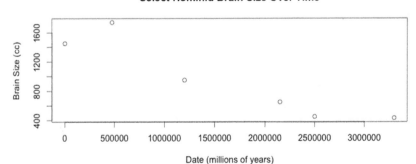

Note: Steps 1 and 2 are all that is needed for a scatter plot. For a line chart, complete the entire tutorial.

If either the *x*- or *y*-axis has labels that look awkward, you can resize the graph by pulling on the bottom right hand corner of the window in which it pops up. This will adjust the graph size and axis labels.

3. To add a line connecting the points and transform the graph into a line chart, use the `lines()` function with the vectors in the order of the *x*-axis and then *y*-axis.

> lines(dates, brains)

The output should be a graph such as the one below.

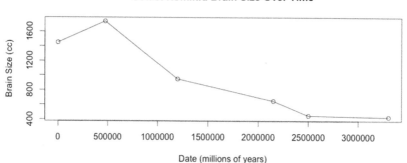

Note: R plots are highly customizable. The help facility and Internet search engines are fantastic resources for learning additional customizations that can be applied.

3.5 Pie Charts

Background

Another way of displaying data is through the use of pie charts. Pie charts allow the reader to visualize the differences between groups as slices (or sectors) in a pie. Rather than graphing the relationships between two groups of data, as we have seen in bar charts and line graphs, pie charts take a unique role in showing the different categories as pieces in relation to the total. In the case of pie charts, the values for each group are converted to percentages of the total, and the size of the pie piece is illustrated in relation to its relative contribution to the whole.

There are several advantages in using pie charts. Pie charts are easy to read and interpret, and allow for large datasets to be summarized and compared in a simple visual format. For these reasons, they are commonly used in business settings and in popular media. However, if oversimplified, pie charts also have the potential to be confusing, and at worst misleading. Without proper labeling, or when shown at odd angles where relative size of the slices is distorted, pie charts may be difficult to interpret. Another common mistake by the novice pie chart maker is in converting the raw data into percentages. If the pieces of the pie do not total to 100%, then the pie chart is meaningless.

The following case study shows the appropriate use of pie charts. Researchers analyzed the fatty acid compositions of olive oil and canola oil. In this case, each oil was shown to contain oleic acid, alpha-linolenic acid, linoleic acid, and saturated fat. The two pie charts show the relative amounts of fatty acids in each oil. A quick glance at the pie charts gives a clear picture of the differences in the fatty acid composition of the oils. For example, it is easy to see that olive oil contains more oleic acid and saturated fat, but less linoleic acid and alpha-linolenic acid than canola oil.

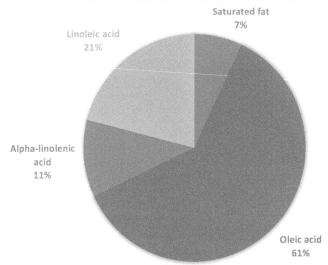

Figure 3.11 Pie chart comparing the fatty acid content (saturated fat, linoleic acid, alpha-linolenic acid, and oleic acid) in standard canola oil.

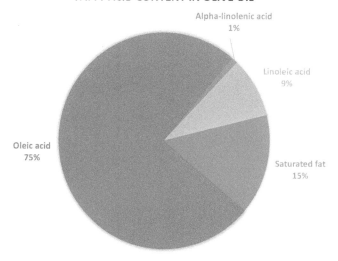

Figure 3.12 Pie chart comparing the fatty acid content (saturated fat, linoleic acid, alpha-linolenic acid, and oleic acid) in standard olive oil.

Tutorials

How to Make a Pie Chart in Excel

The following tutorial will walk you through the construction of a pie chart using Excel.

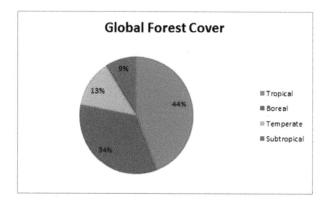

*The following data were gathered from FAO 2011.

Refer to Chapter 12 for tips and tools when using Excel.

1. Arrange the data so that they appear similar to the following.

	A	B
1	Forest Cover	Global Distribution (%)
2	Tropical	44
3	Boreal	34
4	Temperate	13
5	Subtropical	9

2. Make sure the data are not highlighted. If necessary, click on an empty cell.

3. Along the toolbar, select **Insert** and select the pie chart option. Select the first option for a simple pie chart.

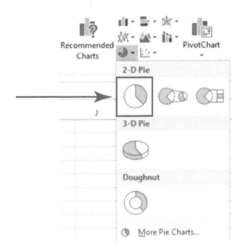

4. A blank canvas will appear.

Showing Your Data | 169

5. Right click on the canvas and click on the **Select Data** option.

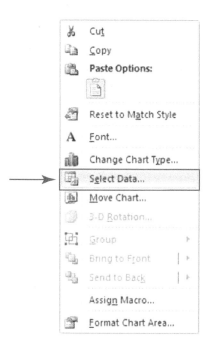

6. The following window will appear. Select the icon to the right of **Chart data range:**.

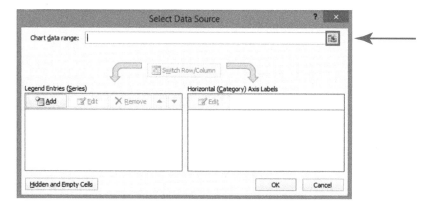

7. The following window will appear. Highlight the labels and the respective values to be included in the pie chart. Then, select the icon to the right of the **Select Data Source** window.

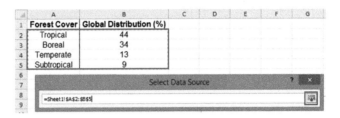

8. Click **OK**.

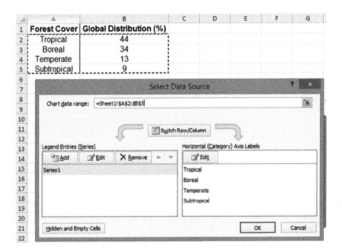

9. A graph similar to the following should appear.

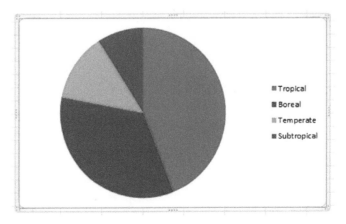

Showing Your Data | 171

10. To format the graph and input a title, make sure that the graph is selected by simply clicking on it.
11. Under **Chart Tools**, select **Design**. Under **Quick Layout**, select the type of layout desired for the pie chart. In this case layout 6, which includes a title, legend, and percentage values, will be used.

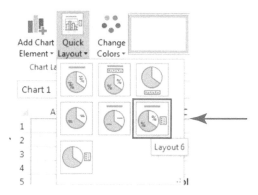

12. The resulting chart will look similar to the following.

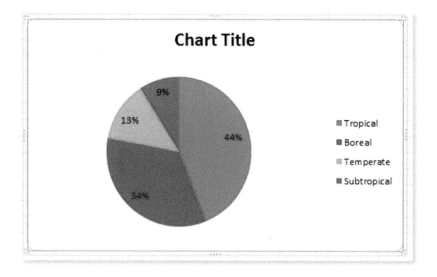

13. To change the title, click on the title box on the chart until it is highlighted.

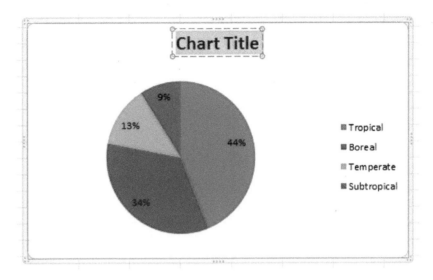

Note: Not all layouts will input a title box. If the title box is not present, go to step 15.

14. Type in an appropriate title for the pie chart that corresponds to the data being represented.

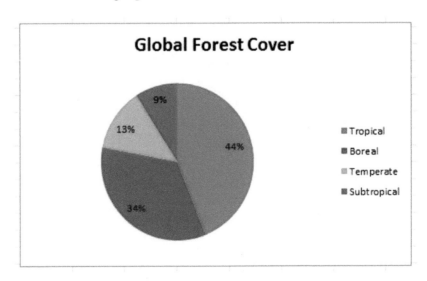

Showing Your Data | 173

15. If the title box is not present on the graph, on the toolbar under **Chart Tools** and **Design**, select **Add Chart Element**, **Chart Title**, then **Above Chart**. The title box will appear and you can now type in the appropriate title.

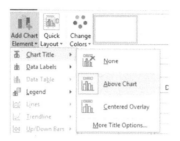

16. To insert any labels or legends, refer to the **Add Chart Element** selection in the **Design Tab**.

17. To change the color of the graph, select the **Design** tab and select **Change Colors**. Scroll over the colors with the mouse to view the changes on the graph. When you find the color you desire, click the option and it will permanently change the graph.

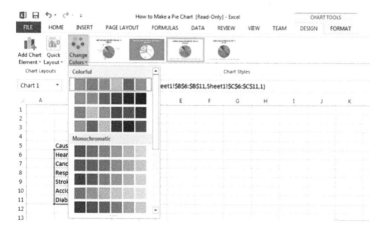

How to Make a Pie Chart in SPSS

The following tutorial will walk you through the construction of a pie chart using SPSS.

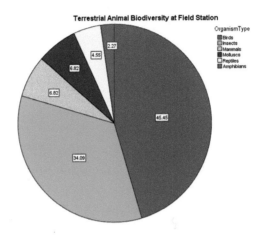

*The data in this example were taken from the research conducted by Drs. Pablo Weaver and Kathleen Weaver.

Refer to Chapter 13 for getting started and understanding SPSS.

1. Arrange the data so that it appears as the following.

	OrganismType	NumberOfOrganisms
1	1	20.00
2	2	15.00
3	3	3.00
4	4	3.00
5	5	2.00
6	6	1.00

2. Along the toolbar at the top, select **Graphs** followed by **Chart Builder**.

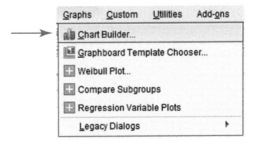

Showing Your Data | 175

3. The following dialog box will appear. Click **OK**.

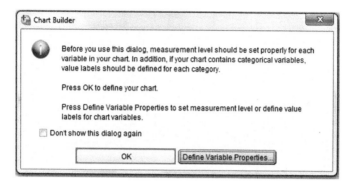

4. The following window will appear. Under the **Gallery** tab, locate **Pie/Polar**.

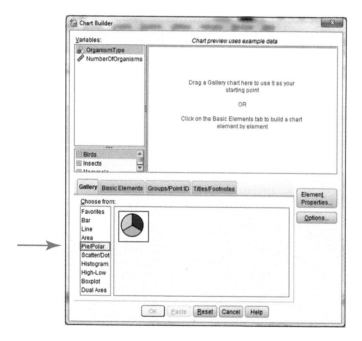

5. Select the **Pie Chart** icon and drag it to the **Chart Preview** window.

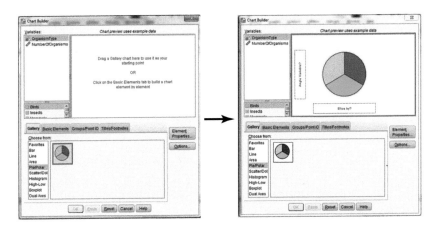

6. Click on the independent variable (**OrganismType**) and drag it to the box labeled **Slice by?** in the chart preview window.

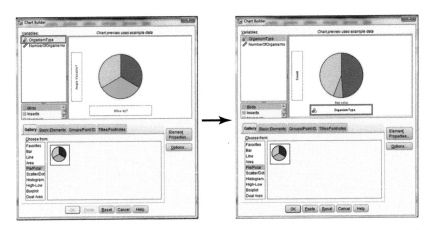

7. Select the dependent variable (**NumberofOrganisms**) and drag it to the box labeled **Count** in the chart preview window.

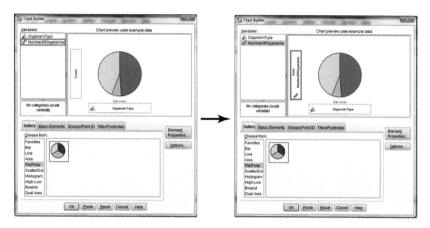

Note: The preview window does not use your data in the preview image. Your graph may look different to the graph in the preview.

8. To give the graph a title, click on **Titles/Footnotes** in the Chart Builder window. Select **Title 1** followed by **Element Properties**.

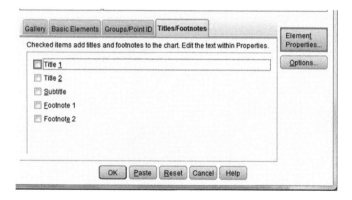

9. In the Element Properties window, type an appropriate title for the graph in the **Content** box. Then, click **Apply** at the bottom of the window.

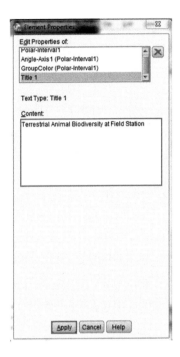

10. Click **OK** at the bottom of the Chart Builder window.

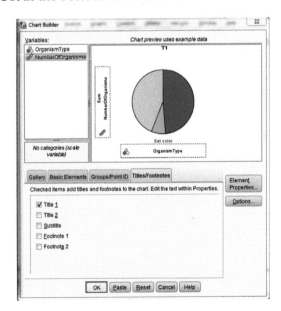

11. The graph will appear as a separate output.

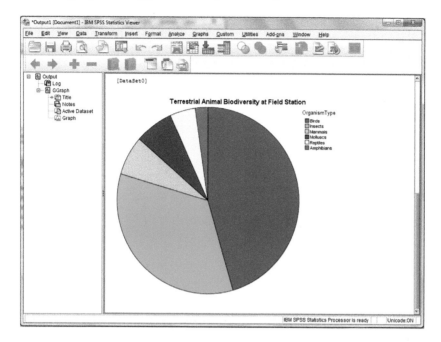

12. To add data labels, double click the chart in the output. The following window will appear.

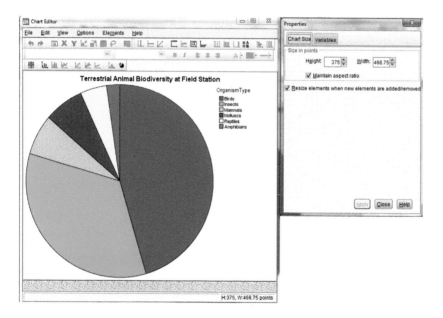

13. Select the **Show Data Labels** icon.

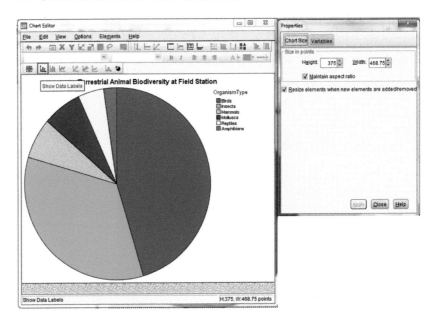

14. SPSS will automatically format the data labels to show the percentage of each slice.

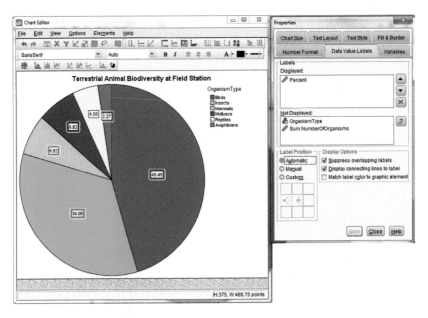

15. When you are done editing the chart, close the **Chart Editor** window. The graph should look similar to the following.

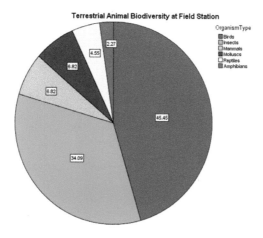

How to Make a Pie Chart in Numbers

The following tutorial will walk you through the construction of a pie chart using Numbers.

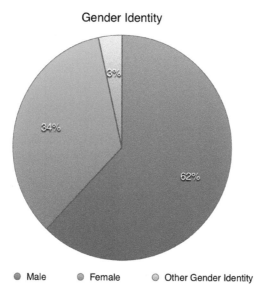

More information on programming in Numbers is found in Chapter 14.

1. Arrange the data so that it appears as the following.

Pie Chart for Gender

C	D
Gender	
1	1= male
2	2= female
2	3= Other gender identity
1	
3	
2	
1	
1	
1	
1	
1	
1	
1	
2	
1	
2	
2	
2	
1	
2	
1	
1	
1	
1	
2	
1	
1	
1	
2	

2. Start by getting a count of how many of your data points fit within each category (see How to make a Histogram in Numbers).

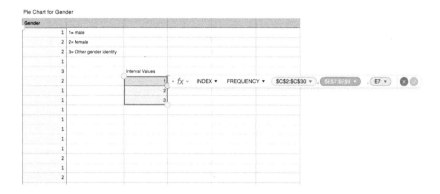

3. The histogram count data will appear like the following.

Interval Values	Counts		
1		18	Male
2		10	Female
3		1	Other Gender Identity

4. You can build a pie chart directly from the count data or you can also calculate the percentage of the total for each category.

Gender	Counts (Histogram)	Percentage
Male	18	62.07
Female	10	34.48
Other Gender Identity	1	3.45
Sum	29	100.00

5. From here, highlight the column you want to build a pie chart from. Click on the chart button in the top toolbar and select pie chart from the drop down menu.

Gender	Counts (Histogram)	Percentage
Male	18	62.07
Female	10	34.48
Other Gender Identity	1	3.45
Sum	29	100.00

6. The following pie chart will pop-up.

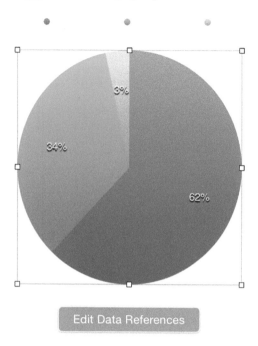

7. If you have the names in the gray area on the left, the names will populate automatically. If not, you will have to add the names to the legend. Click on the pie chart and select one pie wedge. In the bar on the right side, you will see the following menu.

8. Click on **Name**, delete the contents and then click on the cell within the table that has the appropriate name. Repeat this for each slice.

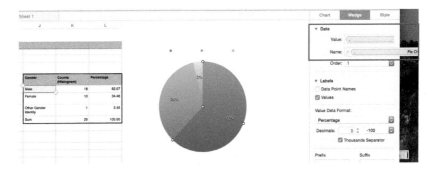

9. After you change the names, the pie chart will look like the following.

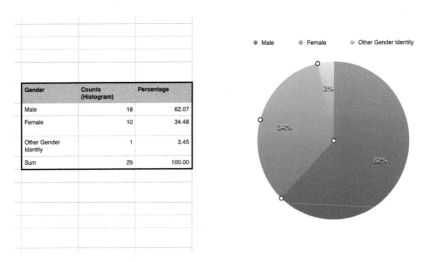

10. Finally, you can adjust the colors and border within the pie chart as desired. With a single slice selected, you can adjust the fill or line within the side format bar. In this case, a 0.75 weight dark grey line is added to each wedge.

Showing Your Data | 187

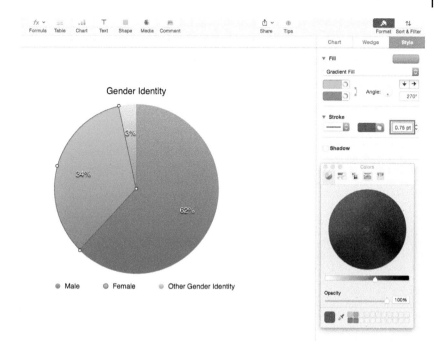

11. The final pie chart will look like this.

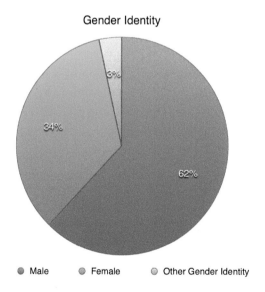

How to Make a Pie Chart in R

The following tutorial will walk you through the construction of a proper pie chart using R.

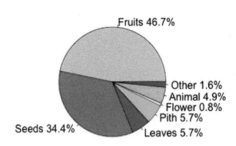

Mandrill Food Consumption

*Data for this example are from Tutin et al. (1999).

Refer to Chapter 15 for R specific terminology and instructions on how to invoke and construct code.

Primate diets can incorporate a variety of foods. A study by Tutin et al. (1999) found the distribution of foods in mandrills (old world monkey) to be composed of fruits, seeds, leaves, pith, flowers, animals, and other items.

1. Create two vectors (one of which contains strings) for the type of food and percentage consumed. The values in **consumed** must be in the same order as the corresponding food type in the **foods** vector to make sure that the percentages align correctly with the type of food. Press enter/return.

```
> consumed<-c(46.7, 34.4, 5.7, 5.7, 0.8, 4.9, 1.6)
> foods<-c("Fruits", "Seeds", "Leaves", "Pith", "Flower", "Animal", "Other")
```

2. Before we plot the actual pie chart, we need to tell R that our numbers are in percentage form and how to express them. The following two lines of code do this. First, we merge **foods** and **consumed** so that they appear next to each other in a single vector using the `paste()` function. Press enter/return.

```
> foods<-paste(foods, consumed)
```

If you want to verify this is the case, you can simply type in **foods** and press enter/return. The following output will appear, showing the merging of the two vectors into one.

```
> foods
[1] "Fruits 47" "Seeds 34" "Leaves 6" "Pith 6"   "Flower 1" "Animal 5" "Other 2"
```

3. Add a "%" after the number and use the `sep=` argument to ensure that there is no space between the number and the symbol. Again, we use the `paste()` function. Press enter/return.

```
> foods<-paste(foods, "%", sep="")
```

4. To verify the vector now contains the "%" symbol, type in **foods** and press enter/return. The following output will appear.

```
> foods
[1] "Fruits 47%" "Seeds 34%"  "Leaves 6%"  "Pith 6%"    "Flower 1%" "Animal 5%"
[7] "Other 2%"
```

5. Finally, we are prepared to plot **consumed** using the `pie()` function. The **consumed** vector that contains the percentages is placed in the function first, and the `labels=` argument is set equal to **foods** for labeling the pie slices. The argument `main=` produces the chart title, while the argument `col=` programs the colors of the pie slices, starting on the top and going counter-clockwise.

```
> pie(consumed, labels=foods, main="Mandrill Food Consumption", col=c("pink", "red", "blue",
"green", "yellow", "orange", "purple"))
```

The output should be a color graph such as the one below.

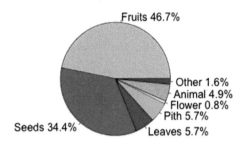

Mandrill Food Consumption

Note: R plots are highly customizable. The help facility and Internet search engines are fantastic resources for learning additional customizations that can be applied.

4

Parametric versus Nonparametric Tests

> **Learning Outcomes**
>
> By the end of this chapter, you should be able to:
>
> 1. Define parametric and nonparametric statistics.
> 2. Determine whether a parametric or nonparametric analysis is most appropriate, based on shape distribution.

In Chapter 2, we covered distribution and examined the various shapes of a sampling distribution (normal, skewed, kurtic, bimodal, and plateau), which form depending on the sampling population and the variables being measured. As we start introducing the statistical tests in this book, a sample's distribution, as well as other parameters, will be critical in helping you to determine the appropriate test to run on a given dataset. This mini chapter will cover parametric and nonparametric tests and considerations for each test. Subsequent chapters will go into the detail on each of these procedures. A few commonly used parametric tests covered in this book include the Student's t-test, the analysis of variance (ANOVA), and Pearson's correlation. A few commonly used nonparametric tests covered in this book include Mann–Whitney U, Wilcoxon signed-rank, Kruskal–Wallis, chi-square, and Spearman's rank order.

Parametric tests (like the t-test and ANOVA) compare means and associated values (e.g., standard deviations). Parametric tests assume that the datasets are on the interval or ratio scale, are normally distributed in some sense (e.g., errors), are randomly sampled from the population, and have equality of variances in select circumstances. In addition, parametric statistics are typically applied to large sampling populations. Nonparametric tests (like the Mann–Whitney U, Wilcoxon signed-rank, and Kruskal–Wallis tests) do not calculate parameters; therefore, they lack any assumptions about the sampling population. Nonparametric tests are also known as distribution-free applications because the data do not need to follow a normal distribution. This characteristic makes these tests more applicable and easy to use on datasets where the distribution is unknown or not made clear.

4.1 Overview

When should you use a parametric or nonparametric test? For each test, evaluate the data type, data distribution, and shape (including skewness and kurtosis), and equality of variance to help guide your decision-making process.

1. **Data type**: Nonparametric statistics are more commonly used when studying populations with variables that are measured as ordinal or nominal. For example, a survey may ask individuals about the effectiveness of a new antihistamine on a scale from extremely ineffective = 1 to extremely effective = 5, making this variable ordinal (see Figure 4.1). In addition, nonparametric tests are also applicable when the sample dataset is small.

    Parametric statistics are only used when the dependent variable is continuous and the sample dataset is sufficiently large. *Note*: A rule of thumb for sample size is that the N should be greater than the number of variables in the study, unless you are running a two-way analysis, when it is best to use a sample size estimator. Before proceeding, you should check with your mentor as sample size can vary by discipline and study subject.

	Extremely ineffective	Moderately ineffective	Neither effective nor ineffective	Moderately effective	Extremely effective
Rate the effectiveness of a new antihistamine	O	O	O	O	O

Figure 4.1 Example of a survey question asking the effectiveness of a new antihistamine in which the response is based on a Likert scale.

2. **Data distribution**: Parametric statistics are only used on data that follow a normal distribution. In order to assess the data for a normal distribution, or normality, you can check for the opposite (non-normal distribution) to determine if the data are in fact not normally distributed. When the data are found to be not normally distributed, they may either be skewed or kurtotic (see Chapter 2).
3. **Equality of variances**: While some tests have an unequal variances version of the test, for example, the Welch t-test, other tests do not, requiring the use of a nonparametric statistic. Testing for equality of variances can be conducted in several ways, of which we will highlight two. The first is by applying Levene's test. In SPSS, Levene's test will run automatically for some statistical procedures (e.g., t-test). For those that require manual selection (e.g., one-way ANOVA), the "Options" menu lists the "Homogeneity of variance test" (see Figure 4.2 for SPSS menu).

The second method for testing if the variances are equal among groups is by examining the standard deviation. If any of the standard deviations of the groups are more than twice that of another, then the variances are not equal. Refer to the Chapter 2 for more information.

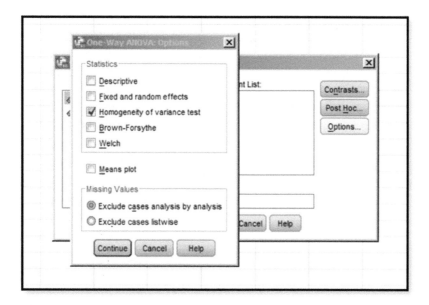

Figure 4.2 Visual representation of the SPSS menu showing how to test for homogeneity of variance.

4.2 Two-Sample and Three-Sample Tests

Previous chapters have discussed the important principles for analyzing a dataset: identifying independent and dependent variables, controlled variables, avoiding biases, and understanding the distribution of your data. The remaining chapters will now delve further into the what, why, and how to apply other statistical analyses to a given dataset. We include parametric and nonparametric two-sample tests and two or more samples tests. The included tests will likely be the ones that you encounter most often.

Two-sample tests allow you to examine a sampling group that has two comparable variables. Whether analyzing the means (parametric) or medians (nonparametric), the resulting output determines the probability value that two samples are from different populations ($p < 0.05$). The two-sample tests to be discussed are the Students t-test, the Mann–Whitney U, and the Wilcoxon signed-rank test. These chapters will dissect each individual test, leaving you with a deeper understanding of how each test is similar, yet unique, in their applications.

In cases where a sampling group has two or more comparable variables, there are several tests, depending on the situation. These chapters will extensively review the main concepts of a parametric (ANOVA) and a nonparametric (Kruskal–Wallis) test. The reading will prepare you to evaluate when it would be appropriate to run an ANOVA and when the conditions would be suitable for a Kruskal–Wallis test.

With parametric and nonparametric two, or two or more, sample tests, you might have samples that are related or paired (e.g., pre- and post-data using the same samples) versus unrelated (e.g., independent groups). The paired t-test, Wilcoxon signed-rank, and rANOVA are paired tests. The independent t-test, Mann–Whitney U test, ANOVA, and Kruskal–Wallis are all run on unrelated or independent samples.

5
t-Test

> **Learning Outcomes**
>
> By the end of this chapter, you should be able to:
>
> 1. Understand the difference between the types of *t*-tests available and know which test to choose for a given dataset.
> 2. Use statistical programs to perform a *t*-test and determine the significance (*p*-value) for your analysis.
> 3. Evaluate the significant difference between two means and construct a logical conclusion for each dataset.
> 4. Use the skills generated to perform, analyze, and evaluate your own dataset from independent research.

5.1 Student's *t*-Test Background

In Chapter 2, we discussed a normal distribution for biological data and the difference between parametric and nonparametric tests. In the next few chapters, we will introduce two types of statistical tests to use with normally distributed data, a Student's *t*-test and an ANOVA. A Student's *t*-test compares two different means, and although an ANOVA has the capability to do the same, an ANOVA is better suited for comparing three or more means.

There are two main types of *t*-tests, one-sample and two-sample. In a one-sample *t*-test, the mean of a sample population or group is compared to a hypothetical mean. In a two-sample *t*-test, the means of two populations or groups are compared against each other to determine if they are different. The two populations or groups can be paired (or related – apples to apples) or unpaired (unrelated – apples to oranges).

To explain the difference between *t*-test options, let us look at the following examples.

5.2 Example *t*-Tests

One-Sample *t*-Test

Canada Geese are a group of migrating birds in the taxonomic family Anatidae. In the fall, many populations migrate south to avoid harsh winter conditions. During migration, geese use a V-shaped flight formation, rotating the front position to share the increased stress and energy cost of flight. The act of migration itself can be physiologically demanding, and individual birds show elevated thyroid hormones to help cope with the needs of the journey, including potential fluctuations in core body temperature. A researcher may be interested in whether the core body temperature of migrating adult geese actually changes compared to the normal, resting temperature (42°C). If we want to determine whether migration causes a significant change from this known body temperature, then we can run a one-sample *t*-test to see if there is a significant deviation from the "normal," 42°C body temperature. See Table 5.1 for a hypothetical dataset of 10 geese.

Normal mean body temperature for an adult goose is 42°C; after migration, the mean body temperature of Canada Geese was 43.2°C. Results from the one-sample *t*-test showed $p < 0.01$. We concluded that migration does result in increased body temperature of geese.

Two-Sample Independent *t*-Test

In North America, Canada Geese can be found from Northern Canada down through the United States and into parts of Northern Mexico. During cold weather, populations at the northern part of

Table 5.1 Body temperature (°C) of 10 geese after migration.

	After Migration (°C)
	43.6
	44.0
	42.8
	43.0
	42.6
	43.3
	42.9
	42.6
	43.8
	43.2
Mean	**43.2**

their range tend to move farther in their migrations than populations from the more southern parts of the range. One question we might ask is whether populations that travel longer distances have a higher mean body temperature immediately after migration than populations that travel shorter distances. In this case, we would perform an unpaired t-test (or independent t-test) because we are looking at two independent populations and are comparing the means of the two different groups.

Let us look at the following data in Table 5.2. The mean body temperature of Canada Geese after a short and a long migration was measured. After a short migration, mean body temperature was 43.2°C, and after a long migration, mean body temperature was 44.7°C. Results from an independent two-sample t-test yielded a $p < 0.01$.

Based on the data in Table 5.2, we can see that mean body temperature was significantly higher after long migrations compared with short migrations. However, a supplementary question we might ask is whether the increase in body temperature is environmentally determined (flight length) or if there is a genetic component. For example, maybe the body temperature of northern groups runs higher than in southern groups. To address this question, a researcher could

Table 5.2 Post migration body temperature (°C) of 10 geese from the Southern group compared to 10 geese from the Northern group.

	Southern Group Short Migration (°C)	Northern Group Long Migration (°C)
	43.6	45.4
	44.0	45.2
	42.8	43.9
	43.0	44.9
	42.6	44.5
	43.3	46.0
	42.9	43.4
	42.6	42.6
	43.8	45.0
	43.2	45.7
Mean	**43.2**	**44.7**

examine the same group of birds in different migration conditions to determine the effect of migration alone.

Two-Sample Paired *t*-Test

In years with significant drought conditions, the migratory route and timing of migration for populations of Canada Geese can change significantly. For example, a population of geese may have to travel longer and farther to find suitable stopover spots along their migration route. The increase in migration travel time may increase the body temperature of these particular geese, in comparison with their body temperature during a previous year's migration. In order to look at this question, we could run a paired *t*-test that looks at mean group body temperatures of the same geese (related) from two time points, 2009 versus 2010, for example.

The data is presented in Table 5.3. Body temperatures of tagged individual Canada Geese were measured in 2009 and 2010. In 2009, mean body temperature was 43.8°C, and in 2010, mean body temperature was 44.7°C. Results from a paired two-sample *t*-test yielded $p > 0.05$. An increase in migration distance due to drought does

Table 5.3 Body temperature (°C) after migration of 10 geese during nondrought or "normal" year compared to 10 geese during a drought year.

Tagged Individuals	Body Temperature During "Normal" Year 2009 (°C)	Body Temperature During Drought Year 2010 (°C)
1	44.0	45.4
2	43.4	45.2
3	45.2	43.9
4	43.7	44.9
5	44.2	44.5
6	43.0	46.0
7	43.6	43.4
8	44.2	42.6
9	42.3	45.0
10	44.0	45.7
Mean	**43.8**	**44.7**

not significantly change the mean body temperature of geese after migration.

Hypotheses

The t-test has a set of hypotheses that addresses the proposed experimental question. For a one-sample t-test, the null hypothesis (H_0) assumes that there is no difference between the "known" mean (A) and the sampled group (B). For a two-sample t-test, the mean of the first sampling group (A) is not statistically different from the mean of the second sampling group (B). The alternative hypotheses assume a difference between groups.

Hypotheses simplified:

$H_0 : A = B$

$H_1 : A > B$ or $A < B$

Output

To report the output from a t-test analysis, you should include the following information: t (df) = t-value, p-value.

Assumptions

The *t*-tests, like most parametric statistics, have five main assumptions:

1. **Data type** – The independent variable is categorical with two groups and the dependent variable is continuous. For one-sample *t*-tests, the variable is continuous.
2. **Distribution of data** – The data fall into a normal distribution. Sample, group, or population size is large.
3. **Independent samples** – The samples are independent of one another (only for Independent *t*-tests).
4. **Homogeneity** – The variances of the populations must be equal, meaning the populations must have an equal spread around the mean.
5. **Random sampling** – The observations were randomly sampled.

Nuts and Bolts

In order to verify if the means from the two populations, samples, or groups are statistically different from each other, the significance level (alpha) must be established and used in reference to the probability value (*p*-value). In biostatistics, the significance level, or the probability that the samples would be different by chance, is generally set to **0.05**. A *p*-value greater than 0.05 indicates that the two compared means are not significantly different. On the other hand, a *p*-value less than or equal to 0.05 indicates that the two compared means are significantly different.

Another consideration is homogeneity of variance. A Levene's test can detect if the variances are equal and statistical programs like SPSS contain this test. Alternatively, using the standard deviations rule as described in Chapter 4 will work in cases where formal tests such as Levene's are not available.

You will need to select the number of tails to test and interpret the relevant output. In order to determine which result to use, you need to understand a little bit more about the biology of your system. Essentially, a one-tailed *t*-test looks for statistical difference in only one direction (Figure 5.1a) and a two-tailed *t*-test looks for statistical difference in two directions (Figure 5.1b).

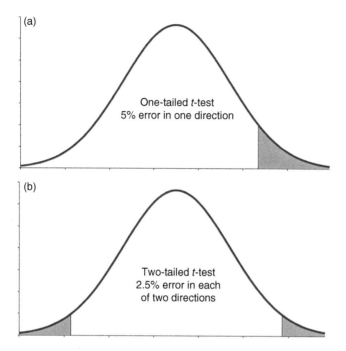

Figure 5.1 Visual representation of the error distribution in a one- versus two-tailed *t*-test. In a one-tailed *t*-test (a), all of the error (5%) is in one direction. In a two-tailed *t*-test (b), the error (5%) is split into the two directions.

In a one-tailed *t*-test, the experimenter must already be familiar with the subject of their research in order to correctly predict the direction of the effect. For this reason, running a one-tail *t*-test is considered less conservative than a two-tail *t*-test. Because the analysis only examines the significance in one direction, this also allows less room for error, which can often produce false results if the direction is in fact unknown or not correctly predicted.

5.3 Case Study

Dr. Gills is interested in studying the pH concentration in Upper Klamath Lake in Central Oregon. She is primarily concerned with whether the pH has significantly changed from the year 2008 to 2009. Past studies have indicated that pH is highest during the month of July and has also increased gradually from year to year. Since prior evidence supports that pH has continually increased from year to year, she could safely run a one-tail t-test and only test for an increase.

As opposed to the one-tail t-test, a two-tail t-test evaluates the statistical difference in both directions. If Dr. Gills is interested in determining whether there was a difference in pH between Upper Klamath and Lower Klamath Lake, where the pH directionality was not already established, then the two-tail t-test would be the most suitable.

Let us look at the Upper Klamath and Lower Klamath Lake example more closely and propose a question, testable hypotheses, and review the data collected (see Table 5.4).

Question: In 2008, was there a difference in the mean pH during the summer months (May–October) between Upper Klamath and Lower Klamath Lake?

Null Hypothesis (H_0): There is no difference in pH between Upper Klamath and Lower Klamath Lake during the summer months.

Alternative Hypothesis (H_1): There is a difference in pH between Upper Klamath and Lower Klamath Lake during the summer months.

Table 5.4 Reported pH levels of Upper and Lower Klamath Lake, OR between May and October 2008.

Collection	Upper Klamath Lake	Lower Klamath Lake
May 08	8.0	7.6
Jun 08	8.2	7.7
July 08	9.7	8.7
Aug 08	9.5	8.6
Sept 08	9.5	8.4
Oct 08	9.3	8.2
Mean	**9.0**	**8.2**

Experimental Data

A preliminary test of equal variances by Levene's test generated a p-value of 0.112 (see Figure 5.2); this exceeds the critical value (0.05), so equal variances are assumed and an independent-samples t-test can proceed. Refer to Figure 5.2 for the experimental SPSS output.

Group Statistics

	LakeLocation	N	Mean	Std. Deviation	Std. Error Mean
pH	1.00	6	9.0333	.73666	.30074
	2.00	6	8.2000	.46043	.18797

Independent Samples Test

		Levene's Test For Equality of Variances		t-test for Equality of Means					
									95% Confidence Interval of the Difference
		F	Sig.	t	df	Sig. (2-tailed)	Mean Difference	Std. Error Difference	Lower	Upper
pH	Equal variances assumed	3.044	.112	2.350	10	.041	.83333	.35465	.04312	1.62355
	Equal variances not assumed			2.350	8.389	.045	.83333	.35465	.02207	1.64460

Figure 5.2 SPSS output showing the results from an independent *t*-test.

> **? Question:** *t*-tests are parametric and evaluate the mean difference between two samples. What type of graph would you use to show your data?
>
> **
>
> A bar graph of the means, using standard deviation as error bars to show the variance of your data around the mean.

Graph

The bar graph in Figure 5.3 is an example of how to illustrate the results from the *t*-test in graphic form.

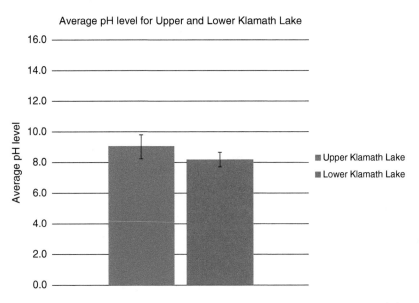

Figure 5.3 Bar graph with standard deviations illustrating the comparison of mean pH levels for Upper and Lower Klamath Lake, OR.

Concluding Statement

As the resulting *p*-value (0.041) is less than the critical value (0.05), you can reject the null hypothesis of no difference and support the alternative hypothesis.

The results indicate a significant difference in the mean pH during the 2008 summer months between Upper Klamath (9.0) versus Lower Klamath (8.2) in 2008 ($t(10) = 2.350, p = 0.041$).

Tutorials

5.4 Excel Tutorial

Coach Riley, the running coach at a local Community College, is preparing for a track meet and wants to determine whether there is a difference between the heart rate in beats per minute (BPM) between male and female runners. Sixteen runners were randomly chosen to report their heart rate immediately after their run. Seven females and nine males were selected. Their pace and distance were controlled, and all members ran at the same intensity. All runners can be considered trained athletes as they have been competing for over a year and exercising five to seven days a week for at least 2 hours each day. All participants measured their heart rate immediately after completing their exercise session.

> **(?)** What type of statistical test should be performed for this type of data?
>
> ***
>
> A two-sample *t*-test is the most suitable statistic for this type of data. We ultimately want to determine whether heart rate is significantly different based on sex. Because the question is generalized to sex and not individuals, the statistical analysis is based on the heart rate for males and females.

> **(R)** With the given information, formulate a question about the data that can be addressed through the use of a *t*-test.
>
> ***
>
> Question: Is the heart rate in beats per minute (BPM) of female runners different than male runners?

> **(H)** Based on the question, formulate the null and alternative hypotheses that address the question proposed.
>
> ***
>
> Null Hypothesis (H_0): There is no difference in the mean heart rate (BPM) values of female and male runners.
>
> Alternative Hypothesis (H_1): There is a difference in the mean heart rate (BPM) of female and male runners.

Now that an appropriate question has been developed, along with a set of testable hypotheses, you can run the statistical analysis.

- This tutorial focuses on running *t*-tests in Excel.
- Refer to Chapter 12 for tips and tools when using Excel.
- Check all assumptions prior to running the test.

t-Test Excel Tutorial

1. Input the following data into Excel with the column labels in cells **A1** and **B1**.

Female BPM	Male BPM
84.5	79.7
88.3	79.5
76.8	78.3
79.9	81.9
81.4	83.7
80.5	80.9
87.2	78.6
	80.7
	81.4

2. In cell **A12**, type in a label for the *p*-value.

	A
1	Female BPM
2	84.5
3	88.3
4	76.8
5	79.9
6	81.4
7	80.5
8	87.2
9	
10	
11	
12	p-value

3. In cell **B12**, use the = **T.TEST** function to calculate the *p*-value. The number of tails and type of *t*-test will be defined in the function.

	A	B
1	Female BPM	Male BPM
2	84.5	79.7
3	88.3	79.5
4	76.8	78.3
5	79.9	81.9
6	81.4	83.7
7	80.5	80.9
8	87.2	78.6
9		80.7
10		81.4
11		
12	p-value	=T.TEST(A2:A8,B2:B10, 2, 3)

Let us look at the function and programming closer so we understand the construction of the code.

= T.TEST(A2:A8,B2:B10,2,3)

The first two sets of cells separated by commas are the two arrays that are being tested. In this case, **female BPM** is located in array A2:A8 and **male BPM** in B2:B10. The next value is the number of tails for the test. Refer to the chart below to code the number of tails you wish to test.

= T.TEST(A2:A8,B2:B10,2,3)

Number of Tails	Excel Code
One tailed	1
Two tailed	2

The last number in the function is the type of *t*-test you want to run. As you can see in the chart below, this function will produce a *p*-value for a two-sample unequal variance *t*-test.

= T.TEST(A2:A8,B2:B10,2,3)

Type of Test	Excel Code
Paired	1
Two-sample equal variance	2
Two-sample unequal variance	3

Thus, the Excel function we constructed calls for a two-tailed, two-sample unequal variance t-test. Press enter/return.

$$= \text{T.TEST}(A2:A8, B2:B10, 2, 3)$$

4. The result is the p-value.

	A	B
1	Female BPM	Male BPM
2	84.5	79.7
3	88.3	79.5
4	76.8	78.3
5	79.9	81.9
6	81.4	83.7
7	80.5	80.9
8	87.2	78.6
9		80.7
10		81.4
11		
12	p-value	0.239754034

The p-value (0.2398) is larger than the significance level we selected (0.05), which means we fail to reject the null hypothesis that the means of **female** and **male BPM** are not different.

Concluding Statement

After measuring the heart rate (BPM) of trained male and female athletes, it was concluded that there was no significant difference in heart rate (BPM) between the sexes (p-value = 0.2398).

Note: Excel does not provide an automated t-value. The equation to calculate a t-value can be programmed into Excel to obtain the value. Otherwise, using a different software, such as SPSS or R, will allow for it to be calculated.

5.5 Paired *t*-Test SPSS Tutorial

The Kinesiology Department at a local university recently became interested in the effect of relaxation simulations on the heart rate of individuals. For the first study, 12 young endurance trained athletes, between 18 and 25 years of age, were asked to participate in a 30-minute audio relaxation demonstration. The audio script focused on body awareness and relaxation techniques. All participants were asked to keep their eyes closed for the entire duration of the audio to get the full effect. Before and after the demonstration, heart rate in beats per minute (BPM) was measured for the participants by the investigator using a heart rate monitor chest strap and watch. After data collection, researchers conducted a paired-samples *t*-test to compare heart rate before and after the simulation.

*The data in this example was taken from the research conducted by Drs. Sarah L. Dunn and Megan Granquist.

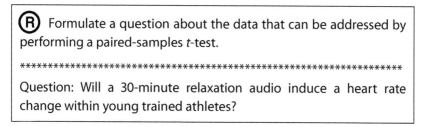

Question: Will a 30-minute relaxation audio induce a heart rate change within young trained athletes?

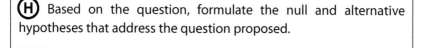

Null Hypothesis (H_0): There is no difference in the heart rate of young trained athletes that listen to the relaxation audio.
Alternative Hypothesis (H_1): There is a difference in the heart rate of young trained athletes that listen to the relaxation audio.

Now that an appropriate testable question has been developed along with a set of testable hypotheses, you can run the statistical analysis.
- Utilize the following tutorial to run a paired-samples *t*-test in SPSS.
- Refer to Chapter 13 for getting started and understanding SPSS.
- Check all assumptions prior to running the test.

Paired-Samples *t*-Test SPSS Tutorial

1. Type the data into the first two columns in SPSS.

BeforeHR	AfterHR
68.00	58.00
88.00	60.00
72.00	65.00
48.00	64.00
55.00	50.00
45.00	55.00
52.00	49.00
63.00	52.00
66.00	53.00
64.00	52.00
58.00	60.00
76.00	59.00

2. Click on the **Analyze** menu and highlight **Compare Means** and then select **Paired-Samples T Test**.

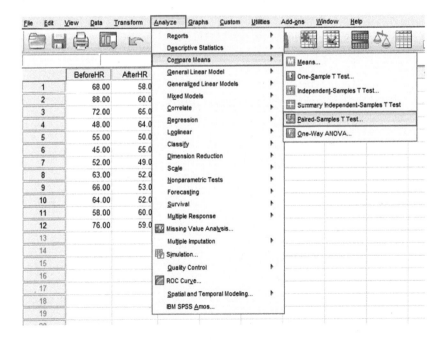

3. The **Paired-Samples T Test** box will pop up. Highlight **BeforeHR** and click on the arrow to the left of the **Paired Variables** box.

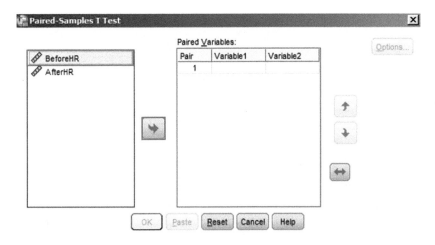

4. Click on **AfterHR** and the arrow next to the **Paired Variables** box.

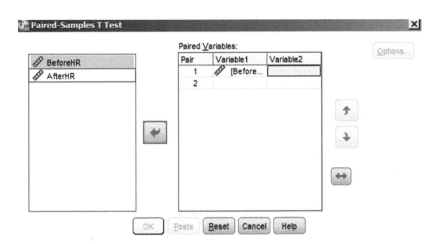

5. Click **OK**.

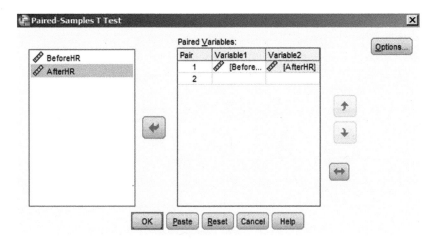

6. A separate document will appear. This is referred to as the output.

T-Test

[DataSet1]

Paired Samples Statistics

		Mean	N	Std. Deviation	Std. Error Mean
Pair 1	BeforeHR	62.9167	12	12.31745	3.55574
	AfterHR	56.4167	12	5.35059	1.54458

Paired Samples Correlations

		N	Correlation	Sig.
Pair 1	BeforeHR & AfterHR	12	.304	.337

Paired Samples Test

		Paired Differences					t	df	Sig. (2-tailed)
		Mean	Std. Deviation	Std. Error Mean	95% Confidence Interval of the Difference				
					Lower	Upper			
Pair 1	BeforeHR - AfterHR	6.50000	11.84368	3.41898	-1.02511	14.02511	1.901	11	.084

In the output we see the t-value is 1.901 (df = 11) with a corresponding p-value of 0.084. The p-value is larger than the significance level we selected (0.05), so we fail to reject the null hypothesis that the means of the BPM tests are the same before and after the relaxation demo.

Concluding Statement

Listening to a relaxation demo for 30 minutes does not result in statistically significant lower heart rate in young trained individuals (t(11) = 1.901, $p > 0.05$).

Note: If you want to save the SPSS file with the inserted data as well as the SPSS output with the results of the statistical analysis performed, then you must save each document separately (see Chapter 13).

5.6 Independent *t*-Test SPSS Tutorial

Example Dataset

After conducting a paired-samples *t*-test on their data (paired *t*-test SPSS tutorial), the Kinesiology Department decided that for their second study, they would compare the change in heart rate of young trained individuals to the change in heart rate of young untrained individuals. The same conditions were applied to this study as the previous study. Participants were asked to listen to a 30-minute relaxation audio and report their heart rates before and after the audio. Because 12 young trained athletes were observed in the first study, 12 young untrained individuals were observed in the second study for comparison.

The change in heart rate was then calculated for both test groups. Given the change in heart rate of young trained individuals and the change in heart rate of young untrained individuals, a *t*-test was used to compare the change in heart rate between the two groups. As the groups are unrelated, a paired-samples *t*-test cannot be used, instead an independent-samples *t*-test is used.

*The data in this example was taken from the research conducted by Drs. Sarah L. Dunn and Megan Granquist.

> **(R)** Formulate a question about the data that can be addressed by performing an independent-samples *t*-test.
>
> ***
>
> Question: Will the change in heart rate induced by a 30-minute relaxation audio simulation be different between young trained individuals and young untrained individuals?

> **(H)** Based on the question, formulate the null and alternative hypotheses that address the question proposed.
>
> ***
>
> Null Hypothesis (H_0): There is no difference between the change in heart rate of young trained individuals and young untrained individuals after a 30-minute relaxation audio simulation.
>
> Alternative Hypothesis (H_1): There is a difference in the heart of trained and untrained individuals after a 30-minute relaxation audio simulation.

Now that an appropriate testable question has been developed, along with a set of testable hypotheses, you can run the statistical analysis.

- To run an independent-samples *t*-test in SPSS, utilize the following tutorial.
- Refer to Chapter 13 for getting started and understanding SPSS.
- Check all assumptions prior to running the test.

Independent *t*–Test SPSS Tutorial

1. Type the data (a categorical variable and the change in heart rate from pre 30 minute audio to post) into the first two columns in SPSS.

    Note: The first column is a categorical vector where:
    1 = trained athletes
    2 = untrained individuals

t-Test

TrainedStat...	ChangeinHR
1.00	10.00
1.00	28.00
1.00	7.00
1.00	16.00
1.00	5.00
1.00	10.00
1.00	3.00
1.00	11.00
1.00	13.00
1.00	12.00
1.00	2.00
1.00	17.00
2.00	4.00
2.00	3.00
2.00	5.00
2.00	8.00
2.00	6.00
2.00	6.00
2.00	7.00
2.00	3.00
2.00	3.00
2.00	2.00
2.00	.00
2.00	3.00

2. Click on the **Analyze** menu and highlight **Compare Means** and then select **Independent-Samples T Test**.

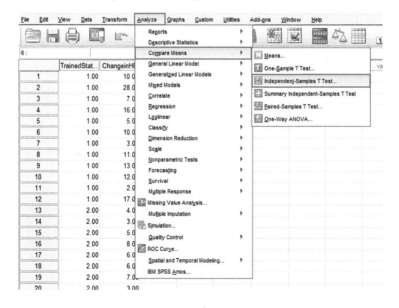

3. The **Independent-Samples T Test** box will pop up. Highlight **ChangeinHR** and click on the arrow to the left of the **Test Variables** box.

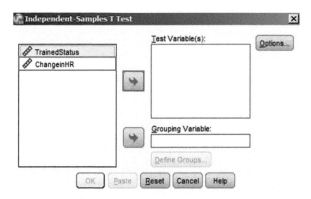

4. Click on **TrainedStatus** and the arrow next to **Grouping Variable.**

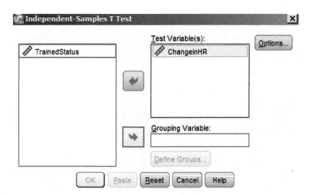

5. Click on **Define Groups**.

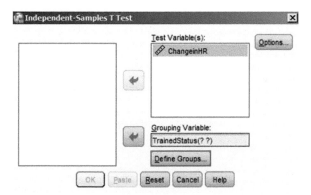

6. Type the group numbers from the **TrainingStatus** variable arranged from top to bottom. Then, click **Continue**.

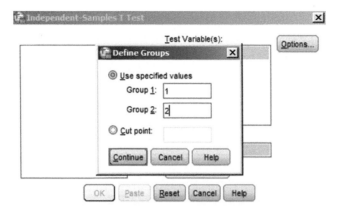

7. Click **OK**.

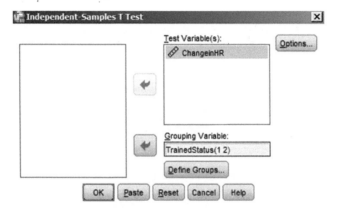

8. A separate document will appear. This is referred to as the output.

T-Test

Group Statistics

	TrainedStatus	N	Mean	Std. Deviation	Std. Error Mean
ChangeinHR	1.00	12	11.1667	7.09460	2.04803
	2.00	12	4.1667	2.28963	.66096

Independent Samples Test

		Levene's Test for Equality of Variances		t-test for Equality of Means					95% Confidence Interval of the Difference	
		F	Sig.	t	df	Sig. (2-tailed)	Mean Difference	Std. Error Difference	Lower	Upper
ChangeinHR	Equal variances assumed	4.968	.036	3.253	22	.004	7.00000	2.15205	2.53692	11.46308
	Equal variances not assumed			3.253	13.267	.006	7.00000	2.15205	2.36027	11.63973

Using the unequal variances test, the *t*-value is 3.253 and has a corresponding *p*-value of 0.006, which is less than the significance level that we selected (0.05). Thus, we reject the null hypothesis that the means of the change in heart rate between the two groups of trained and untrained young individuals is the same.

Note: If your assumptions show the variances are equal, then the equal variances section of the output can be interpreted similarly to the unequal variances portion.

Concluding Statement

The change in heart rate before and after listening to a 30-minute relaxation demo is significantly different between young trained athletes and young untrained individuals (t(13.267) = (3.253), *p*-value = 0.006).

Note: If you want to save the SPSS file with the inserted data as well as the SPSS output with the results of the statistical analysis performed, then you must save each document separately (see Chapter 13).

5.7 Numbers Tutorial

Researchers discovered two species of fish (*Limia* species 1 and *Limia* species 2) living in the same lake on the Caribbean island of Hispaniola. Genetic tests revealed the two fish to be of different species, which was consistent with some general observations about their appearance. Another factor used to differentiate species is their behavior. The researchers were particularly interested in evaluating differences in shoaling behavior (like "schooling" and thought to be related to antipredatory defense), as one of the species seemed to be found more often in groups during its collection.

*Data taken from Dr. Pablo Weaver's lab. Some numbers were changed; however, the trend was maintained.

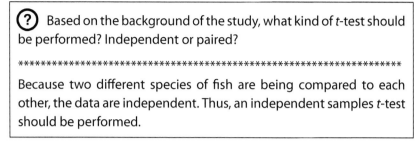

> **?** Based on the background of the study, what kind of *t*-test should be performed? Independent or paired?
>
> ***
>
> Because two different species of fish are being compared to each other, the data are independent. Thus, an independent samples *t*-test should be performed.

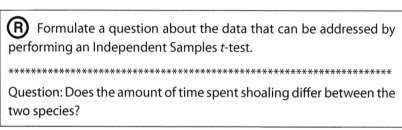

> **R** Formulate a question about the data that can be addressed by performing an Independent Samples *t*-test.
>
> ***
>
> Question: Does the amount of time spent shoaling differ between the two species?

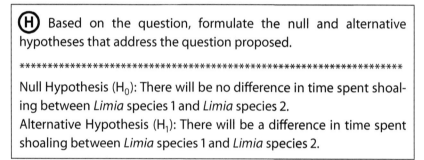

> **H** Based on the question, formulate the null and alternative hypotheses that address the question proposed.
>
> ***
>
> Null Hypothesis (H_0): There will be no difference in time spent shoaling between *Limia* species 1 and *Limia* species 2.
> Alternative Hypothesis (H_1): There will be a difference in time spent shoaling between *Limia* species 1 and *Limia* species 2.

Now that an appropriate testable question has been developed, along with a set of testable hypotheses, you can run the statistical analysis.

- To run an independent samples *t*-test in Numbers, utilize the following tutorial.
- More information on programming in Numbers is found in Chapter 14.
- Check all assumptions prior to running the test.

t-Test Numbers Tutorial

1. Arrange your data into two columns, ***Limia* species 1** and ***Limia* species 2**.

A	B	C	D
	Time spent shoaling		
	Limia species 1	*Limia* species 2	
	15	22	
	15	22	
	16	19	
	16	20	
	17	20	
	17	21	
	18	21	
	18	21	
	18	21	
	18	21	
	19	21	
	19	22	
	19	22	
	19	22	
	19	22	
	19	22	
	19	22	
	19	22	
	19	22	
	19	22	
	19	22	
	19	22	
	19	22	
	19	22	
	19	22	
	19	22	
	19	22	
	19	22	
	19	22	
	19	22	
	19	22	
	19	22	
	19	22	
	19	22	
	19	22	
	19	22	
	19	22	
	19	22	
	19	23	
	19	23	
	19	23	
	20	23	
	20	23	
	20	23	
	21	23	
	21	24	
	22	24	

2. Next to the column of data enter the *t*-test function **=ttest**.

Time spent shoaling							
Limia species 1	Limia species 2						
15	22						
15	22						
16	19						
16	20						
17	20						
17		*fx* ˅ TTEST ▼	sample-1-values	sample-2-values	tails ▼	test-type ▼	⊗ ⊘
18	21						
18	21						
18	21						

3. Click on **sample-1-values**, and highlight the data for *Limia* **species 1**.

Limia species 1	Limia species 2						
15	22						
15	22						
16	19						
16	20						
17	20						
17		*fx* ˅ TTEST ▼	B2:B50 ▼	sample-2-values	tails ▼	test-type ▼	⊗ ⊘
18	21						
18	21						
18	21						

4. Click on **sample-2-values**, and highlight the data for *Limia* **species 2**.

Limia species 1	Limia species 2						
15	22						
15	22						
16	19						
16	20						
17	20						
17		*fx* ˅ TTEST ▼	B2:B50 ▼	C2:C50 ▼	tails ▼	test-type ▼	⊗ ⊘
18	21						
18	21						
18	21						

5. Click on the arrow for tails, select **two tails (2)**.

6. Click on the arrow for test-type, select **two-sample equal (2)**.

7. Here is the full equation. Click the check box or press enter/return.

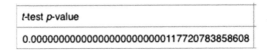

8. The following is the *t*-test output and averages for the two *Limia* species.

t-test p-value
0.00000000000000000000000117720783858608

	Limia species 1	Limia species 2
Average	18.8	22.0

Concluding Statement

The mean time spent shoaling by *Limia* species 1 (18.8 seconds) is significantly ($p < 0.01$) less than the mean time spent shoaling by *Limia* species 2 (22.0 seconds).

Note: In Numbers, we can calculate the *p*-value for the *t*-test but not the *t*-value; therefore, this output looks different than the ones in SPSS and R.

5.8 R Independent/Paired-Samples *t*-Test Tutorial

Anthropologists are often employed to take measurements from living populations, the study of which is called anthropometry. Measurements obtained range from height and body weight to skin fold thickness and head circumference. Height has been shown to be reflective of nutrition, among other factors. Currently, the Dutch are the tallest living population in the world.

Anthropometric height data from two living populations (Americans and Dutch) were measured by a researcher. The researcher wanted to test if there were significant differences in the height of Dutch versus American males. She selected an independent *t*-test as a means to examine her data.

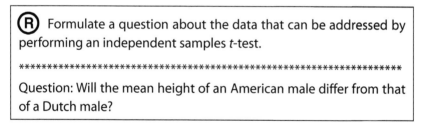

Question: Will the mean height of an American male differ from that of a Dutch male?

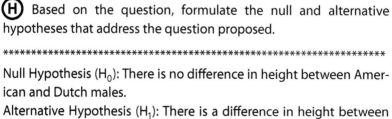

Null Hypothesis (H_0): There is no difference in height between American and Dutch males.
Alternative Hypothesis (H_1): There is a difference in height between American and Dutch males.

Now that an appropriate testable question has been developed along with a set of testable hypotheses, you can run the statistical analysis.

- This tutorial focuses on running an independent samples and paired samples *t*-test in R.
- Refer to Chapter 15 for R specific terminology and instructions on how to invoke and construct code.
- Check all assumptions prior to running the test.

Independent-Samples *t*-Test R Tutorial

1. Type the following heights (reported in centimeters) into a vector and store it by the population name (e.g., American) at the command prompt. Press enter/return.

```
> American<-c(167.61, 172.71, 177.81, 182.90, 182.90, 177.81, 193.03, 190.52, 162.58, 177.81, 180.32)
```

2. Type in a second vector of heights (in centimeters) at the prompt, this time with the name "Dutch." Press enter/return.

```
> Dutch<-c(185.42, 182.9, 187.94, 187.94, 190.52, 185.42, 180.32, 182.90, 193.03, 182.9, 182.9)
```

3. Run the **t.test()** function by first listing the two vectors separated by a comma. As an independent *t*-test is being used, add the argument **paired=** and set it to **FALSE** (for a paired *t*-test the argument is set to **TRUE**). Type in the code below and press enter/return.

```
> t.test(American, Dutch, paired=FALSE)
```

The default *t*-test assumes unequal variances (a Welch test), but if you find the variances are equal when you check your assumptions, you can add the argument **var.equal=TRUE**.

Note: The default is a two-tailed test, although one-tailed tests can be specified (by adding the argument **alternative=c()** and writing either "**less**" or "**greater**"). For example, **t.test (American, Dutch, paired=FALSE, alternative=c**

("less")) or t.test(American, Dutch, paired=FALSE, alternative=c("greater")).
4. The following screen will appear. This is known as the output.

```
        Welch Two Sample t-test

data:  American and Dutch
t = -2.355, df = 13.548, p-value = 0.03418
alternative hypothesis: true difference in means is not equal to 0
95 percent confidence interval:
 -13.2541416  -0.5985857
sample estimates:
mean of x mean of y
 178.7273  185.6536
```

The p-value (0.03418) is significant ($p < 0.05$), indicating the researcher can reject the null hypothesis. Thus, there are differences in mean height between American and Dutch males. A quick glance at the means in the output shows that Dutch males are taller.

Concluding Statement

Mean Dutch male height (185.65 cm) is significantly (t (13.548) = −2.355, $p < 0.05$) greater than mean American male height (178.73 cm) in this sample.

6
ANOVA

> **Learning Outcomes:**
>
> By the end of this chapter, you should be able to:
>
> 1. Understand the scenarios in which a one-way, two-way, and repeated measures ANOVA (rANOVA) are applicable and why.
> 2. Use statistical programs to perform a one-way and rANOVA and determine the significance (p-value) for your analysis.
> 3. Evaluate the significant difference between groups based on the chosen analysis and post hoc measures.
> 4. Use the skills generated to perform, analyze, and evaluate your own dataset from independent research and construct a logical conclusion.

6.1 ANOVA Background

Analysis of variance, otherwise known as an ANOVA, is similar to a Student's t-test in that it compares means across groups of data. In the t-test, we compared the means of two sampling groups, or two variables, to determine if there was a significant difference between groups. ANOVA is similar in concept, but allows us to compare the means of three or more sampling groups at the same time. This type of ANOVA is known as the one-way ANOVA. In essence, if we have three groups (A, B, and C), a one-way ANOVA allows us to compare means from all of the groups at once with a single statistic. There

An Introduction to Statistical Analysis in Research: With Applications in the Biological and Life Sciences, First Edition. Kathleen F. Weaver, Vanessa C. Morales, Sarah L. Dunn, Kanya Godde and Pablo F. Weaver.
© 2018 John Wiley & Sons, Inc. Published 2018 by John Wiley & Sons, Inc.
Companion Website: www.wiley/com/go/weaver/statistical_analysis_in_research

are variations to the ANOVA (one-way, two-way, repeated measures, etc.), some of which will be explained in more detail below.

One-Way ANOVA

The one-way ANOVA allows us to compare the means of three or more sampling groups. Results from an ANOVA analysis reveal whether there is a significant difference between groups, but do not give information about the significance of pairwise comparisons (or comparisons of each group to the other group). Thus, if a significant difference is detected, it is best to run post hoc analyses to further inform the pairwise comparisons between groups. Additional information on post hoc analyses is presented later in this chapter. It is important to note that using Student's t-test multiple times (one t-test per pair) is not an appropriate statistical practice when analyzing multiple groups or time points. By running multiple t-test comparisons, the chances of committing a type I error (false positive, rejecting the null when it is true) are high. Therefore, an ANOVA is the most appropriate method for assessing multiple groups or time points.

For example, a certified athletic trainer is concerned about the number of concussions reported across the Division III National Collegiate Athletic Association (NCAA) fall sporting season. A concussion is a severe blow to the head that can impact brain function. To compare the number of concussions reported across multiple sports, a one-way ANOVA would be appropriate to use for the analysis (see Figure 6.1 for a visual example of a one-way ANOVA protocol).

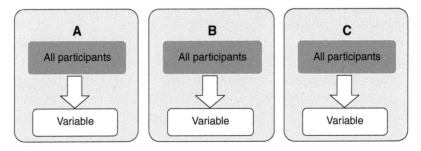

Figure 6.1 One-way ANOVA example protocol using three groups (A, B, and C).

Two-Way ANOVA

An extension of the one-way ANOVA is the two-way ANOVA. As with the one-way ANOVA, the two-way ANOVA allows us to perform an analytical comparison of means; however, we also compare the means of two or more subgroups within multiple comparison groups. In the previous case of three groups (A, B, and C), the two-way ANOVA would allow us to compare multiple independent variables within each group, for example, variables 1 and 2 within groups A, B, and C. Another way to think of this is if you have variables A, B, C and a variable of 1 and 2 assigned to the participants where either an A, a B, or a C and a 1 and a 2 is attributed to each participant (e.g., the first participant is assigned an *A* and a 1). Refer to Figure 6.2 for a visual example of a two-way ANOVA protocol.

If the athletic trainer wanted to assess the concussion rates amongst all sports by sex, the two-way ANOVA would allow us to make this comparison by analyzing sports and sex at the same time. After running the two-way ANOVA, the athletic trainer would know if there was a significant difference in the number of concussions reported between males and females for all sports in the fall season.

Two-way ANOVAs include interaction terms where the two main effects (the two independent variables) are examined along with a

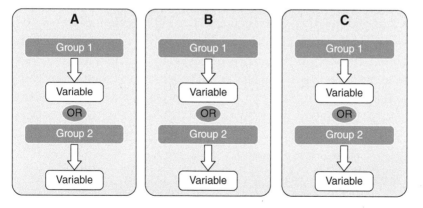

Figure 6.2 Two-way ANOVA example protocol using three groups (A, B, and C) with subgroups (1 and 2).

variable that is the combination of the two main effects, where the level of one main effect depends on the level of the other main effect (the interaction variable). The best models have significance across the main effects and interaction term. If the interaction term is significant, both main effects are left in the model, regardless of significance. If the interaction effect is not significant, it can be dropped and only the two main effects are analyzed. In this book, we will focus on the full model, employing both main effects and the interaction term.

Repeated Measures ANOVA

The rANOVA is the analysis of samples similar to those examined in the paired *t*-test (where the samples are related), applied to the framework of the one-way ANOVA. rANOVA is an extension of the ANOVA whereby the dependent variable lacks independence because it reflects multiple measures (different time points) from the same groups.

In the following example, the strength and conditioning coach at a local university was interested in strength gains for a soccer team. By incorporating a strengthening program across the season involving resistance-based exercises, in addition to regular training at practice, the student-athletes should see positive benefits that would translate into performance gains on competition day. In order to assess strength gains across a season, the strength and conditioning coach measured the student-athletes for gains in muscle strength, using a power output test. The test was implemented the week before the season started (day 1 of the resistance training program or pre-season) and then again at mid- and post-season (three time points). All the student-athletes on the team were expected to complete the resistance training program unless an injury prevented them from doing so. Refer to Figure 6.3 for a visual example of a one-way repeated

Figure 6.3 One-way repeated measures ANOVA study protocol for the measurement of muscle power output at pre-, mid-, and post-season.

measures ANOVA protocol. Because the measurements of strength are linked to a single student, the appropriate test is the one-way rANOVA, which would allow us to test for significant changes in strength for the pre-, mid-, and post-assessments.

Two-Way rANOVA

If you were interested in the difference between time points 1, 2, or 3 for three groups (X, Y, or Z), then a two-way rANOVA would be the most appropriate to use. This type of analysis determines both the differences between each group (X, Y, and Z) and within the group (X, time points 1, 2, and 3; Y, time points 1, 2, and 3; and Z, time points 1, 2, and 3). Figure 6.4 illustrates a two-way repeated measures ANOVA protocol.

In our strength training example, a two-way rANOVA would be needed when analyzing the difference between the same dependent variable (power output of the muscle) across the season (pre-, mid-, and post-season) with an additional element. In the following example, that additional element is the time of day (morning, mid-day, or evening) that the student-athletes participated in their training.

The strength and conditioning coach heard at a recent conference that "evening" may be the best time for muscle building because of

Figure 6.4 Two-way repeated measures ANOVA study protocol for the measurement of muscle power output at pre-, mid-, and post-season for three resistance training groups (morning, mid-day, and evening).

increased protein synthesis during that time. The coach decided to test this idea by dividing the student-athletes into three groups:

Group 1 – Morning Resistance Training
Group 2 – Mid-day Resistance Training
Group 3 – Evening Resistance Training

The two-way rANOVA would allow the coach to determine if there is a significant difference in the times of training across multiple points in the season. Additional post hoc analyses would then inform the coach which time of the day led to the best strength gains. The two-way rANOVA would help to determine the best time of the day for a muscle strengthening program in relation to the greatest gains in muscle strength. By testing three groups, the coach could determine whether time of day (morning, mid-day, or evening) led to better strength gains. Figure 6.5 demonstrates an intervention to compare the effects of time of day of strength training on a muscle power output test across a season.

One- or Two-Way ANOVA or rANOVA Nuts and Bolts

The one- or two-way ANOVA analyses test the differences in means between multiple groups and possibly subgroups. The commonly used parametric analyses to compare multiple groups with related time points are one- or two-way rANOVA. All analyses provide the researcher with a *p*-value (significance). When deciding on whether

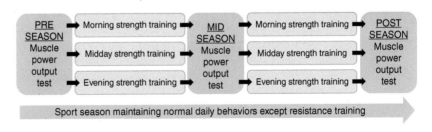

Figure 6.5 An intervention design layout to compare the effects of time of day for strength training (morning, mid-day, and evening) on muscle power output across a season (pre-, mid-, and post-season).

to use the parametric or nonparametric analysis, the decision is based on the assumptions similar to those made in previous chapters.

Assumptions

The one- or two-way ANOVA or rANOVA is most appropriate when certain conditions and assumptions are fulfilled involving multiple groups or variables. The following six assumptions must be satisfied before running these tests.

1. **Data type** – The dependent variable must be interval or ratio. Additionally, the independent variable should have two or more categorical groups.
2. **Distribution of data** – The data follow a normal distribution. This includes the dependent variable's distribution within each category (group) of the independent variable.
3. **Independent samples** – The samples are independent (with the exception of the rANOVA); independent samples have no effect on one another.
4. **Homogeneity** – The variances of the populations must be equal, meaning the populations must have an equal spread around the mean.
5. **Random sampling** – The observations are randomly sampled and are independent from one another.
6. **Sphericity** – rANOVA is sensitive to issues of sphericity. Sphericity is calculated by comparing each level, or group, of the factor variable to each other. The assumption of sphericity is met when the variance of the differences between the levels is equal. Please note that this assumption is easily measured in some software programs such as SPSS and R, while others do not allow for this option to perform the calculation.

In the case that one or more of these assumptions are not fully satisfied, then a nonparametric version of the one- or two-way ANOVA or rANOVA tests would be more appropriate. For example, the nonparametric equivalent of the one-way ANOVA is the Kruskal–Wallis test, which we cover in Chapter 8.

Hypotheses

As mentioned in previous chapters, the null hypothesis (H_0) states that the means of the samples are not statistically different. The alternative hypotheses (H_1, H_2, etc.) state that at least one inequality exists between means.

Hypotheses simplified:

$H_0 : A = B = C$

$H_1 : A \neq B \neq C$

The researcher running an ANOVA can compare the resulting probability value (p) with the same critical value (0.05) previously applied in other analytical tests to determine whether the compared means are statistically different.

$p \leq 0.05$ (statistically significant)

$p > 0.05$ (NOT statistically significant)

Output

To report the output from an ANOVA analysis, you should include the following information: $F(\text{df}, \text{df}) = F\text{-value}, p\text{-value}$.

Post Hoc Analyses

One of the limitations of the ANOVA test is that it will not determine which group is statistically different from the other. For instance, in the concussion example provided above, the certified athletic trainer would use the results from the one-way ANOVA to determine if there was a significant difference between the concussion rates of different sports. However, following the initial analysis, if a significant difference was detected, the athletic trainer would not know which sport had more or less concussions, just that there was a difference found in the reported rate of concussions. To compare each sport, an additional test would need to be performed. For more specific pairwise comparisons within the dataset, post hoc tests are used to determine those differences.

If the ANOVA yields a significant result, a post hoc analysis can be applied to determine where the differences lie between groups (A > B > C or B > C > A, etc.). Post hoc analyses account for multiple comparisons and minimize the probability of making a type I error. If, for instance, the athletic trainer finds a significant p-value from an ANOVA test, a post hoc analysis then allows for tests of significant differences between each of the groups (A with B, A with C, and B with C). As mentioned above, a series of pairwise t-tests would be inappropriate because the type I error compounds with each test, decreasing the level of confidence in the analysis. There are many post hoc analyses to consider. For this brief introduction to ANOVA and post hoc analysis, we include the least significant differences (LSD) and Tukey honestly significant difference (HSD).

When running post hoc analyses, both the LSD and Tukey HSD will provide pairwise comparisons of significant differences. Current accepted protocol is to use both measures and compare the results, looking specifically at the confidence bands (confidence intervals). The smaller the band, the greater the confidence in the data, and the better the choice of post hoc test. For a visual representation of post hoc analysis, refer to Figure 6.6.

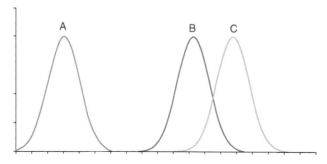

Figure 6.6 Diagram illustrating the relationship between distribution curves where groups B and C are similar but A is significantly different.

> The confidence bands are also known as the confidence intervals (see Chapter 2). The lower confidence limit to the interval is the lower bound, while the higher limit is the upper bound. The narrowest range of these bounds among the post hoc tests is the most appropriate selection. For example, if the range of the bounds in LSD is smaller than the range in Tukey HSD, we would select the LSD as the appropriate post hoc measure and report the significance associated with it.

> While the Bonferroni correction is listed as a post hoc test in some analysis software, it is a correction method and not an additional test. Thus, it is not appropriate to refer to Bonferroni as a post hoc analysis. When used appropriately, the Bonferroni correction allows for an adjustment to the *p*-value with a post hoc analysis, using the number of groups, in order to reduce a type I error (false positive, rejecting the null when it is true).

In situations when significance is found ($p < 0.05$) for the ANOVA, a post hoc is needed in order to differentiate and more closely examine pairwise comparisons. A significant post hoc test indicates there is at least one significant difference among the groups. For example, in Figure 6.6, we might expect that A is significantly less than B and C; however, B and C may not be different from each other. The case study presented next explains the one-way ANOVA in greater detail using a real example.

6.2 Case Study

Let us try the following case study using another exercise example.

The Cori cycle is a physiological metabolic process that recycles and reuses lactate for future energy (glucose). Lactate is a by-product of glucose metabolism (glycolysis) in working muscles. In anaerobic conditions (when no oxygen is present), glucose is broken down to create energy for the muscle to move and lactate is eventually produced via pyruvate. The accumulation of lactate in the working muscle is shuttled into the liver (via the systemic circulation) and eventually the fate of the lactate in the liver is glucose formation (gluconeogenesis).

A student researcher in human exercise physiology is interested in testing for the difference in blood lactate levels between three different exercise testing sessions: control (resting), moderate intensity steady state continuous exercise (SSE), and high intensity interval exercise (HIIE). Under varying levels of exercise intensity (different workloads placed on the muscles), different levels of lactate will be produced and accumulated. It could be expected that with no exercise, there would be low levels of lactate. With moderate intensity SSE, there would be some, but mostly low levels of lactate, and with HIIE, there would be high levels of lactate produced.

Participants in this study were asked to rest or exercise (either SSE or HIIE) and then rest again; blood lactate levels were measured 1 hour post intervention (control, SSE, or HIIE). Because of the need to compare more than two groups with a single dependent variable, a one-way ANOVA would be the appropriate analysis in this case. For this example, participants were randomly assigned into separate groups for exercise testing (control, SSE, or HIIE) as indicated in Figure 6.7.

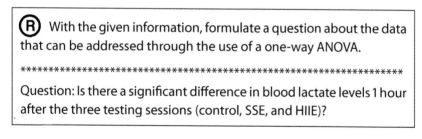

> **Ⓡ** With the given information, formulate a question about the data that can be addressed through the use of a one-way ANOVA.
>
> **
>
> Question: Is there a significant difference in blood lactate levels 1 hour after the three testing sessions (control, SSE, and HIIE)?

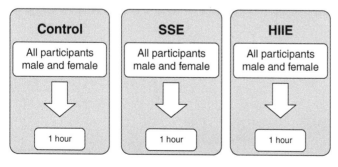

Figure 6.7 One-way ANOVA case study experimental design diagram.

> **(H)** Based on the question, formulate the null and alternative hypotheses that address the question proposed.
>
> ***
>
> Null Hypothesis (H_0): There is no significant difference among the blood lactate levels for the control, SSE, and HIIE sessions.
>
> Alternative Hypothesis (H_1): There is a significant difference among the lactate levels for the control, SSE, and HIIE sessions.

Experimental Result

We can use the results from the one-way ANOVA to determine if there was a significant difference in blood lactate levels among the three exercise sessions. Figure 6.8 provides the experimental SPSS output obtained by the ANOVA. The results indicate a p-value of 0.001, which is less than 0.05; therefore, we can reject the null hypothesis and support the alternative hypothesis. Our results, with degrees of freedom from between the groups (2) and within each group (57), support that there is a significant difference between the average blood lactate levels following the control, SSE, or HIIE sessions. See Figure 6.9 for a graphic illustration of the ANOVA results.

Now that we know there is a significant difference in blood lactate levels (control, SSE, and HIIE), let us look at the following results from the one-way ANOVA and run a post hoc analysis. For this situation, we will use an LSD analysis for post hoc comparisons within the one-way ANOVA as that post hoc measure has the smallest confidence band reported (see Figure 6.10 for a list of available post hoc tests

ANOVA

Lactate1hour

	Sum of Squares	df	Mean Square	F	Sig.
Between Groups	2.715	2	1.358	7.812	.001
Within Groups	9.907	57	.174		
Total	12.622	59			

Figure 6.8 One-way ANOVA SPSS output.

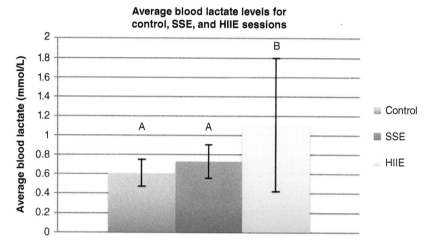

Figure 6.9 Bar graph illustrating the average blood lactate levels (A significantly different than B) for the control and experimental groups (SSE and HIIE).

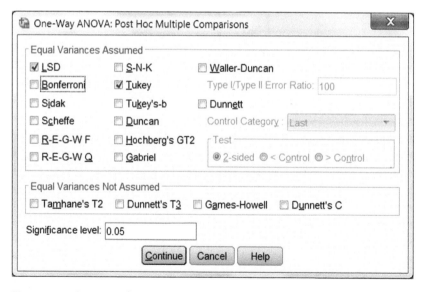

Figure 6.10 SPSS post hoc options when analyzing data for multiple comparisons.

Multiple comparisons

Dependent variable: lactate 1 hour

	(I) Session CON1SSE2HIIE3	(J) Session CON1SSE2HIIE3	Mean Difference (I-J)	Std. Error	Sig.	95% Confidence interval	
						Lower bound	Upper bound
Tukey HSD	1.00	2.00	-.12050	.13184	.634	-.4378	.1968
		3.00	-.49930*	.13184	.001	-.8166	-.1820
	2.00	1.00	.12050	.13184	.634	-.1968	.4378
		3.00	-.37880*	.13184	.015	-.6961	-.0615
	3.00	1.00	.49930*	.13184	.001	.1820	.8166
		2.00	.37880*	.13184	.015	.0615	.6961
LSD	1.00	2.00	-.12050	.13184	.365	-.3845	.1435
		3.00	-.49930*	.13184	.000	-.7633	-.2353
	2.00	1.00	.12050	.13184	.365	-.1435	.3845
		3.00	-.37880*	.13184	.006	-.6428	-.1148
	3.00	1.00	.49930*	.13184	.000	.2353	.7633
		2.00	.37880*	.13184	.006	.1148	.6428

*.The mean difference is significant at the 0.05 level.

Figure 6.11 Post hoc multiple comparison SPSS output.

in SPSS). The LSD analysis provides a pairwise comparison of the means for the rest sessions (1), the SSE (2), and the HIIE (3) as indicated in Figure 6.11.

Pay close attention to the significance levels. After analyzing our output with post hoc measures (LSD), we can see that there is a significant difference between the blood lactate level for control (1) and HIIE (3), with a p-value of 0.000 (reported as less than 0.0001) and between SSE (2) and HIIE (3) with a p-value of 0.006. In addition to determining where the significance lies, the LSD results also indicate where there is no significant difference. For example, there is no significant difference in the blood lactate levels in the control (1) and SSE (2) sessions (p-value = 0.365). The results from the one-way ANOVA indicate a significant difference and the post hoc LSD helps to determine which specific pairs were significantly different from one another.

Concluding Statement

For the current case study, an example of an appropriate concluding statement would read: the average blood lactate levels of the control, SSE, and HIIE were statistically different, $F(2, 57) = 7.812$, $p < 0.05$. From the post hoc comparisons, it was determined that the blood lactate levels in the HIIE group were significantly elevated compared to both the control and SSE sessions.

Tutorials

6.3 One-Way ANOVA Excel Tutorial

Suppose a local candy company was trying to market a new candy bar. Employees want the wrapper to be the most aesthetically appealing as possible to increase sales. Executives decided to gather some preliminary data and distribute the candy to their five best selling stores. Wrappers had similar patterns but differed in color. The colors included: silver, gold, red, green, and blue.

Candy bars were placed on store shelves for 5 days. After the fifth day, total sales for each wrapper was reported. The color with the highest sales would be the color chosen for commercial distribution.

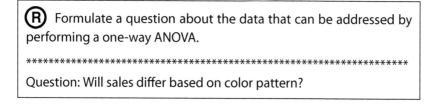

(R) Formulate a question about the data that can be addressed by performing a one-way ANOVA.

****

Question: Will sales differ based on color pattern?

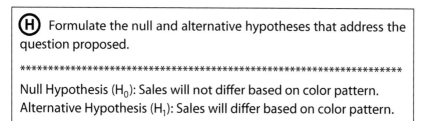

(H) Formulate the null and alternative hypotheses that address the question proposed.

****

Null Hypothesis (H_0): Sales will not differ based on color pattern.
Alternative Hypothesis (H_1): Sales will differ based on color pattern.

Now that an appropriate question has been developed, along with a set of testable hypotheses, you can run the statistical analysis.

- This tutorial focuses on running an ANOVA in Excel.
- Refer to Chapter 12 for tips and tools when using Excel.
- Check all assumptions prior to running the test.

ANOVA Excel Tutorial

1. Arrange the data table so that it appears similar to the following.

	A	B	C	D	E
1	Silver Design	Gold Design	Red Design	Green Design	Blue Design
2	18	18	24	18	20
3	21	12	20	17	28
4	20	17	21	19	20
5	18	14	25	16	26
6	15	14	22	15	27

2. Under the **Data** tab located along the toolbar, select **Data Analysis** (if the Data Analysis option does not appear, refer to **Installing Analysis ToolPak for Excel 2013** tutorial for installation).

3. Select **ANOVA: Single Factor** and click **OK**.

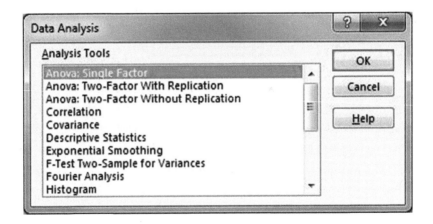

4. Click on the icon corresponding to the **Input Range**.

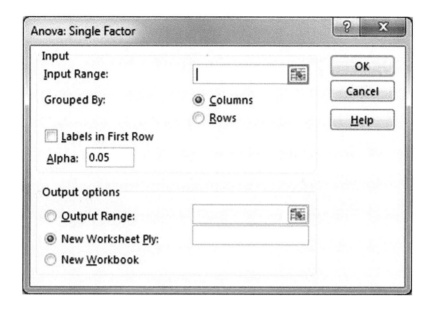

5. Select all data including labels.

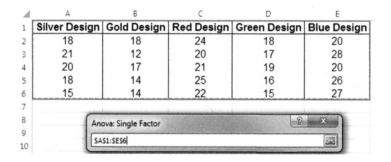

6. If data are grouped by columns, make sure the data are grouped by **Columns,** otherwise check **Rows.**

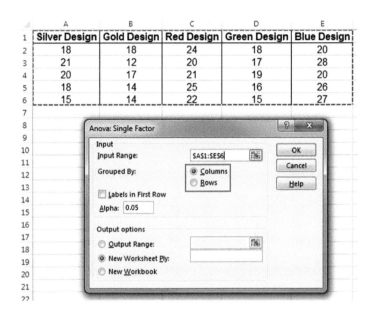

7. Check the box to include the **Labels in the first row**, so that the labels are not confused for actual numerical data.

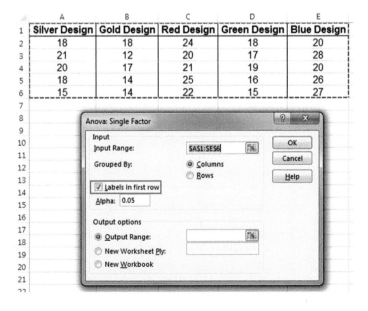

8. If you want the output to appear in the same spreadsheet as the data, check the circle for **Output Range**. Then click on the corresponding icon.

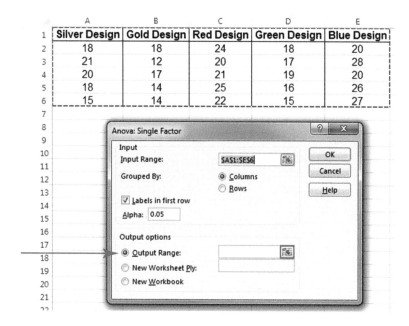

9. Select a space in the spreadsheet for the output to be placed. Then click the small icon on the right.

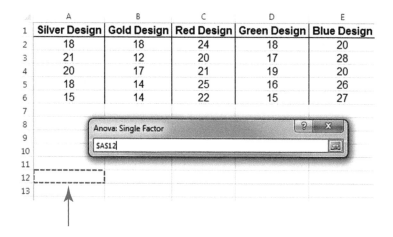

10. Click **OK**.

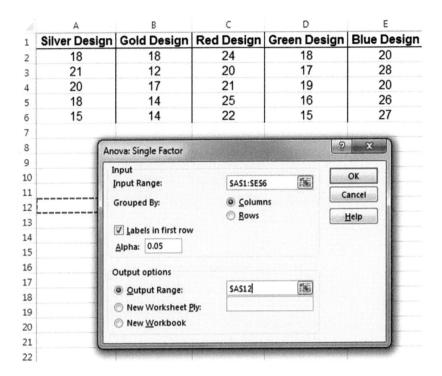

	A	B	C	D	E
1	Silver Design	Gold Design	Red Design	Green Design	Blue Design
2	18	18	24	18	20
3	21	12	20	17	28
4	20	17	21	19	20
5	18	14	25	16	26
6	15	14	22	15	27

11. The output for the **ANOVA: Single Factor** will look similar to:

Anova: Single Factor

SUMMARY

Groups	Count	Sum	Average	Variance
Silver Design	5	92	18.4	5.3
Gold Design	5	75	15	6
Red Design	5	112	22.4	4.3
Green Design	5	85	17	2.5
Blue Design	5	121	24.2	15.2

ANOVA

Source of Variation	SS	df	MS	F	P-value	F crit
Between Groups	290.8	4	72.7	10.91591592	7.36E-05	2.866081
Within Groups	133.2	20	6.66			
Total	424	24				

Concluding Statement

The number of sales significantly differed among the five color patterns, $F(4, 20) = 10.916$, $p < 0.05$; therefore, the null hypothesis can be rejected.

Note: When a *p*-value indicates statistical significance, a post hoc statistical analysis is normally performed. However, Excel is not capable of performing a post hoc analysis. To run a post hoc test, consider using SPSS, R, or another statistical program. In addition, Excel does not provide an automated feature for calculating rANOVA. Other software, such as SPSS, have automated functions that are easy to use. Thus, it is suggested another program be used for rANOVA other than Excel.

6.4 One-Way ANOVA SPSS Tutorial

Obesity rates are alarmingly high in westernized societies. Exercise programs targeting fat loss, the breakdown of the triglyceride (lipolysis) into a glycerol and three fatty acids, have been somewhat unsuccessful long term, although HIIE has been shown to elicit positive effects on overall body fat compared to SSE. More specifically, glycerol levels present in circulating blood represent the breakdown of the triglyceride into free glycerol and the three individual fatty acids. In order to assess if greater amounts of triglycerides are being broken down after one type of exercise compared to another, an ANOVA would be the appropriate statistical analysis to use. Comparing the plasma or serum glycerol levels at 1 hour after each of no exercise (CON for a control session), SSE, and HIIE using a one-way ANOVA will allow the researcher to determine the greatest amount of fat lost after each exercise session.

*The data in this example were taken from the research conducted by Dr. Sarah L. Dunn.

> (R) Formulate a question about the data that can be addressed by performing a one-way ANOVA.
>
> **
>
> Question: Do glycerol levels change based on the type of exercise performed (control, steady state exercise, and high intensity interval exercise)?

> **(H)** Based on the question, formulate the null and alternative hypotheses that address the question proposed.
>
> **
>
> Null Hypothesis (H_0): Glycerol levels do not change based on the type of exercise (control, steady state exercise, and high intensity interval exercise).
>
> Alternative Hypothesis (H_1): Glycerol levels do change based on the type of exercise (control, steady state exercise, and high intensity interval exercise).

Now that an appropriate testable question has been developed along with a set of testable hypotheses, you can run the statistical analysis.

- Utilize the following tutorial to run an ANOVA in SPSS.
- Refer to Chapter 13 for getting started and understanding SPSS.
- Check all assumptions prior to running the test.

ANOVA SPSS Tutorial

1. The data should now look similar to the following.

2. After all data have been inserted, click on the **Analyze** tab. Scroll down to the **Compare Means** option and select **One-Way ANOVA**.

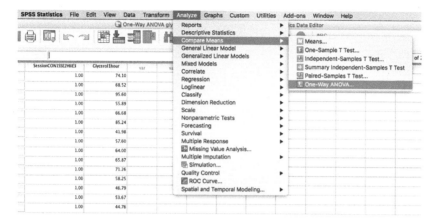

3. The following screen will appear.

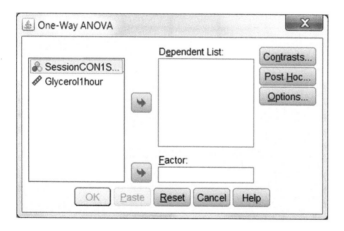

4. Click on the independent variable (**SessionCON1S...**) that appears in the box to the left, then click on the corresponding arrow to move the variable over to the **Factor** list.

5. Next click on the dependent variable remaining in the box to the left (**Glycerol1hour**) and click the corresponding arrow to move the variable over to the **Dependent List**.

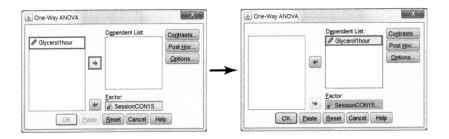

6. Select the **Post Hoc…** option on the right side.

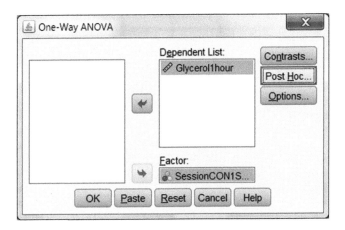

7. Under the **Equal Variances Assumed** heading, select the box for **LSD** and/or **Tukey HSD**, depending on which post hoc test you prefer. Then click Continue followed by **OK**.

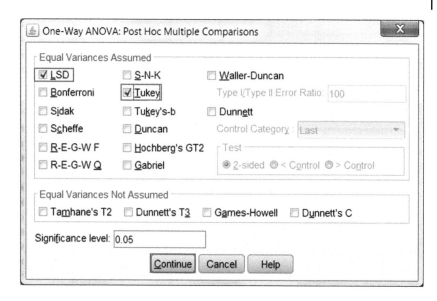

8. A separate document will appear. This is referred to as the output.

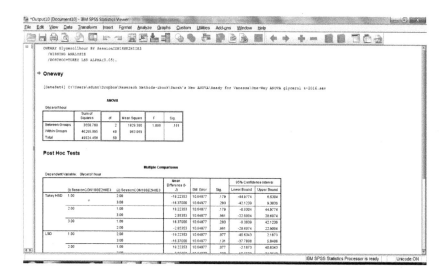

The following is a close view of the one-way ANOVA output.

ANOVA

Glycerol 1 hour

	Sum of Squares	df	Mean Square	F	Sig.
Between Groups	3658.760	2	1829.380	1.898	.161
Within Groups	46265.695	48	963.869		
Total	49924.456	50			

According to the output, the p-value resulting from the one-way ANOVA was not statistically significant ($p = 0.161$), suggesting that circulating plasma or serum glycerol in the blood 1 hour after each session was not due to varying intensity: HIIE, SSE, or CON. As the ANOVA is not significant, the post hoc analysis does not need to be reviewed.

Concluding Statement

Circulating plasma or serum glycerol at 1 hour post exercise was not significantly different, $F(2, 48) = 1.898$, $p = 0.161$, based on the type of exercise session (control, steady state, or HIIE); therefore, the null hypothesis cannot be rejected.

Note: If you want to save the SPSS file with the inserted data as well as the SPSS output with the results of the statistical analysis performed, then you must save each document separately (see Chapter 13).

6.5 One-Way Repeated Measures ANOVA SPSS TUTORIAL

For consistency, a similar example to the one presented in the previous tutorial (one-way ANOVA) is used here to help explain the difference between ANOVA and one-way rANOVA. With the one-way ANOVA, three groups were analyzed for the differences between them. In the one-way repeated measures, three time points following a single exercise session will be assessed.

Obesity rates are a public health concern in westernized societies. To assess the issue, exercise physiologists may examine exercise programs to specifically investigate the rate of fat loss either with different types of exercise or following a specific type of exercise.

In order to determine fat loss, researchers examine glycerol levels present in circulating blood as that can represent the breakdown of the triglyceride into a glycerol and three fatty acids. Measuring plasma or serum glycerol levels at baseline (pre-HIIE), 1 hour, and 2 hours post exercise may represent fat oxidation and fat usage as a fuel source during and following exercise to replenish what was lost (glycogen stores used). In order to analyze the plasma or serum glycerol levels following HIIE at the related time points, a one-way rANOVA is the preferred analysis to perform.

Again, the difference between the one-way ANOVA and the one-way rANOVA is the factors used. The factors used for the one-way in the previous tutorial were the types of exercise (HIIE, steady state exercise, or no exercise) and for this tutorial, the time points before and after one type of exercise (HIIE) will be used.

*The data in this example were taken from the research conducted by Dr. Sarah Dunn

---

(R) Formulate a question about the data that can be addressed by performing a one-way rANOVA.

****

Question: Do glycerol levels change in the short term after high intensity interval training (baseline, 1 hour after, and 2 hours after)?

(H) Based on the question, formulate the null and alternative hypotheses that address the question proposed.

****

Null Hypothesis (H_0): Glycerol levels do not change after high intensity interval training (baseline, 1 hour after, and 2 hours after).

Alternative Hypothesis (H_1): Glycerol levels do change after high intensity interval training (baseline, 1 hour after, and 2 hours after).

Now that an appropriate testable question has been developed along with a set of testable hypotheses, you can run the statistical analysis.

- Utilize the following tutorial to run an ANOVA in SPSS.
- Refer to Chapter 13 for getting started and understanding SPSS.
- Check all assumptions prior to running the test.

One-Way ANOVA with Repeated Measures SPSS Tutorial

1. The data should now look similar to the following.

	HIIEGlycerolatbaseline	HIIEGlycerolat1hour	HIIEGlycerolat2hours
1	145.12	136.21	144.02
2	87.18	67.06	92.87
3	86.68	80.49	81.64
4	73.77	68.11	95.24
5	61.92	60.36	107.30
6	49.53	115.65	125.26
7	60.36	60.36	102.05
8	95.24	68.11	107.30
9	107.30	88.44	102.87
10	34.02	73.77	127.42
11	40.82	27.21	81.64
12	74.30	59.37	76.18
13	60.36	80.47	92.69
14	40.82	47.62	53.65
15	93.89	145.12	118.95
16	120.71	111.81	132.78
17	65.97	52.77	107.30

2. After all data have been inserted, click on the **Analyze** tab. Scroll down to the **General Linear Model** option and select **Repeated Measures**.

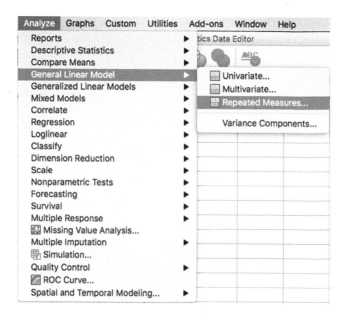

3. The following screen will appear.

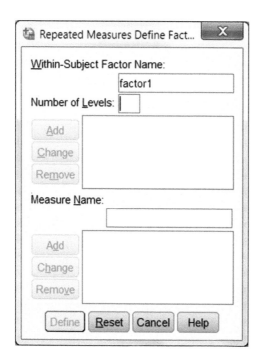

4. Type in a **Within-Subject Factor Name** corresponding to your dataset. In this case **TIME** was used. Next, insert the **Number of Levels** corresponding to the dataset, in this case **3** time points were examined so 3 will be the number of levels used. Click **Add** and then click **Define**.

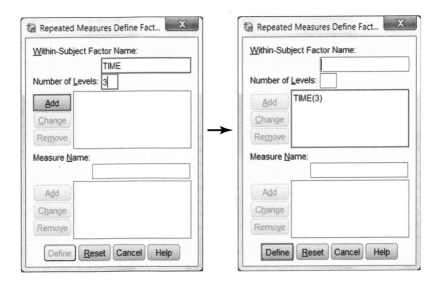

5. Click on each dependent variable remaining in the box and click the corresponding arrow to move the variable over to the **Within-Subjects Variables (TIME)** box and match it up to the 3 time points listed. Select the **Options** tab along the right side.

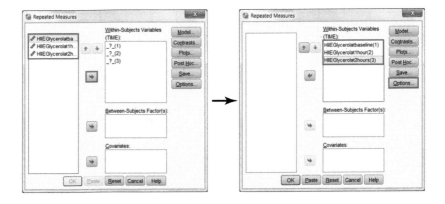

6. Click on each **Factor** (**OVERALL** and the independent variable you named, **TIME**) remaining in the box to the left and click the corresponding arrow to move the variable over to the **Display Means for:** box.

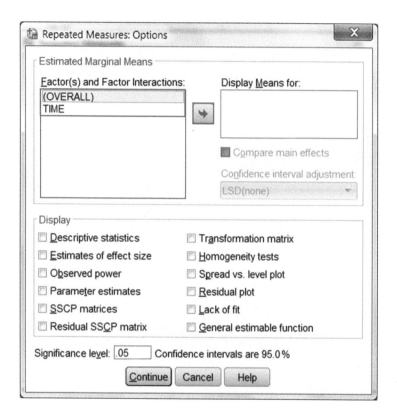

7. Make sure to check the box just below the **Display Means for:** box labeled **Compare main effects** and use the **LSD** as the **Confidence interval adjustment**. Check the box for **Descriptive Statistics** located under **Display**. Also, make sure that the **Significance level** is set to **0.05** if not already set by default. Then click **Continue** followed by **OK**.

258 | An Introduction to Statistical Analysis in Research

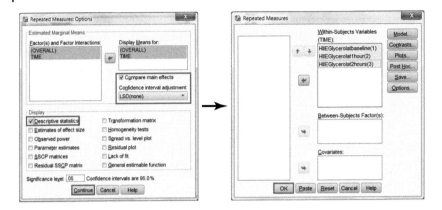

8. A separate document will appear. This is referred to as the output.

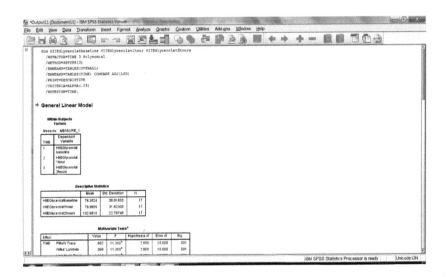

The following is a close view of the one-way rANOVA output.

Mauchly's Test of Sphericity[a]

Measure: MEASURE_1

Within Subjects Effect	Mauchly's W	Approx. Chi-Square	df	Sig.	Epsilon[b]		
					Greenhouse-Geisser	Huynh-Feldt	Lower-bound
TIME	.937	.975	2	.614	.941	1.000	.500

Tests the null hypothesis that the error covariance matrix of the orthonormalized transformed dependent variables is proportional to an identity matrix.

a. Design: Intercept
 Within Subjects Design: TIME

b. May be used to adjust the degrees of freedom for the averaged tests of significance. Corrected tests are displayed in the Tests of Within-Subjects Effects table.

To begin analyzing the results, first look at the significance value generated from **Mauchly's Test of Sphericity**. If significance is observed, then the assumptions have been violated and the data must be analyzed using **Greenhouse-Geisser** or **Huynh-Feldt**. According to Mauchly's test, no significance was observed; therefore, the output generated for within subjects should be reviewed for the correct analysis, **Sphericity Assumed**.

According to the **Tests of Within-Subjects Effects**, significance was observed. All p-values are less than 0.05. This suggests that HIIE does affect circulating plasma or serum glycerol in the blood in the short term (baseline, 1 hour post HIIE, 2 hours post HIIE). Because the rANOVA is significant, the post hoc analysis should be reviewed to determine where the significant difference is within the time points. The results indicate the following:

Tests of Within-Subjects Effects

Measure: MEASURE_1

Source		Type III Sum of Squares	df	Mean Square	F	Sig.
TIME	Sphericity Assumed	7266.604	2	3633.302	11.077	.000
	Greenhouse-Geisser	7266.604	1.882	3861.971	11.077	.000
	Huynh-Feldt	7266.604	2.000	3633.302	11.077	.000
	Lower-bound	7266.604	1.000	7266.604	11.077	.004
Error(TIME)	Sphericity Assumed	10496.601	32	328.019		
	Greenhouse-Geisser	10496.601	30.105	348.663		
	Huynh-Feldt	10496.601	32.000	328.019		
	Lower-bound	10496.601	16.000	656.038		

Tests of Within-Subjects Contrasts

Measure: MEASURE_1

Source	TIME	Type III Sum of Squares	df	Mean Square	F	Sig.
TIME	Linear	5986.893	1	5986.893	15.539	.001
	Quadratic	1279.710	1	1279.710	4.726	.045
Error(TIME)	Linear	6164.363	16	385.273		
	Quadratic	4332.238	16	270.765		

2. TIME

Estimates

Measure: MEASURE_1

TIME	Mean	Std. Error	95% Confidence Interval	
			Lower Bound	Upper Bound
1	76.352	7.281	60.918	91.786
2	78.996	7.670	62.737	95.255
3	102.892	5.527	91.176	114.608

Pairwise Comparisons

Measure: MEASURE_1

(I) TIME	(J) TIME	Mean Difference (I-J)	Std. Error	Sig.[b]	95% Confidence Interval for Difference[b]	
					Lower Bound	Upper Bound
1	2	-2.644	6.422	.686	-16.258	10.971
	3	-26.539*	6.732	.001	-40.812	-12.267
2	1	2.644	6.422	.686	-10.971	16.258
	3	-23.896*	5.404	.000	-35.351	-12.440
3	1	26.539*	6.732	.001	12.267	40.812
	2	23.896*	5.404	.000	12.440	35.351

No significant difference ($p = .686$) between baseline (pre-HIIE) and 1 hour post HIIE

A significant difference ($p = .001$) between baseline (pre-HIIE) and 2 hours post HIIE

A significant difference ($p = .000$) between 1 hour post HIIE and 2 hours post HIIE

Therefore, significant differences in plasma glycerol levels were found in the 2 hours post HIIE. You can conclude that glycerol significantly increased following HIIE between 1 and 2 hours post HIIE and that led to a significant increase between baseline and 2 hours post HIIE.

Concluding Statement

Fasting plasma glycerol levels were significantly different at two hours post HIIE, $F(2, 32) = 11.077$, $p < 0.0001$ compared to baseline with the greatest increase found between 1 and 2 hours post HIIE. The null hypothesis can be rejected.

6.6 Two-Way Repeated Measures ANOVA SPSS Tutorial

For consistency, a similar example to the ones presented in the previous ANOVA tutorials (one-way ANOVA and one-way rANOVA) is used here.

To assess if different types of exercise (HIIE and SSE) and the time points following those types of exercise show a difference in fat loss, a two-way rANOVA should be used. In order to determine fat loss, researchers examine glycerol levels present in circulating blood representing the breakdown of the triglyceride into a glycerol and three fatty acids. Measuring plasma or serum glycerol levels at baseline, 1 hour, and 2 hours post different types of exercise (HIIE or SSE) may represent fat oxidation and fat usage as a fuel source during and following that specific exercise session to replenish what was lost (glycogen stores used). For this tutorial, it is important to understand other variables related to exercise testing. The following steps will demonstrate how to complete the two-way rANOVA using time as a variable with the same SSE and HIIE examples previously mentioned.

*The data in this example were taken from the research conducted by Dr. Sarah Dunn

---

(R) Formulate a question about the data that can be addressed by performing a two-way rANOVA.

****

Question: Do glycerol levels change in the short term (baseline, 1 hour after, and 2 hours after) after two different types of exercise (steady state or high intensity interval exercise)?

(H) Based on the question, formulate the null and alternative hypotheses that address the question proposed.

****

Null Hypothesis (H_0): Fasting plasma glycerol levels do not vary significantly between the two exercise types, three time points, or with the interaction of exercise type and time.

Alternative Hypothesis (H_1): Fasting plasma glycerol levels vary significantly between the two exercise types, three time points, and/or with the interaction of exercise type and time.

Now that an appropriate testable question has been developed along with a set of testable hypotheses, you can run the statistical analysis.

- Utilize the following tutorial to run an ANOVA in SPSS.
- Refer to Chapter 13 for getting started and understanding SPSS.
- Check all assumptions prior to running the test.

Two-Way ANOVA with Repeated Measures SPSS Tutorial

1. The data should now look similar to the following.

	HIIEGlycerolatbaseline	HIIEGlycerolat1hour	HIIEGlycerolat2hours	SSEGlycerolatbaseline	SSEGlycerolat1hour	SSEGlycerolat2hours
1	145.12	136.21	144.02	130.88	76.87	104.83
2	87.18	67.06	92.87	125.79	39.58	46.18
3	86.68	80.49	81.64	146.76	125.79	125.76
4	73.77	68.11	95.24	52.77	46.18	74.80
5	61.92	60.36	107.30	111.81	139.77	131.93
6	49.53	115.65	125.26	98.95	48.92	55.91
7	60.36	60.36	102.05	55.91	52.77	46.18
8	95.24	68.11	107.30	48.92	34.94	41.93
9	107.30	88.44	102.87	46.18	92.35	69.79
10	34.02	73.77	127.42	118.80	170.39	116.21
11	40.82	27.21	81.64	49.53	53.65	74.84
12	74.30	59.37	76.18	13.19	85.76	52.77
13	60.36	80.47	92.69	76.80	43.34	80.49
14	40.82	47.62	53.65	103.77	74.84	105.65
15	93.89	145.12	118.95	80.49	136.21	108.85
16	120.71	111.81	132.78	47.62	61.23	50.49
17	65.97	52.77	107.30	111.45	108.85	129.67

2. Now you can start the statistical analysis. Click on the **Analyze** tab. Scroll down to the **General Linear Model** option and select **Repeated Measures**.

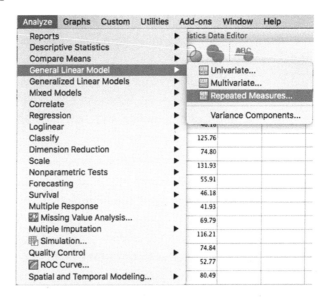

ANOVA | 263

3. The following screen will appear.

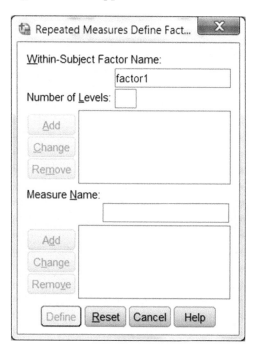

4. Because we are interested in glycerol levels, under the **Within-Subject Factor Name** type in **ExerciseType**. Under the **Number of Levels** type in **2**. Click **Add**.

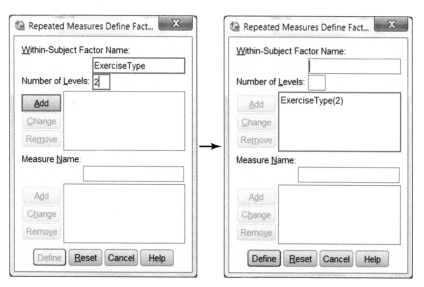

5. Under **Within-Subject Factor Name** type in **TimeInHrs**. Under the **Number of Levels** type in **3**, as the samples were taken at three different time points (baseline, 1 hour, and 2 hours following each exercise session). Click **Add**.

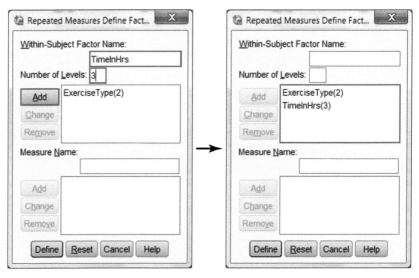

6. Click **Define**.

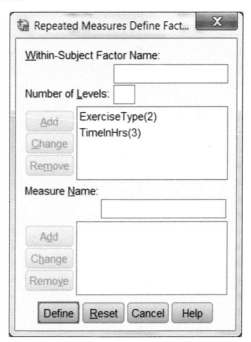

7. The following window will appear.

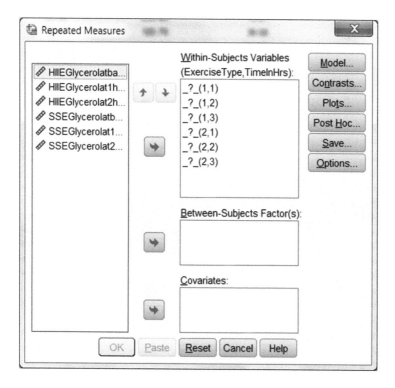

Now you have to move each of the groups over to the **Within-Subjects Variables** box. To do so, click on the first group (HIIEGlycerolbaseline). Then click on the arrow to the right corresponding to the **Within-Subjects Variables** box. Do the same for all the dependent variables listed in the correct order (based on the timing of the variables).

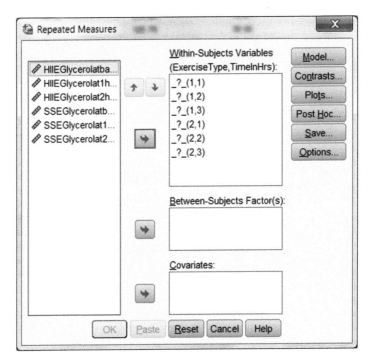

8. Select **Options** on the right side of the window.

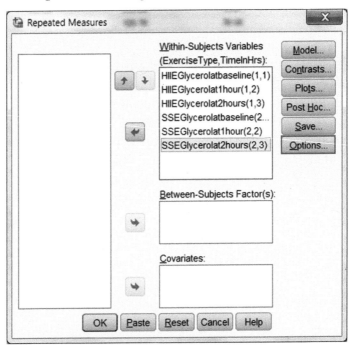

9. Click on each **Factor** (**OVERALL** and the independent variable you named, **TIMEInHrs**) remaining in the box to the left and click the corresponding arrow to move the variable over to the **Display Means for:** box.

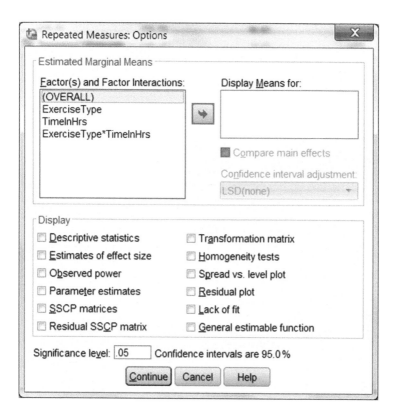

10. Make sure to check the box just below the **Display Means for:** box labeled **Compare main effects** and use the **LSD** as the **Confidence interval adjustment**. Check the box for **Descriptive Statistics** located under **Display**. Also, make sure that the **Significance level** is set to **0.05,** if not already set by default. Then click **Continue**.

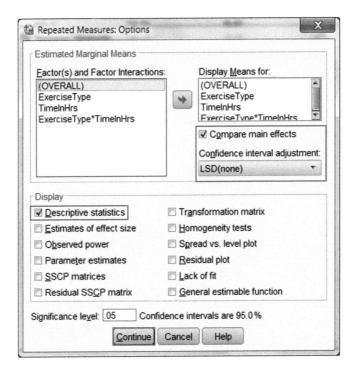

11. Click **OK**.

12. A separate document will appear. This is the output.

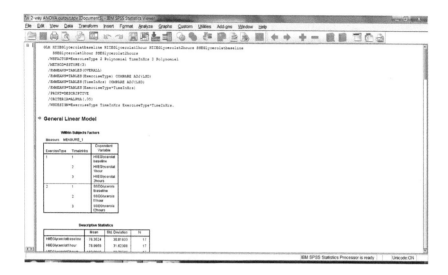

The following is a close view of the two-way rANOVA output.

Mauchly's Test of Sphericity[a]

Measure: MEASURE_1

Within Subjects Effect	Mauchly's W	Approx. Chi-Square	df	Sig.	Epsilon[b]		
					Greenhouse-Geisser	Huynh-Feldt	Lower-bound
ExerciseType	1.000	.000	0	.	1.000	1.000	1.000
TimeInHrs	.737	4.578	2	.101	.792	.864	.500
ExerciseType * TimeInHrs	.837	2.671	2	.263	.860	.953	.500

Tests the null hypothesis that the error covariance matrix of the orthonormalized transformed dependent variables is proportional to an identity matrix.

a. Design: Intercept
 Within Subjects Design: ExerciseType + TimeInHrs + ExerciseType * TimeInHrs

b. May be used to adjust the degrees of freedom for the averaged tests of significance. Corrected tests are displayed in the Tests of Within-Subjects Effects table.

To begin analyzing the results, first look at the significance value generated from **Mauchly's Test of Sphericity**. If significance is observed, then the assumptions have been violated and the data must be analyzed using **Greenhouse-Geisser** or **Huynh-Feldt**. According to Mauchly's test, no significance was observed; therefore, the output generated for within subjects should be reviewed for the correct analysis, **Sphericity Assumed**. The output will look similar to the following.

Tests of Within-Subjects Effects

Measure: MEASURE_1

Source		Type III Sum of Squares	df	Mean Square	F	Sig.
ExerciseType	Sphericity Assumed	259.650	1	259.650	.115	.739
	Greenhouse-Geisser	259.650	1.000	259.650	.115	.739
	Huynh-Feldt	259.650	1.000	259.650	.115	.739
	Lower-bound	259.650	1.000	259.650	.115	.739
Error(ExerciseType)	Sphericity Assumed	36168.776	16	2260.549		
	Greenhouse-Geisser	36168.776	16.000	2260.549		
	Huynh-Feldt	36168.776	16.000	2260.549		
	Lower-bound	36168.776	16.000	2260.549		
TimeInHrs	Sphericity Assumed	3790.725	2	1895.363	4.132	.025
	Greenhouse-Geisser	3790.725	1.584	2393.879	4.132	.036
	Huynh-Feldt	3790.725	1.729	2193.009	4.132	.032
	Lower-bound	3790.725	1.000	3790.725	4.132	.059
Error(TimeInHrs)	Sphericity Assumed	14676.924	32	458.654		
	Greenhouse-Geisser	14676.924	25.336	579.289		
	Huynh-Feldt	14676.924	27.657	530.681		
	Lower-bound	14676.924	16.000	917.308		
ExerciseType * TimeInHrs	Sphericity Assumed	3503.767	2	1751.883	4.027	.028
	Greenhouse-Geisser	3503.767	1.719	2037.683	4.027	.035
	Huynh-Feldt	3503.767	1.907	1837.432	4.027	.030
	Lower-bound	3503.767	1.000	3503.767	4.027	.062
Error(ExerciseType*TimeInHrs)	Sphericity Assumed	13919.887	32	434.996		
	Greenhouse-Geisser	13919.887	27.512	505.961		
	Huynh-Feldt	13919.887	30.510	456.238		
	Lower-bound	13919.887	16.000	869.993		

According to the **Tests of Within-Subjects Effects**, no significance was observed for the exercise type; the p-value was greater than 0.05. Significant differences were found across time and when type was combined with time; both p-values were less than 0.05. Because the purpose was to compare the different time points (baseline, 1 hour, and 2 hours) with the exercise type, the significant values to take note of are the p-values reported with **Sphericity Assumed** within the section allotted for time.

Summary of Results:

No significant difference ($p = 0.739$) between exercise type (HIIE and SSE)

A significant difference ($p = 0.025$) between time in hours (baseline, 1 hour, and 2 hours)

A significant difference ($p = 0.028$) within the time points between exercise type (HIIE and SSE)

No significant differences in plasma glycerol levels were found between the different types of exercise (SSE and HIIE) although significant differences were found among the time points (baseline, 1 hour, and 2 hours) and between the groups when comparing the time points. You can conclude from this dataset that the differences in plasma glycerol at different time points do depend on the exercise type (HIIE or SSE).

Pairwise Comparisons

Measure: MEASURE_1

(I) TimeInHrs	(J) TimeInHrs	Mean Difference (I-J)	Std. Error	Sig.[b]	95% Confidence Interval for Difference[b]	
					Lower Bound	Upper Bound
1	2	-.493	6.336	.939	-13.924	12.938
	3	-13.171*	4.909	.016	-23.579	-2.764
2	1	.493	6.336	.939	-12.938	13.924
	3	-12.679*	4.086	.007	-21.341	-4.016
3	1	13.171*	4.909	.016	2.764	23.579
	2	12.679*	4.086	.007	4.016	21.341

Based on estimated marginal means
*. The mean difference is significant at the .05 level.
b. Adjustment for multiple comparisons: Least Significant Difference (equivalent to no adjustments).

To determine the differences in time points, check the pairwise comparisons for significance:

No significant difference ($p = 0.939$) between baseline and 1 hour post exercise

A significant difference ($p = 0.016$) between baseline and 2 hours post exercise

A significant difference ($p = 0.007$) between 1 and 2 hours post exercise

A significant difference ($p = 0.028$) within the time points between exercise type (HIIE and SSE)

Concluding Statement

Although the fasting plasma glycerol levels were not significantly different between the exercise types (SSE or HIIE), $F(1, 16) = 0.115$, $p = 0.739$; they were significantly different between the three time points, $F(2, 32) = 4.132$, $p = 0.025$, and with the interaction of exercise type and time, $F(2, 32) = 4.027, p = 0.028$. The null hypothesis can be rejected.

Note: If you want to save the SPSS file with the inserted data as well as the SPSS output with the results of the statistical analysis performed, then you must save each document separately (see Chapter 13).

6.7 One-Way ANOVA Numbers Tutorial

"Climate change can be seen within the petal colors of the Hydrangea flower"

The hydrangea plant, shrub, or woody vine found in Asia and North and South America comes from the Hydrangeaceae family. The flower of the hydrangea produces a variety of petal colors with the most common being white, pink, purple, or blue. The vibrant differences in the flower petal color are due to soil makeup and the nutritional content of the hydrangea environment.

If the soil is more acidic (greater exposure to certain minerals and acids with a lower pH level) the flower petals are more blue or purple in appearance, whereas a neutral soil will produce a cream or white flower petal color. The more basic the environment (increased alkalinity and greater pH level), the greater the likelihood that hydrangea flower petal will be pink or fuchsia in color.

Hydrangea flowers in a series of mountains were historically white in coloration. However, recent reports show that flowers from different regions now vary in coloration, and are purple, blue, salmon pink, or bright pink. Alterations in soil content from climate change (such as acid rain) could be responsible for changing the soil chemistry in these areas and causing the regional differentiation of petal color. In this example, the amount of acid precipitation (rainfall) was measured at different sites over a period of 14 days in four regions of different petal colors. To assess the regional differences of acid rain, a one-way ANOVA can be used to determine the differences in rainfall between regions.

Note: the following data set is a hypothetical situation and created solely for teaching purposes.

ANOVA | 273

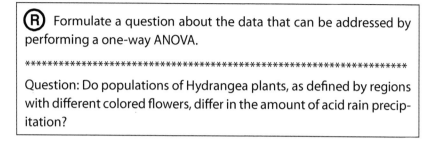

Question: Do populations of Hydrangea plants, as defined by regions with different colored flowers, differ in the amount of acid rain precipitation?

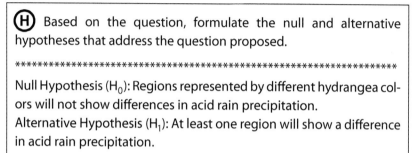

Null Hypothesis (H_0): Regions represented by different hydrangea colors will not show differences in acid rain precipitation.
Alternative Hypothesis (H_1): At least one region will show a difference in acid rain precipitation.

Now that an appropriate testable question has been developed, along with a set of testable hypotheses, you can run the statistical analysis.

- To run a one-way ANOVA in Numbers, utilize the following tutorial.
- More information on programming in Numbers is found in Chapter 14.
- Check all assumptions prior to running the test.

After running your assumptions, you might have noticed one or more violations. Complete the exercise assuming a parametric test is appropriate and revisit this example when working through nonparametric ANOVAs.

ANOVA Numbers Tutorial

1. Start by inputting the labels for your data and creating a chart similar to the one below. Note that the samples were taken at different, independent locations each day.

Table 1

	Day	Purple Hydrangea Plant 1	Blue Hydrangea Plant 2	Salmon Pink Hydrangea Plant 3	Bright Pink Hydrangea Plant 4
Daily Acid Precipitation "Rain Fall" Level Within a Two-Week Period (inches)	1	0.30	0.46	0.20	0.03
	2	1.21	0.61	0.22	0.22
	3	0.44	0.34	0.18	0.21
	4	0.65	0.59	0.01	0.08
	5	0.22	0.75	0.44	0.19
	6	1.28	0.22	0.26	0.15
	7	0.88	0.43	0.32	0.11
	8	0.43	0.21	0.02	0.08
	9	0.58	0.98	0.43	0.05
	10	1.02	0.58	0.05	0.10
	11	1.55	0.45	0.21	0.12
	12	0.57	0.55	0.09	0.24
	13	0.94	0.54	0.18	0.17
	14	0.42	0.45	0.11	0.19
AVG					
SD					
N					

2. Before starting the calculations for an ANOVA analysis, you need to first calculate the mean (average), standard deviation, and N for each group. To calculate the mean, type **=average(cell range)** and select the cell range (e.g., **C3:C16**) that you want calculated.

Table 1

	Day	Purple Hydrangea Plant 1	Blue Hydrangea Plant 2	Salmon Pink Hydrangea Plant 3	Bright Pink Hydrangea Plant 4
Daily Acid Precipitation "Rain Fall" Level Within a Two-Week Period (inches)	1	0.30	0.46	0.20	0.03
	2	1.21	0.61	0.22	0.22
	3	0.44	0.34	0.18	0.21
	4	0.65	0.59	0.01	0.08
	5	0.22	0.75	0.44	0.19
	6	1.28	0.22	0.26	0.15
	7	0.88	0.43	0.32	0.11
	8	0.43	0.21	0.02	0.08
	9	0.58	0.98	0.43	0.05
	10	1.02	0.58	0.05	0.10
	11	1.55	0.45	0.21	0.12
	12	0.57	0.55	0.09	0.24
	13	0.94	0.54	0.18	0.17
	14	0.42	0.45	0.11	0.19
AVG					
SD					
N					

ANOVA | 275

	13	0.94	0.54
	14	0.42	0.45
AVG	· fx ∨ AVERAGE ▾ C3:C16 ▾		1

3. Click on the yellow circle and drag to the right to fill in the remaining mean values for the groups.

	A	B	C	D	E	F
					Table 1	
1						
2		Day	Purple Hydrangea Plant 1	Blue Hydrangea Plant 2	Salmon Pink Hydrangea Plant 3	Bright Pink Hydrangea Plant 4
3		1	0.30	0.46	0.20	0.03
4		2	1.21	0.61	0.22	0.22
5		3	0.44	0.34	0.18	0.21
6		4	0.65	0.59	0.01	0.08
7	Daily Acid Precipitation "Rain Fall" Level Within a Two-Week Period (inches)	5	0.22	0.75	0.44	0.19
8		6	1.28	0.22	0.26	0.15
9		7	0.88	0.43	0.32	0.11
10		8	0.43	0.21	0.02	0.08
11		9	0.58	0.98	0.43	0.05
12		10	1.02	0.58	0.05	0.10
13		11	1.55	0.45	0.21	0.12
14		12	0.57	0.55	0.09	0.24
15		13	0.94	0.54	0.18	0.17
16		14	0.42	0.45	0.11	0.19
17	AVG		0.7493			
18	SD					
19	N					
20						

4. To calculate the standard deviation, type **=stdev(cell range)** and select the cell range, similar to the mean already calculated in the previous steps.

Table 1

	Day	Purple Hydrangea Plant 1	Blue Hydrangea Plant 2	Salmon Pink Hydrangea Plant 3	Bright Pink Hydrangea Plant 4
Daily Acid Precipitation "Rain Fall" Level Within a Two-Week Period (inches)	1	0.30	0.46	0.20	0.03
	2	1.21	0.61	0.22	0.22
	3	0.44	0.34	0.18	0.21
	4	0.65	0.59	0.01	0.08
	5	0.22	0.75	0.44	0.19
	6	1.28	0.22	0.26	0.15
	7	0.88	0.43	0.32	0.11
	8	0.43	0.21	0.02	0.08
	9	0.58	0.98	0.43	0.05
	10	1.02	0.58	0.05	0.10
	11	1.55	0.45	0.21	0.12
	12	0.57	0.55	0.09	0.24
	13	0.94	0.54	0.18	0.17
	14	0.42	0.45	0.11	0.19
AVG		0.7493	0.5114	0.1943	0.1386
SD		*fx* STDEV C3:C16			
N					

5. Click on the yellow circle and drag to the right to fill in the remaining standard deviation values for the groups.

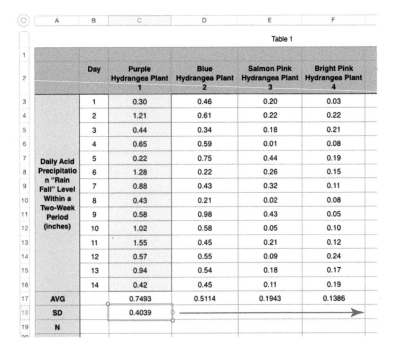

ANOVA

6. Next fill in the N for each group. Type a 14 (based on example data) in the empty cell corresponding to the first treatment and drag the cell to the right to copy the N to the remaining groups.

Table 1

	Day	Purple Hydrangea Plant 1	Blue Hydrangea Plant 2	Salmon Pink Hydrangea Plant 3	Bright Pink Hydrangea Plant 4
Daily Acid Precipitation "Rain Fall" Level Within a Two-Week Period (inches)	5	0.22	0.75	0.44	0.19
	6	1.28	0.22	0.26	0.15
	7	0.88	0.43	0.32	0.11
	8	0.43	0.21	0.02	0.08
	9	0.58	0.98	0.43	0.05
	10	1.02	0.58	0.05	0.10
	11	1.55	0.45	0.21	0.12
	12	0.57	0.55	0.09	0.24
	13	0.94	0.54	0.18	0.17
	14	0.42	0.45	0.11	0.19
AVG		0.7493	0.5114	0.1943	0.1386
SD		0.4039	0.2003	0.1360	0.0665
N		14			

Table 1

	Day	Purple Hydrangea Plant 1	Blue Hydrangea Plant 2	Salmon Pink Hydrangea Plant 3	Bright Pink Hydrangea Plant 4
Daily Acid Precipitation "Rain Fall" Level Within a Two-Week Period (inches)	5	0.22	0.75	0.44	0.19
	6	1.28	0.22	0.26	0.15
	7	0.88	0.43	0.32	0.11
	8	0.43	0.21	0.02	0.08
	9	0.58	0.98	0.43	0.05
	10	1.02	0.58	0.05	0.10
	11	1.55	0.45	0.21	0.12
	12	0.57	0.55	0.09	0.24
	13	0.94	0.54	0.18	0.17
	14	0.42	0.45	0.11	0.19
AVG		0.7493	0.5114	0.1943	0.1386
SD		0.4039	0.2003	0.1360	0.0665
N		14	14	14	14

7. Now create the following "results" table just below your data on the same data sheet.

Source of Variation	Sum of Squares	df	Mean Squares	F-ratio	F-critical	P-value
Groups (treatment)						
Error						
Total						

8. Create a second table "calculations" just below your newly created results table.

Sum of Squares (error):	
MS error:	
Grand Mean:	
Sum of Squares (group):	
MS groups:	
F:	
df (MSgroups):	
df (MSerror):	
F-crit:	

9. To perform a one-way ANOVA, start by calculating the sum of squares and within group variance, also referred to as the error mean square (MS_{error}). The equations for the sum of squares (error) and MS_{error} are shown below. Notice that the sum of squares is just the value of the numerator to the MS_{error} equation.

$$\text{Sum of squares (error)} = \sum s_i^2(n_i - 1)$$

$$MS_{error} = \frac{\sum s_i^2(n_i - 1)}{N - k}$$

s_i^2 = Sample standard deviation from each group
n_i = Sample size
N = Total number of datapoints in all groups combined
k = Number of groups

10. To calculate the sum of squares (error) in the designated cell, type in an equal sign. An equation box will appear.

Table 1

	Day	Purple Hydrangea Plant 1	Blue Hydrangea Plant 2	Salmon Pink Hydrangea Plant 3	Bright Pink Hydrangea Plant 4			
	11	1.55	0.45	0.21	0.12			
	12	0.57	0.55	0.09	0.24			
	13	0.94	0.54	0.18	0.17			
	14	0.42	0.45	0.11	0.19			
AVG		0.7493	0.5114	0.1943	0.1386			
SD		0.4039	0.2003	0.1360	0.0665			
N		14	14	14	14			

Source of Variation	Sum of Squares	df	Mean Squares	F-ratio	F-critical	P-value
Groups (treatment)						
Error						
Total						

Sum of Squares (error):	$fx =$ (((Purple Hydrangea Plant 1 SD ▾)× (Purple Hydrangea Plant 1 SD ▾))×(14−1))+
MS error:	((Blue Hydrangea Plant 2 SD ▾)×(Blue Hydrangea Plant 2 SD ▾))×(14−1))+
Grand Mean:	(((Salmon Pink Hydrangea Plant 3 SD ▾)×(Salmon Pink Hydrangea Plant 3 SD ▾))×(14−1))+
Sum of Squares (group):	(((Bright Pink Hydrangea Plant 4 SD ▾)×(Bright Pink Hydrangea Plant 4 SD ▾))×(14−1)))

11. Next, follow the equation shown. You can either type in the values by hand or scroll over and click the actual cells containing the data to be included in the equation. As you insert your values, your equation should look similar to the one displayed in the figure. After you have entered all values into the equation, press enter/return. The resulting value in the designated cell is the sum of squares (error).

12. Type that value into the results table in the sum of the squares cell corresponding to error, as shown below.

Source of Variation	Sum of Squares	df	Mean Squares	F-ratio	F-critical	P-value
Groups (treatment)						
Error	2.9400					
Total						

Sum of Squares (error):	2.9400
MS error:	
Grand Mean:	
Sum of Squares (group):	
MS groups:	
F:	
df (MSgroups):	
df(MSerror):	
F-crit:	

13. Next calculate **MS error**. Because we already calculated the numerator of the equation, we do not need to enter each individual value again. First type in an equal sign, followed by a dollar sign.

$$MS_{error} = \frac{\sum s_i^2(n_i - 1)}{N - k}$$

14. Then, select the value for the sum of squares (error) calculated previously, followed by the remaining values for the equation. Then click the green check or press enter/return.

Source of Variation	Sum of Squares	df	Mean Squares	F-ratio	F-critical	P-value
Groups (treatment)						
Error	2.9400					
Total						

Sum of Squares (error):	2.9400
MS error:	• fx ⌄ $D23 ⌄ ÷(56−4)
Grand Mean:	
Sum of Squares (group):	
MS groups:	
F:	
df (MSgroups):	
df(MSerror):	
F-crit:	

15. Type the calculated value into the results table in the mean squares cell corresponding to error, as shown below.

Source of Variation	Sum of Squares	df	Mean Squares	F-ratio	F-critical	P-value
Groups (treatment)						
Error	2.9400		0.0565			
Total						

Sum of Squares (error):	2.9400
MS error:	0.0565
Grand Mean:	
Sum of Squares (group):	
MS groups:	
F:	
df (MSgroups):	
df(MSerror):	
F-crit:	

16. Now, calculate the grand mean using the following equation.

$$\bar{Y} = \frac{\sum n_i \bar{Y}_i}{N}$$

n_i = Sample size
N = Total number of datapoints in all groups combined
$\bar{Y}_i$ = Group mean

17. Once again, start by typing in an equal sign into the empty cell in the calculations table. Then select the values from the data table (mean and N) to complete the equation. A finished equation should look similar to the one below.

	Day	Purple Hydrangea Plant 1	Blue Hydrangea Plant 2	Salmon Pink Hydrangea Plant 3	Bright Pink Hydrangea Plant 4			
	12	0.57	0.55	0.09	0.24			
	13	0.94	0.54	0.18	0.17			
	14	0.42	0.45	0.11	0.19			
AVG		0.7493	0.5114	0.1943	0.1386			
SD		0.4039	0.2003	0.1360	0.0665			
N		14	14	14	14			
	Source of Variation	Sum of Squares	df	Mean Squares	F-ratio	F-critical	P-value
	Groups (treatment)						
	Error	2.9400		0.0565			
	Total						
	Sum of Squares (error):	2.9400					
	MS error:	0.0565					
	Grand Mean:	• fx − (((Purple Hydrangea Plant 1 AVG ▼)×(Purple Hydrangea Plant 1 N ▼))+					
	Sum of Squares (group):	((Blue Hydrangea Plant 2 AVG ▼)×(Blue Hydrangea Plant 2 N ▼))+ (Salmon Pink Hydrangea Plant 3 AVG ▼)×(Salmon Pink Hydrangea Plant 3 N ▼))+					
	MS groups:	(Bright Pink Hydrangea Plant 4 AVG ▼)×(Bright Pink Hydrangea Plant 4 N ▼))÷56					
	F:						
	df (MSgroups):						

18. Now that you have calculated the grand mean from the data table (mean and N), you can calculate the sum of squares for the groups from the data table (mean, N, and grand mean).
19. The finished equation should look similar to the one below.

Day	Purple Hydrangea Plant 1	Blue Hydrangea Plant 2	Salmon Pink Hydrangea Plant 3	Bright Pink Hydrangea Plant 4
14	0.42	0.45	0.11	0.19
AVG	0.7493	0.5114	0.1943	0.1386
SD	0.4039	0.2003	0.1360	0.0665
N	14	14	14	14

Source of Variation	Sum of Squares	df	Mean Squares	F-ratio	F-critical	P-value
Groups (treatment)						
Error	2.9400		0.0565			
Total						

Sum of Squares (error):	2.9400
MS error:	0.0565
Grand Mean:	0.3984
Sum of Squares (group):	fx ((Purple Hydrangea Plant 1 N)×((Purple Hydrangea Plant 1 AVG) − (D28))) + ((Blue Hydrangea Plant 2 N)×((Blue Hydrangea Plant 2 AVG) − (D28))) + ((Salmon Pink Hydrangea Plant 3 N)×((Salmon Pink Hydrangea Plant 3 AVG) − (D28))) + ((Bright Pink Hydrangea Plant 4 N)×((Bright Pink Hydrangea Plant 4 AVG) − (D28)))
MS groups:	
F:	
df (MSgroups):	
df(MSerror):	
F-crit:	

20. Insert the computed value into the results table.

Source of Variation	Sum of Squares	df	Mean Squares	F-ratio	F-critical	P-value
Groups (treatment)	3.4310					
Error	2.9400		0.0565			
Total						

Sum of Squares (error):	2.9400
MS error:	0.0565
Grand Mean:	0.3984
Sum of Squares (group):	3.4310
MS groups:	
F:	
df (MSgroups):	
df(MSerror):	
F-crit:	

21. The sum of squares of the groups, just calculated, is the numerator of the **MSgroups** equation which is the next calculation.

Sum of squares (group) $= \sum n_i(\bar{Y}_i - \bar{Y})^2$

n_i = Sample size
$\bar{Y}_i$ = Group mean
$\bar{Y}$ = Grand mean

22. Next calculate **MSgroups**. Because we already calculated the numerator (sum of squares) of the equation in the calculations table, we do not need to enter each individual value again. The equation we will use is given below.

$$MS_{groups} = \frac{\sum n_i(\bar{Y}_i - \bar{Y})^2}{k-1}$$

n_i = Sample size
$\bar{Y}_i$ = Group size
$\bar{Y}$ = Grand mean
k = Number of groups

23. First type in an equal sign in the calculations table. Then insert the values as you have done before. Your finished equation should look similar to the one below.

Source of Variation	Sum of Squares	df	Mean Squares	F-ratio	F-critical	P-value
Groups (treatment)	3.4310					
Error	2.9400		0.0565			
Total						

Sum of Squares (error):	2.9400
MS error:	0.0565
Grand Mean:	0.3984
Sum of Squares (group):	3.4310
MS groups:	= fx ~ D29 ▼ ÷(4−1)
F:	
df (MSgroups):	
df(MSerror):	
F-crit:	

24. Insert the calculated **MSgroups** value into the results table.

Source of Variation	Sum of Squares	df	Mean Squares	F-ratio	F-critical	P-value
Groups (treatment)	3.4310		1.1437			
Error	2.9400		0.0565			
Total						

Sum of Squares (error):	2.9400
MS error:	0.0565
Grand Mean:	0.3984
Sum of Squares (group):	3.4310
MS groups:	1.1437
F:	
df (MSgroups):	
df(MSerror):	
F-crit:	

25. Now you can calculate the F ratio using the following equation in the calculations table.

$$F = \frac{MS_{groups}}{MS_{error}}$$

26. Once again, type in an equal sign into the empty cell in the calculations table and input the values for the variables in the equation.

Source of Variation	Sum of Squares	df	Mean Squares	F-ratio	F-critical	P-value
Groups (treatment)	3.4310		1.1437			
Error	2.9400		0.0565			
Total						

Sum of Squares (error):	2.9400
MS error:	0.0565
Grand Mean:	0.3984
Sum of Squares (group):	3.4310
MS groups:	1.1437
F:	= fx ˅ (D30 ˅) ÷ (D27 ˅)
df (MSgroups):	
df(MSerror):	
F-crit:	

27. Insert the calculated value into the results table.

Source of Variation	Sum of Squares	df	Mean Squares	F-ratio	F-critical	P-value
Groups (treatment)	3.4310		1.1437	20.2281		
Error	2.9400		0.0565			
Total						

Sum of Squares (error):	2.9400
MS error:	0.0565
Grand Mean:	0.3984
Sum of Squares (group):	3.4310
MS groups:	1.1437
F:	20.2281
df (MSgroups):	
df(MSerror):	
F-crit:	

28. In order to calculate the *p*-value, the *F* distribution has to be determined based on the degrees of freedom calculated from the numerator and denominator of the *F* ratio. To determine the degrees of freedom for **MSgroups**, use the following equation.

    $df(\text{MSgroups}) = k - 1$
    $df(\text{MSgroups}) = 4 - 1 = 3$
    N = Total number of datapoints in all groups combined
    k = Number of groups

29. To determine the degrees of freedom for **MSerror**, use the following equation.

    $df(\text{MSerror}) = N - k$
    $df(\text{MSerror}) = 56 - 4 = 52$

30. Insert both of the calculated degrees of freedom into the results table.

Source of Variation	Sum of Squares	df	Mean Squares	F-ratio	F-critical	P-value
Groups (treatment)	3.4310		1.1437	20.2281		
Error	2.9400		0.0565			
Total						

Sum of Squares (error):	2.9400
MS error:	0.0565
Grand Mean:	0.3984
Sum of Squares (group):	3.4310
MS groups:	1.1437
F:	20.2281
df (MSgroups):	3
df(MSerror):	52
F-crit:	

Source of Variation	Sum of Squares	df	Mean Squares	F-ratio	F-critical	P-value
Groups (treatment)	3.4310	3	1.1437	20.2281		
Error	2.9400	52	0.0565			
Total						

Sum of Squares (error):	2.9400
MS error:	0.0565
Grand Mean:	0.3984
Sum of Squares (group):	3.4310
MS groups:	1.1437
F:	20.2281
df (MSgroups):	3
df(MSerror):	52
F-crit:	

31. Within the results table, calculate the sum of each of the following categories in the total cell: sum of squares, degrees of freedom (df), and mean squares.

Source of Variation	Sum of Squares	df	Mean Squares	F-ratio	F-critical	P-value
Groups (treatment)	3.4310	3	1.1437	20.2281		
Error	2.9400	52	0.0565			
Total	• fx ~ SUM ▼ (D22:D23 ▼)					

Sum of Squares (error):	2.9400
MS error:	0.0565
Grand Mean:	0.3984
Sum of Squares (group):	3.4310
MS groups:	1.1437
F:	20.2281
df (MSgroups):	3
df(MSerror):	52
F-crit:	

32. Click on the yellow circle and drag to the right to fill in the remaining sum values for the sum of squares, degrees of freedom (df), and mean squares.

ANOVA | 287

Source of Variation	Sum of Squares	df	Mean Squares	F-ratio	F-critical	P-value
Groups (treatment)	3.4310	3	1.1437	20.2281		
Error	2.9400	52	0.0565			
Total	6.3710					

Sum of Squares (error):	2.9400
MS error:	0.0565
Grand Mean:	0.3984
Sum of Squares (group):	3.4310
MS groups:	1.1437
F:	20.2281
df (MSgroups):	3
df(MSerror):	52
F-crit:	

33. Using the calculated degrees of freedom (**3, 52**), refer to an F distribution table, based on your proposed alpha level (e.g., 0.05) in order to determine the critical value F.

$\alpha = 0.05$ df	Numerator Degrees of Freedom (df)		
	1	2	3
1	161.45	199.50	215.71
2	18.51	19.00	19.16
3	10.13	9.55	9.28
4	7.71	6.94	6.59
5	6.61	5.79	5.41
//			
50	4.03	3.18	2.79
51	4.03	3.18	2.79
52	4.03	3.18	2.78

(Denominator Degrees of Freedom (df))

34. Insert the F-critical value into the results table based on the distribution table referenced.

Source of Variation	Sum of Squares	df	Mean Squares	F-ratio	F-critical	P-value
Groups (treatment)	3.4310	3	1.1437	20.2281	2.78	
Error	2.9400	52	0.0565			
Total	6.3710	55	1.2002			

Sum of Squares (error):	2.9400
MS error:	0.0565
Grand Mean:	0.3984
Sum of Squares (group):	3.4310
MS groups:	1.1437
F:	20.2281
df (MSgroups):	3
df(MSerror):	52
F-crit:	2.78

35. In order to determine the exact p-value, reference a different statistical program such as Excel or SPSS. However, based on the F ratio and F critical (derived from an F table), you can still determine whether the data are significant or not. Because the observed value of F(20.2281) is greater than the critical value of F(2.78), the p-value is less than 0.05. Insert the observed F-value into the results table.

Source of Variation	Sum of Squares	df	Mean Squares	F-ratio	F-critical	P-value
Groups (treatment)	3.4310	3	1.1437	20.2281	2.78	< 0.05
Error	2.9400	52	0.0565			
Total	6.3710	55	1.2002			

Concluding Statement

With the data collected over a 2-week period, it was determined that the various regions with different flower colors had different acid rain fall levels, $F(3, 52) = 20.2281$, $p < 0.05$; therefore, the null hypothesis can be rejected.

6.8 One-Way R Tutorial

Physical anthropologists have difficulty estimating body mass index (BMI) from skeletal remains, with their best efforts little better than guesses. Many methods have been put forth, with the most recent emphasis on cross-sectional (measurements from within the bone) properties from long bones (e.g., arm and selected leg bones). External dimensions (measurements on the outside of the bone) have been shown to be excellent proxies for cross-sectional dimensions in the absence of radiographic images.

The data for this example come from Wilson Taylor and Godde (2011), who examined selected measurements and how they vary across different BMI categories defined by the World Health Organization (Obese I (30–34.99), Obese II (35–35.99), and Obese III (40+)). In this example, we will focus on the anterior–posterior (front–back) measurement of the radius (a forearm bone) to see if the measurements differ among the three obesity categories.

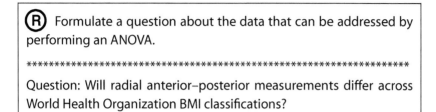

Question: Will radial anterior–posterior measurements differ across World Health Organization BMI classifications?

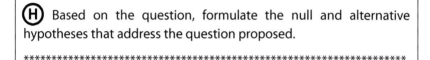

Null Hypothesis (H_0): Radial anterior–posterior measurements will not differ across BMI classifications.
Alternative Hypothesis (H_1): Radial anterior–posterior measurements will differ across BMI classifications.

Now that an appropriate testable question has been developed along with a set of testable hypotheses, you can run the statistical analysis.

- This tutorial focuses on running an ANOVA in R.
- Refer to Chapter 15 for R specific terminology and instructions on how to invoke and construct code.
- Check all assumptions prior to running the test.

One-Way ANOVA R Tutorial

1. Type the following data into a vector and store it with the measurement name, **radialap**, at the command prompt "`>`." Press enter/return.

```
> radialap<-c(9, 11, 9, 10, 11, 11, 11, 11, 10, 10, 10, 11, 10, 9, 11, 11)
```

2. Create a second vector with the name **obeseclass** using the values below. Press enter/return.

```
> obeseclass<-c(1,1,1,1,1,1,2,2,2,2,2,3,3,3,3,3)
```

3. Ensure **obeseclass** is recognized in R as a categorical variable by inquiring if it is a factor using the `is.factor()` function. Type in the code below and press enter/return.

```
> is.factor(obeseclass)
[1] FALSE
```

An answer of **FALSE** indicates it is not a categorical variable, while **TRUE** means it is.

4. If **obeseclass** is not a factor, convert it to a factor by applying the `factor()` function to **obeseclass** and saving it to a variable of the same name. Type in the programming below and press enter/return.

```
> obeseclass<-factor(obeseclass)
```

5. Invoke the `aov()` function by using the formula Y~X or Dependent~Independent variable. Store the function to the name of your choice. In this case, we will use **result**. Type in the script below and press enter/return.

```
> result<-aov(radialap~obeseclass)
```

6. Process the ANOVA results by using the `summary()` function and the name of the results (**result**). Type in the following code and press enter/return.

```
> summary(result)
```

7. Analyze the output below.

	Df	Sum Sq	Mean Sq	F value	Pr(>F)
obeseclass	2	0.204	0.1021	0.144	0.867
Residuals	13	9.233	0.7103		

The df is calculated as $k - 1$, which in this case is the number of groups (3) minus 1, or 2. The F-value (0.144) and corresponding

p-value (0.867) indicate there is no difference between the three obesity classes.

Concluding Statement

There is no significant difference ($F(2, 13) = 0.144$, p-value $= 0.867$) in anterior–posterior radial measurements among the three obese classifications (I, II, and III) as defined by the World Health Organization.

Note: rANOVA is complex to complete in R. Other software, such as SPSS, have automated functions that are easy to use. Thus, it is suggested another program be used for rANOVA other than R.

6.9 Two-Way ANOVA R Tutorial

Musculoskeletal stress markers (MSM) are alterations to the bone originally thought to be the result of muscle pulling on bone. Thus, it is thought that the more active someone is, and the more muscle they build in areas relating to these activities, the greater the pull on the bone, resulting in a larger MSM. Recent research conflicts with this idea, indicating that there are more factors contributing to the expression of MSM, which blurs the cause and effect relationship originally postulated.

As presented in the one-way ANOVA tutorial for R, cross-sectional measurements of the bone are associated with physical activity and BMI, which is an indicator of body composition. The femur has been identified as the best indicator of physical activity in the literature and new research suggests BMI may also play a role. The data from Godde and Wilson Taylor (2011) and Wilson Taylor and Godde (2011) will be examined to assess the cross-sectional properties of long bones and their association with BMI in American males; although for this tutorial, the variable MSM expression will be examined in addition to the others. MSM robusticity is a categorical variable that is visually scored for degree of expression using a 0–3 scale, with 0 reflecting no changes to the bone and 3 indicating extreme changes. The cross-sectional measurements and MSM in this study are from the left femur of American males.

> **(R)** Formulate a question about the data that can be addressed by performing an ANOVA.
>
> ***
>
> Question: Do cross-sectional measurements of the femur change with expression of MSM and BMI categories based on WHO criteria?

> **(H)** Based on the question, formulate the null and alternative hypotheses that address the question proposed.
>
> ***
>
> Null Hypothesis (H_0): Femoral cross-sectional measurements do not vary with MSM expression, BMI, or the interaction of MSM and BMI.
> Alternative Hypothesis (H_1): Femoral cross-sectional measurements vary with MSM expression, BMI, or the interaction of MSM and BMI.

Now that an appropriate testable question has been developed along with a set of testable hypotheses, you can run the statistical analysis.

- This tutorial focuses on running an ANOVA in R.
- Refer to Chapter 15 for R specific terminology and instructions on how to invoke and construct code.
- Check all assumptions prior to running the test.

Two-Way ANOVA R Tutorial

1. At the command prompt, type in the following 3 vectors of variables: (1) **BMI** (using WHO classifications described in the one-way ANOVA R tutorial: **1** = Normal, **2** = Obese I, **3** = Obese II, **4** = Obese III), (2) **GLMLR** (gluteus maximus MSM robusticity expression on a scale of 0–3), and (3) **FMS** (anterioposterior measurement/mediolateral measurement taken at midshaft). Input the appropriate data formatted in Excel into each vector. Press enter/return.

	A	B	C
1	BMI	FMS_L	GLML R
2	1	1.00	0
3	1	1.00	0
4	1	1.00	0
5	1	0.90	1
6	1	1.07	1
7	1	1.03	1
8	1	1.03	1
9	1	1.04	1
10	2	1.08	1
11	2	1.06	1
12	2	1.00	1
13	2	0.93	1
14	2	1.21	1
15	3	1.03	1
16	3	1.25	1
17	3	1.14	1
18	3	1.03	1
19	4	1.00	1
20	4	1.10	1
21	4	0.94	1
22	1	1.23	2
23	1	1.18	2
24	1	1.18	2
25	1	1.11	2
26	1	1.07	2
27	2	0.88	2
28	2	1.00	2
29	2	1.26	2
30	2	1.03	2
31	2	1.14	2
32	3	1.28	2
33	3	1.03	2
34	3	1.03	2
35	3	1.07	2
36	3	1.03	2
37	4	1.19	2
38	4	0.94	2
39	4	1.19	2
40	2	1.11	3
41	3	1.00	3
42	3	1.00	3
43	4	1.03	3
44	4	1.10	3
45	4	1.06	3
46	4	1.00	3

```
> BMI<-c(1,1,1,1,1,1,1,1,2,2,2,2,2,3,3,3,3,4,4,4,1,1,1,1,1,2,2,2,2,2,3,3,3,
3,3,4,4,4,2,3,3,4,4,4,4)
```

```
> GLMLR<-c(0,0,0,1,1,1,1,1,1,1,1,1,1,1,1,1,1,1,1,1,2,2,2,2,2,2,2,2,2,2,2,2,
2,2,2,2,2,2,3,3,3,3,3,3,3)
```

```
> FMS<-c(1.00,1.00,1.00,0.90,1.07,1.03,1.03,1.04,1.08,1.06,1.00,0.93,1.21,
1.03,1.25,1.14,1.03,1.00,1.10,0.94,1.23,1.18,1.18,1.11,1.07,0.88,1.00,
1.26,1.03,1.14,1.28,1.03,1.03,1.07,1.03,1.19,0.94,1.19,1.11,1.00,1.00,
1.03,1.10,1.06,1.00)
```

2. Like in the one-way ANOVA R tutorial, ensure that **GLMLR** and **BMI** are recognized by R as a factor variable by using the `factor()` function. Press enter/return.

```
> is.factor(GLMLR)
[1] FALSE
> is.factor(BMI)
[1] FALSE
```

An answer of FALSE indicates it is not a categorical variable, while TRUE means it is.

3. If one or both of the variables are not considered a factor variable, convert it to a factor by applying the `factor()` function to each vector name and saving it to a variable of the same name. Type in the programming below for each vector and press enter/return.

```
> GLMLR<-factor(GLMLR)
> BMI<-factor(BMI)
```

4. The `aov()` or `lm()` function can be utilized for a two-way ANOVA by multiplying a second independent variable against the first. The order the vectors should appear in the function is Y~X1*X2, or FMS ~ GLMLR * BMI. As an R script, the programming would be typed in at the command prompt as the following.

```
> femur<-lm(FMS~GLMLR*BMI)
```

Press enter/return.

5. Analyze the output below.

```
> anova(femur)
Analysis of Variance Table

Response: FMS
           Df   Sum Sq   Mean Sq  F value  Pr(>F)
GLMLR       3  0.045110 0.0150366  1.6496  0.1968
BMI         3  0.003112 0.0010375  0.1138  0.9514
GLMLR:BMI   5  0.054032 0.0108065  1.1855  0.3374
Residuals  33  0.300803 0.0091153
```

From the output, we see that none of the variables are significant at an alpha of 0.05, as they all exceed this cutoff. Thus, we should fail to reject the null hypothesis and understand that FMS does not vary in relation to the main effects (independent variables) of **GLMLR** and **BMI**, nor the interaction of these two variables **GLMLR:BMI**.

6. The significance of the main effects can change with the addition of the interaction variable. If you would like to ascertain how well they do in the model alone, or are not interested in interactions, change the "*" to a "+."

```
> femur<-lm(FMS~GLMLR+BMI)
```

7. Analyze the output below.

```
> anova(femur)
Analysis of Variance Table

Response: FMS
          Df  Sum Sq   Mean Sq  F value Pr(>F)
GLMLR      3 0.04511 0.0150366  1.6103 0.2030
BMI        3 0.00311 0.0010375  0.1111 0.9531
Residuals 38 0.35484 0.0093378
```

The *p*-values are all greater than 0.05 for the independent variables, indicating that **FMS** does not vary with **GLMLR** or **BMI**; therefore, we would fail to reject the null hypothesis.

Concluding Statement

Steps 4 and 5: There is no significant difference in anterior–posterior femoral measurements among normal weight and the three obese classifications (I, II, and III) as defined by the World Health Organization ($F(3) = 1.6496$, *p*-value $= 0.1968$), among gluteus maximus MSM expression scores ($F(3) = 0.1138$, *p*-value $= 0.9514$), or the interaction of these two variables, ($F(5) = 1.1855$, *p*-value $= 0.3374$).

Steps 6 and 7: There is no significant difference in anterior–posterior femoral measurements among normal weight and the three obese classifications (I, II, and III) as defined by the World Health

Organization ($F(3) = 1.6103$, p-value $= 0.2030$) or among gluteus maximus MSM expression scores ($F(3) = 0.1111$, p-value $= 0.9531$).

Note: rANOVA is complex to complete in R. Other software, such as SPSS, have automated functions that are easy to use. Thus, it is suggested another program be used for rANOVA other than R.

7

Mann–Whitney *U* and Wilcoxon Signed-Rank

> **Learning Outcomes**
>
> By the end of this chapter, you should be able to:
>
> 1. Determine when and why a Mann–Whitney *U* or Wilcoxon signed-rank test is suitable for a particular dataset.
> 2. Use statistical programs to run a Mann–Whitney *U* and a Wilcoxon signed-rank test and determine the significance, or *p*-value, for an analysis.
> 3. Evaluate the significant difference between two medians and construct a logical conclusion for each dataset.
> 4. Use the skills generated to run and evaluate your own dataset from independent research.

7.1 Mann–Whitney *U* and Wilcoxon Signed-Rank Background

The Mann–Whitney *U* test and the Wilcoxon signed-rank test are both commonly used two-sample nonparametric statistical tests. Nonparametric tests are useful for data that do not follow a normal distribution or for datasets with a small *N*.

The Mann–Whitney *U* test is the nonparametric equivalent to the parametric, independent *t*-test and can be most useful when determining whether there is a difference between two unrelated **independent** variables. The Wilcoxon signed-rank differs only slightly, as it is characterized as the nonparametric equivalent to the parametric, paired *t*-test and is applicable when determining the difference between two **related** variables. Whether the two variables are unrelated, as in the Mann–Whitney U test, or related, as in the Wilcoxon signed-rank test, both tests evaluate potential differences between two variables, based on their sampling groups. The following chapter will elaborate on the main principles for both tests.

7.2 Assumptions

The Mann–Whitney U test and Wilcoxon signed-rank test are most applicable when the following assumptions are fulfilled:

1. **Data type** – The Mann–Whitney U test can be used with a dependent variable that is ordinal, interval, or ratio. We commonly see dependent variables reported in the form of ordinal variables. For example, subjects using a Likert scale to rate the intensity of their back pain may respond with rankings between 0 and 10 (0 = no back pain at all, 10 = severe, intense back pain). The independent variable must be categorical with two groups.

   Note: Disciplines vary on how they analyze data on a Likert scale, sometimes classifying them as ordinal and sometimes as continuous. Check with your advisor for the appropriate way to treat the data for your discipline.

2. **Distribution of data** – The Mann–Whitney U test and Wilcoxon signed-rank test are nonparametric applications; therefore, the data can fall into a non-normal distribution (see Chapter 2). The evaluation of medians allows us to assess the overall data, despite the presence of outliers. The distribution of the groups is important. For Mann–Whitney U, if the shape of the distribution is the same across the groups, then the medians can be used in interpretations. If the distributions are of different shapes, then the test interprets if there are differences in the shapes of the distributions. The Wilcoxon signed-rank test is slightly different; the shape of the differences between the groups must be symmetrical. If they are not symmetrical, the data need to be transformed or a different statistical test applied.

3. **Sampling groups and observations** – Both the Mann–Whitney *U* test and Wilcoxon signed-rank test require two sampling groups but differ in the relationship between the two variables. The Mann–Whitney *U* test assumes the two variables are independent. If the data are independent, then the observations gathered from one group will not influence or have any effect on the second sampling group. In other words, the observations are gathered from separate subjects/groups/samples rather than the same one. The Wilcoxon signed-rank assumes that the observations within each group are paired. In this case, the repeated observations are derived from the same subject/group/sample.
4. **Equal sample sizes** – Wilcoxon signed-rank test requires the two variables to be of equal sample size.
5. **Random sampling** – The observations are collected at random.

Generalized Hypotheses

The Mann–Whitney *U* and the Wilcoxon signed-rank tests share similar hypotheses. In general, the hypotheses compare the median for population A to that of population B and evaluate differences between the populations. The null hypothesis (H_0) states that the two samples come from the same population, indicating no difference. The first alternative hypothesis (H_1) states that the two independent samples come from different populations.

Null Hypothesis (H_0): The ranks of A = the ranks of B.

Alternative Hypothesis (H_1): The ranks of A > the ranks of B or the ranks of A < the ranks of B.

7.3 Case Study – Mann—Whitney *U* Test

A common practice when conserving biodiversity is to preserve natural areas in the form of parks. A valid question is whether the park designation actually mitigates anthropogenic threats, leading to the preservation of biodiversity. Bruner et al. (2001) compared anthropogenic threats to wildlife in 93 parks and 93 nonpark areas in 22 tropical countries.

Note: In the Bruner et al. (2001) study, the question was proposed: "Do parks help to protect tropical biodiversity?" The original data have been revised for teaching purposes; however, the statistical outcome was kept the same in order to reinforce the findings of the study. Although the authors looked at five anthropogenic threats (land clearing, hunting, logging, fire, and grazing), for the sake of this example, we will provide the output for only the effects of park designation on land clearing.

> **(H)** Based on the question above, formulate the null and alternative hypotheses that address the question proposed.
>
> **
>
> Null Hypothesis (H_0): There will be no difference in land clearing between the park and the unmanaged surrounding area.
> Alternative Hypothesis (H_1): There will be a difference in land clearing between the park and the unmanaged surrounding area.

Experimental Data

Looking at the collected data in Table 7.1 (percent of land cleared in each area), several outliers are present. In addition, the mean and

Table 7.1 Mann–Whitney U experimental data showing the percent of land cleared in the unprotected surrounding area and the park area.

	Percent of Land Cleared in Surrounding Area ($N = 10$)	Percent of Land Cleared in Park Area ($N = 10$)
	8	0
	20	8
	16	2
	11	3
	11	2
	9	4
	9	5
	6	12
	3	0
	7	5
Median	9.0	3.5
Mean	10.0	4.1

median are different. Outliers typically indicate a non-normal distribution; therefore, a Mann–Whitney U test is appropriate for this particular dataset.

Experimental Results

The Mann–Whitney U test yielded a p-value of 0.008 (see Figure 7.1). A p-value less than or equal to the critical value (0.05) indicates a significant difference. The histograms (not shown) also indicate different distributions; therefore, we can proceed with interpreting the mean ranks.

Mann-Whitney Test

Ranks

	LandType	N	Mean Rank	Sum of Ranks
PercentLandClear	1	10	14.00	140.00
	2	10	7.00	70.00
	Total	20		

Test Statistics[a]

	PercentLandClear
Mann-Whitney U	15.000
Wilcoxon W	70.000
Z	−2.653
Asymp. Sig. (2-tailed)	.008
Exact Sig. [2*(1-tailed Sig.)]	.007[b]

a. Grouping Variable: LandType
b. Not corrected for ties.

Figure 7.1 Mann–Whitney U SPSS output.

Graph

For a graphic illustration of the experimental results, see Figure 7.2.

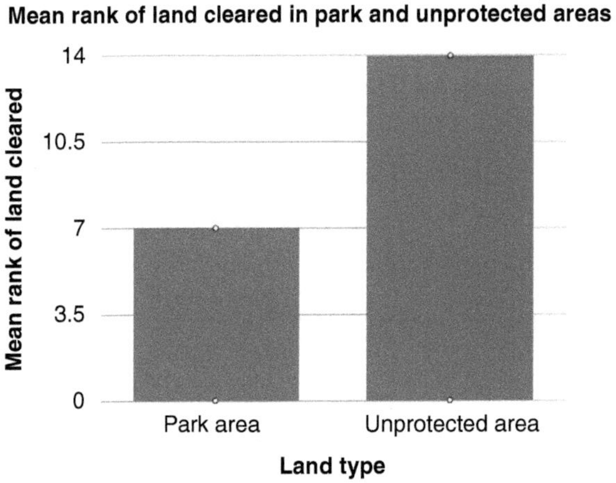

Mean rank of land cleared in park and unprotected areas

(Bar graph: Park area ≈ 7, Unprotected area ≈ 14; y-axis "Mean rank of land cleared" with gridlines at 0, 3.5, 7, 10.5, 14; x-axis "Land type")

Figure 7.2 Bar graph illustrating the mean ranks of land cleared for the unprotected surrounding areas and park areas.

Concluding Statement

Protected park areas had significantly less land clearing than the surrounding unmanaged areas ($U = 15.0$, $p < 0.05$). The null hypothesis can be rejected. The output shows the mean ranks for each area. The mean rank for "park area" ($N_2 = 10$) is 7.00, and the mean rank for "unprotected area" ($N_1 = 10$) is 14.00.

Note: The mean rank is calculated by rank ordering each number and assigning ties a midpoint rank.

7.4 Case Study – Wilcoxon Signed-Rank

The Madagascar hissing cockroach, *Gromphadorhina portentosa*, has been the focus of many studies, including those in the Contreras Laboratory at the University of La Verne. Dr. Contreras is interested in looking at respiratory patterns induced by changes in the metabolic rate (MR) of hissing cockroaches. Metabolic rate is a measurement of energy (calories) usage by the body at rest. Cockroaches are fed a low protein diet and their metabolic rate is then determined by measuring the CO_2 output (ppm) and mass (g) of the individuals.

Suppose a student in the Contreras lab was interested in looking specifically at the roaches' metabolic rate. The student would propose the following question: "Does dietary intake induce changes in metabolic rate?" The roaches were starved for three consecutive days and their resting metabolic rate was then collected (MR_{pre}). The roaches were fed and their metabolic rate was collected immediately following the meal (MR_{post}). The observations accumulated for MR_{pre} and for MR_{post} are considered matched pairs or related groups because they stem from the same cockroaches. Because the data contain matched "related" pairs that are not independent from one another, it is necessary to use Wilcoxon signed-rank test rather than the Mann–Whitney U test.

> **(H)** Based on the question above, formulate the null and alternative hypotheses that address the question proposed.
>
> **
>
> Null Hypothesis (H_0): There will be no difference in metabolic rate before and after feeding.
>
> Alternative Hypothesis (H_1): There will be a difference in metabolic rate before and after feeding.

*The data in this example were taken from the research conducted by Dr. Heidy Contreras. Data were modified for teaching purposes.

Experimental Data

By looking at the data in Table 7.2 (metabolic rate before and after a meal), we can see that the mean and median are different. In addition, it is a relatively small dataset; therefore, a Wilcoxon signed-rank test is an appropriate analysis.

Experimental Results

The z score is -0.052 and has a corresponding p-value of 0.958 (refer to Figure 7.3 and Figure 7.4). The p-value exceeds the critical value (0.05), and therefore, we fail to reject the null hypothesis.

Table 7.2 Wilcoxon signed-rank experimental data showing the metabolic rate of *Gromphadorhina portentosa* before and after a meal.

Cockroach	MR_{pre}	MR_{post}
1	0.024	0.004
2	0.005	0.025
3	0.012	0.002
4	0.017	0.017
5	0.002	0.022
6	0.048	0.008
7	0.024	0.034
8	0.001	0.021
9	0.024	0.004
10	0.005	0.025
11	0.012	0.002
12	0.017	0.017
Median	**0.015**	**0.017**
Mean	**0.016**	**0.015**

Wilcoxon Signed Ranks Test

Ranks

		N	Mean Rank	Sum of Ranks
VAR00002 − VAR00001	Negative Ranks	5[a]	5.40	27.00
	Positive Ranks	5[b]	5.60	28.00
	Ties	2[c]		
	Total	12		

a. VAR00002 < VAR00001
b. VAR00002 > VAR00001
c. VAR00002 = VAR00001

Test Statistics[a]

	VAR00002 − VAR00001
Z	−.052[b]
Asymp. Sig. (2-tailed)	.958

a. Wilcoxon Signed Ranks Test
b. Based on negative ranks.

Figure 7.3 Wilcoxon signed-rank SPSS output.

Graph

Figure 7.4 provides a graphic illustration of the experimental results.

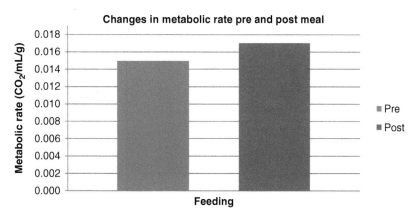

Figure 7.4 Bar graph illustrating the median changes in metabolic rate (CO_2/mL/g) pre and post meal of *Gromphadorhina portentosa*.

Concluding Statement

> **?** Given the *p*-value, what might a concluding sentence look like for a Wilcoxon signed-rank test?
>
> ***
>
> No significant difference was found in the metabolic rate before and after a meal ($Z = -0.052, p = 0.958$); therefore, the null hypothesis fails to be rejected.

Tutorials

7.5 Mann–Whitney *U* Excel Tutorial

Extreme environments have severe impacts on the physiology of fishes. Fishes in the genus *Cyprinodon* are found in both freshwater habitats and hypersaline environments (sometimes over three times the salinity of seawater) throughout the island of Hispaniola. The mechanism to explain how these fish are able to adapt and thrive in both environments remains unknown. In an attempt to investigate adaptations by different populations, a pair of students wanted to study the upregulation of genes associated with oxidative stress, a potential aid in dealing with extreme environments. One particular gene of interest was phospholipid-hydroperoxide glutathione peroxidase (PHGP), which functions in defending cells in a stressful environment. The goal of the study was to determine if relative gene expression of PHGP differed between freshwater and hypersaline fish.

> (R) Formulate a question about the data that can be addressed by performing a Mann–Whitney U test.
>
> **
>
> Question: Will hypersaline fish differ from freshwater fish based on PHGP relative gene expression.

> (H) Based on the question, formulate the null and alternative hypotheses that address the question proposed.
>
> **
>
> Null Hypothesis (H_0): Hypersaline fish will not differ from freshwater fish based on PHGP relative gene expression.
>
> Alternative Hypothesis (H_1): Hypersaline fish will differ from freshwater fish based on PHGP relative gene expression.

Now that an appropriate question has been developed, along with a set of testable hypotheses, you can run the statistical analysis.

- This tutorial focuses on running Mann–Whitney U in Excel.
- Refer to Chapter 12 for tips and tools when using Excel.
- Check all assumptions prior to running the test.

Mann–Whitney U Test Excel Tutorial

1. Arrange the data so that they look similar to the following. (*Note*: **Raq** = La Raqueta, the freshwater environment, **Sal** = Las Salinas, the hypersaline environment)

	A	B
1	Sample	PHGP ΔCt
2	Raq 1	4.85
3	Raq 5	4.17
4	Raq 11	5.97
5	Raq 13	5.17
6	Raq 15	5.25
7	Sal 5	4.40
8	Sal 8	4.43
9	Sal 12	5.81
10	Sal 13	6.95
11	Sal 14	4.76

2. Sort the **PHGP ΔCt** column from lowest to highest. Label **C1** as **Rank**. Assign each value a rank, you can do this manually or using the = RANK function.

	A	B	C
1	Sample	PHGP ΔCt	Rank
2	Raq 5	4.17	1
3	Sal 5	4.40	2
4	Sal 8	4.43	3
5	Sal 14	4.76	4
6	Raq 1	4.85	5
7	Raq 13	5.17	6
8	Raq 15	5.25	7
9	Sal 12	5.81	8
10	Raq 11	5.97	9
11	Sal 13	6.95	10

3. Separate the data based on the categorical variable (e.g., **Raq** and **Sal**). Label **B8** and **F8** as **R1** and **R2**, respectively.

	A	B	C	D	E	F	G
1	Sample	PHGP ΔCt	Rank		Sample	PHGP ΔCt	Rank
2	Raq 5	4.17	1		Sal 5	4.40	2
3	Raq 1	4.85	5		Sal 8	4.43	3
4	Raq 13	5.17	6		Sal 14	4.76	4
5	Raq 15	5.25	7		Sal 12	5.81	8
6	Raq 11	5.97	9		Sal 13	6.95	10
7							
8		R1 =				R2 =	

4. Determine the sum of the ranks for both variables (e.g., **Rag** and **Sal**). To start, type in =**sum** in **C8**. Select the sum function.

R1 = =SUM

- SUM
- SUMIF
- SUMIFS
- SUMPRODUCT
- SUMSQ
- SUMX2MY2
- SUMX2PY2
- SUMXMY2

5. Highlight cells **C2:C6**. Press enter/return.

	A	B	C
1	Sample	PHGP ΔCt	Rank
2	Raq 5	4.17	1
3	Raq 1	4.85	5
4	Raq 13	5.17	6
5	Raq 15	5.25	7
6	Raq 11	5.97	9
7			
8		F	=SUM(C2:C6

6. The following is the calculated sum of the ranks (**R1**) for the first categorical variable (**Raq**).

	A	B	C
1	Sample	PHGP ΔCt	Rank
2	Raq 5	4.17	1
3	Raq 1	4.85	5
4	Raq 13	5.17	6
5	Raq 15	5.25	7
6	Raq 11	5.97	9
7			
8		R1 =	28

7. Repeat steps 4–6 to determine the sum of ranks (**R2**) for the second categorical variable (**Sal**). The following is the calculated sum of the ranks for **Sal**.

E	F	G
Sample	PHGP ΔCt	Rank
Sal 5	4.40	2
Sal 8	4.43	3
Sal 14	4.76	4
Sal 12	5.81	8
Sal 13	6.95	10
	R2 =	27

8. Determine the sample number for each category. Label as **N1** and **N2** for **Raq** and **Sal**, respectively. There are five samples per group; therefore, **N1** and **N2** will equal 5.

	A	B	C	D	E	F	G
1	Sample	PHGP ΔCt	Rank		Sample	PHGP ΔCt	Rank
2	Raq 5	4.17	1		Sal 5	4.40	2
3	Raq 1	4.85	5		Sal 8	4.43	3
4	Raq 13	5.17	6		Sal 14	4.76	4
5	Raq 15	5.25	7		Sal 12	5.81	8
6	Raq 11	5.97	9		Sal 13	6.95	10
7							
8		R1 =	28			R2 =	27
9		N1 =	5			N2 =	5

9. Arrange **R1, N1, R2,** and **N2** in a more accessible manner as seen below.

R1 =	28
N1 =	5
R2 =	27
N2 =	5

10. Compute the **U1** using the equation below.

$$U(1) = N1 \cdot N2 + \left(\frac{N1 \cdot (N1 + 1)}{2}\right) - R1$$

11. In Excel, the equation should be like the following:

R1 =	28	
N1 =	5	=J3*J5+((J3*(J3+1))/2)-J2
R2 =	27	
N2 =	5	

12. Press enter/return. Label this value as **U1**.

R1 =	28
N1 =	5
R2 =	27
N2 =	5
U1 =	12

13. Compute the **U2** using the equation below.

$$U(2) = N1 \cdot N2 + \left(\frac{N2 \cdot (N2 + 1)}{2}\right) - R2$$

14. Press enter/return. Label this value as **U2**.

R1 =	28
N1 =	5
R2 =	27
N2 =	5
U1 =	12
U2 =	13

15. Calculate μu, using the following equation:

$$\mu u = \frac{N1 * N2}{2}$$

16. The final µu value should be equal to the calculated value below.

R1 =	28
N1 =	5
R2 =	27
N2 =	5
U1 =	12
U2 =	13
µu =	12.5

17. Calculate the standard deviation (σU) using the following equation.

$$\sigma u = SQRT\left(\frac{N1 \cdot N2 \cdot (N1 + N2 + 1)}{12}\right)$$

18. To access the square root function, type in = **SQRT** and select the **SQRT** function. Then continue to input the values. Press enter/return.

19. Label value as σU.

R1 =	28
N1 =	5
R2 =	27
N2 =	5
U1 =	12
U2 =	13
µu =	12.5
σu =	4.78714

20. Lastly, the **z score** can be calculated using the smaller U value. In this case, we will use **U1** to compute the **z score**. Use the equation:

$$Z\ Score = \frac{U - \mu u}{\sigma u}$$

21. The formula should look similar to the one below. Press enter/return.

R1 =	28
N1 =	5
R2 =	27
N2 =	5
U1 =	12
U2 =	13
μu =	12.5
σu =	4.78714
	=(J6-J8)/J9

22. The result is the **z score** value.

R1 =	28
N1 =	5
R2 =	27
N2 =	5
U1 =	12
U2 =	13
μu =	12.5
σu =	4.78714
z score	-0.1044

Note: When examining the z table, the area between −1.96 and 1.96 under the z distribution curve indicates no statistical significance. However, the area outside these parameters, below −1.96 or above 1.96, indicates statistical significance. Therefore, the calculated z score can reference these critical values to determine significance. The calculated z score for this example is −0.1044, which falls in the area between −1.96 and 1.96 under the z distribution curve; therefore, there is no significance between the two groups ($p > 0.05$).

Concluding Statement

There is no significant difference between the relative expression of PGHP in fish from the freshwater and the hypersaline populations. ($U = 12, p > 0.05$). The null hypothesis cannot be rejected.

7.6 Wilcoxon Signed-Rank Excel Tutorial

The Centers for Disease and Control and Prevention (CDC) reported 73.5 million adults in the United States have high LDL (130 mg/dL or more). Elevated cholesterol, specifically elevated low-density lipoprotein (LDL) is considered the "bad cholesterol." Those with high LDL levels have approximately twice the risk for heart disease than those with ideal levels.

With 31.7% of Americans struggling with elevated cholesterol, regular exercise is a general recommendation to help lower cholesterol and increase overall good health. Other than regular exercise, prescribed medication is another alternative for reducing LDL and increasing the good cholesterol known as high-density lipoproteins (HDL).

The following hypothetical study recruited 10 males to participate in a research study focused on testing a new drug. Males were of similar demographics and reported similar laboratory results confirming high cholesterol. The research aimed to determine if the new drug would serve as a cholesterol reducing agent. Participants were asked to participate in a 90-day drug treatment trial. Blood was drawn before and after the trial.

> (R) Formulate a question about the data that can be addressed by performing a Wilcoxon signed-rank test.
>
> **
>
> Question: Will pre LDL levels (mg/dL) differ from post LDL levels (mg/dL) after 90 days of treatment?

> **(H)** Based on the question, formulate the null and alternative hypotheses that address the question proposed.
>
> **
>
> Null Hypothesis (H_0): There will be no difference in pre and post LDL levels (mg/dL).
>
> Alternative Hypothesis (H_1): There will be a difference in pre and post LDL levels (mg/dL).

Now that an appropriate question has been developed, along with a set of testable hypotheses, you can run the statistical analysis.

- This tutorial focuses on running Wilcoxon signed-rank test in Excel.
- Refer to Chapter 12 for tips and tools when using Excel.
- Check all assumptions prior to running the test.

Wilcoxon Signed-Rank Excel Tutorial

1. Input data onto an Excel spreadsheet with distinct labels.

	A	B	C
1	Patient	pre LDL (mg/dl)	post LDL (mg/dl)
2	1	132	130
3	2	147	151
4	3	161	158
5	4	143	139
6	5	155	159
7	6	140	147
8	7	156	149
9	8	139	130
10	9	143	134
11	10	157	168

2. Label a new column as **Difference**. Calculate the difference between the **pre LDL (mg/dL)** values and the **post LDL (mg/dL)** values by subtracting the pre from the post using the = ABS function.

	A	B	C	D
1	Patient	pre LDL (mg/dl)	post LDL (mg/dl)	Difference
2	1	132	130	=ABS(B2-C2)
3	2	147	151	
4	3	161	158	
5	4	143	139	
6	5	155	159	
7	6	140	147	
8	7	156	149	
9	8	139	130	
10	9	143	134	
11	10	157	168	

Note: Mathematically, it does not matter the order in which the difference is calculated (pre minus post or post minus pre), as the smaller value is used from the two calculated sums (sum of positive ranks and sum of negative ranks) in determination of the T value in step #9.

3. Press enter/return. The resulting value is the difference between the **pre** and **post LDL** levels. Apply the command to the rest of the column with the fill handle by clicking on the bottom right hand corner of the cell and dragging down.

	A	B	C	D
1	Patient	pre LDL (mg/dl)	post LDL (mg/dl)	Difference
2	1	132	130	2
3	2	147	151	4
4	3	161	158	3
5	4	143	139	4
6	5	155	159	4
7	6	140	147	7
8	7	156	149	7
9	8	139	130	9
10	9	143	134	9
11	10	157	168	11

4. Label a new column as **Ranks**. Calculate the rank of the difference by using the = **RANK** function. The programming within the parentheses follows this order: cell of the difference that you want to compare other differences (in this case **D2**), the array of differences you want to compare it to (in this case **D$2:D$11**),

and a 1, which instructs Excel to rank in ascending order. Press enter/return.

	A	B	C	D	E
1	Patient	pre LDL (mg/dl)	post LDL (mg/dl)	Difference	Ranks
2	1	132	130	2	=RANK(D2, D$2:D$11,1)
3	2	147	151	4	RANK(number, ref, [orde
4	3	161	158	3	
5	4	143	139	4	
6	5	155	159	4	
7	6	140	147	7	
8	7	156	149	7	
9	8	139	130	9	
10	9	143	134	9	
11	10	157	168	11	

Note: Absolute references are used to prevent the row numbers from changing when using the fill handle.

5. Use the fill handle to drag the equation in cell **E2** to cell **E11**.

	A	B	C	D	E
1	Patient	pre LDL (mg/dl)	post LDL (mg/dl)	Difference	Ranks
2	1	132	130	2	1
3	2	147	151	4	3
4	3	161	158	3	2
5	4	143	139	4	3
6	5	155	159	4	3
7	6	140	147	7	6
8	7	156	149	7	6
9	8	139	130	9	8
10	9	143	134	9	8
11	10	157	168	11	10

6. Add two new columns and label one as **Positive Ranks** and the other as **Negative Ranks**. Input the rank values for all data that experienced a positive difference under the **Positive Rank** and a negative difference under **Negative Rank**. A positive difference arises when the cells in the **B** column are larger than the cells in the **C** column. The opposite is also true.

	A	B	C	D	E	F	G
1	Patient	pre LDL (mg/dl)	post LDL (mg/dl)	Difference	Ranks	Positive Ranks	Negative Ranks
2	1	132	130	2	1	1	
3	2	147	151	4	3		4
4	3	161	158	3	2	2	
5	4	143	139	4	3	4	
6	5	155	159	4	3		4
7	6	140	147	7	6		6.5
8	7	156	149	7	6	6.5	
9	8	139	130	9	8	8.5	
10	9	143	134	9	8	8.5	
11	10	157	168	11	10		10

Note: In our example, several of the numbers have the same value. For instance, there are three numbers with a rank value of 3. In this case, you would take the average of 3, 4, and 5 (2) to obtain the positive rank. The next rank value would start at 6; however, we again see there are two numbers with the same value (6). Take the average of 6 and 7 (6.5). Repeat for the remaining numbers.

7. Calculate the sum of ranks for **Positive Ranks** and the **Negative Ranks**. To begin, type in an equals sign underneath the Positive Ranks column. Use the =SUM function to add all of the numbers in the array for **Positive Ranks**. Do the same for **Negative Ranks**.

F Positive Ranks	G Negative Ranks
1	
	4
2	
4	
	4
	6.5
6.5	
8.5	
8.5	
	10
=SUM(F2:F11)	=SUM(G2:G11)

8. Organize your results so they appear like below.

Positive Rank	30.5
Negative Rank	24.5
T=	
N=	
Z=	

9. Of the two calculated sums, whichever is the smaller value will be the **T** value that will be used to calculate z later. Therefore, insert the smaller value into the table as **T**. Also calculate **N**. There are 10 patients that participated in the study, therefore the **N** is 10.

Positive Rank	30.5
Negative Rank	24.5
T=	24.5
N=	10
Z=	

10. Calculate z using the following equation.

$$z = \frac{T - \frac{n(n+1)}{4}}{\sqrt{\frac{n(n+1)(2n+1)}{24}}}$$

11. In Excel, the final z equation should look similar to the one below. Press enter/return.

Positive Rank	30.5
Negative Rank	24.5
T=	24.5
N=	10
Z=	=(G16-(10*(10+1))/4)/(SQRT((10*(10+1))*((2*10)+1)/24))

12. The result is the z score.

Positive Rank	30.5
Negative Rank	24.5
T=	24.5
N=	10
Z=	-0.305788315

Note: When examining the z table, the area between -1.96 and 1.96 under the z distribution curve indicates no statistical significance. However, the area outside these parameters, below -1.96 or above 1.96, indicates statistical significance. Therefore, the calculated z score can reference these critical values to determine significance.

The calculated z score for this example is -0.3058, which does not fall between -1.96 and 1.96 under the z distribution curve; therefore there is a significant difference between the two groups ($p < 0.05$).

Concluding Statement

There was a significant difference in LDL (mg/dL) before and after the 90-day treatment ($Z = -0.3058$, p-value < 0.05). The null hypothesis can be rejected.

7.7 Mann–Whitney *U* SPSS Tutorial

Cranial capacity can tell us much about how brains evolved in the hominid line. Studies of hominids are primarily descriptive as preserved material is sparse and fragmentary. A student is interested in applying a quantitative approach to examine differences between early and late members of our genus *Homo*. He collected a representative sample of later *Homo* brain sizes from the published literature to have an equal sample size to the entirety of early *Homo* material available.

*Data on cranial capacities are from Antón (2012).

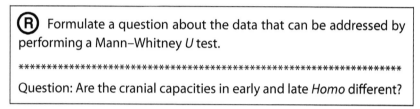

Question: Are the cranial capacities in early and late *Homo* different?

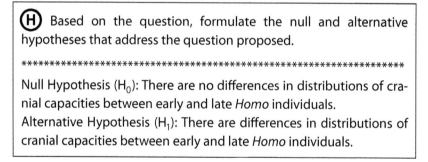

Null Hypothesis (H_0): There are no differences in distributions of cranial capacities between early and late *Homo* individuals.
Alternative Hypothesis (H_1): There are differences in distributions of cranial capacities between early and late *Homo* individuals.

Now that an appropriate question has been developed, along with a set of testable hypotheses, you can run the statistical analysis.

- To run a Mann–Whitney *U* in SPSS, utilize the following tutorial.
- Refer to Chapter 13 for getting started and understanding SPSS.
- Check all assumptions prior to running the test.

Mann–Whitney U Test SPSS Tutorial

1. Type the data into the first two columns in SPSS.

	Homocc	EarlyLate
1	638.00	2.00
2	655.00	2.00
3	690.00	2.00
4	727.00	2.00
5	909.00	2.00
6	995.00	2.00
7	1067.00	2.00
8	510.00	1.00
9	580.00	1.00
10	595.00	1.00
11	631.50	1.00
12	662.50	1.00
13	668.50	1.00
14	750.00	1.00
15		
16		
17		
18		
19		
20		
21		
22		
23		

Note: The first column is a vector of brain sizes and the second column assigns the brain sizes to early (1) or late (2) *Homo*.

2. Click on the **Analyze** menu and select **Nonparametric**, then **Legacy Dialogs,** and then click on **2 Independent Samples.**

Mann–Whitney U and Wilcoxon Signed-Rank | 321

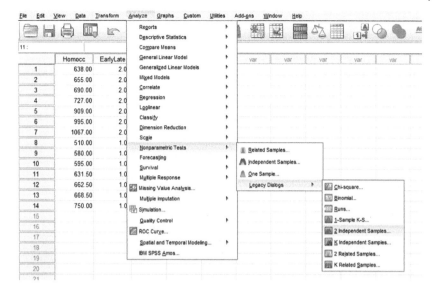

3. The **Two Independent-Samples Tests** box will appear similar to the picture below. Click on **Homocc** to highlight it and click on the arrow next to the **Test Variable List** box.

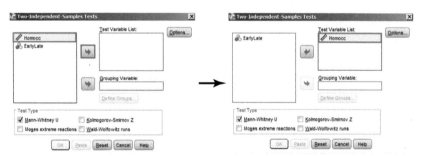

4. Click on the variable for **EarlyLate** and click on the arrow next to the **Grouping Variable** box.

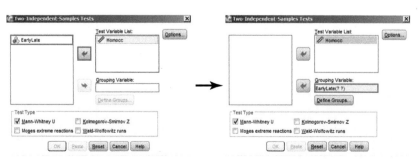

5. Once the two variables are in their respective places in the **Test Variable List** and **Grouping Variable** box, click **Define Groups**.

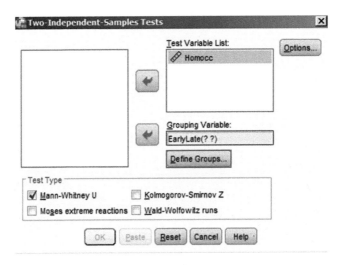

6. In the **Define Groups** box, assign the group numbers in the **EarlyLate** variable in the order they appear in the spreadsheet. Then click **Continue**.

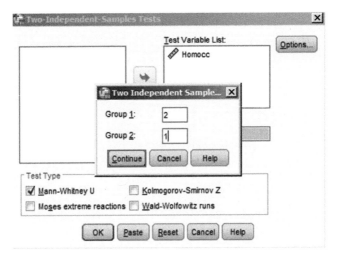

7. SPSS will go back to the **Two Independent-Samples Tests** screen. Make sure the **Test Type** is marked as **Mann-Whitney U**. Click **OK**.

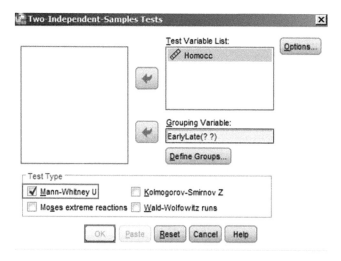

8. A separate document will appear. This is referred to as the output.

Mann-Whitney Test

Ranks

	EarlyLate	N	Mean Rank	Sum of Ranks
Homocc	1.00	7	5.14	36.00
	2.00	7	9.86	69.00
	Total	14		

Test Statistics[a]

	Homocc
Mann-Whitney U	8.000
Wilcoxon W	36.000
Z	-2.108
Asymp. Sig. (2-tailed)	.035
Exact Sig. [2*(1-tailed Sig.)]	.038[b]

a. Grouping Variable: EarlyLate
b. Not corrected for ties.

The Mann–Whitney U value is 8 and the corresponding p-value is low ($p = 0.035$) indicating the null hypothesis should be rejected and there are different distributions between Early and Late *Homo*.

Concluding Statement

Early and Late *Homo* display significantly different distributions of cranial capacities ($U = 8$, p-value $= 0.035$).

Note: If you want to save the SPSS file with the inserted data as well as the SPSS output with the results of the statistical analysis performed, then you must save each document separately (see Chapter 13).

7.8 Wilcoxon Signed-Rank SPSS Tutorial

The following tutorial utilizes some data from the previous Mann–Whitney *U* test in SPSS; it was modified for teaching purposes.

The student was further interested in the variations of different reconstructions of hominid skulls. So he tested the cranial capacities derived from an alternative reconstruction of each hominid's brain size in his study of early and late *Homo*, in order to quantify the disparities between reconstructions.

> **Ⓡ** Formulate a question about the data that can be addressed by performing a Wilcoxon signed-rank test.
>
> **
>
> Question: What is the relationship between two different reconstructions of cranial capacity?

> **Ⓗ** Based on the question, formulate the null and alternative hypotheses that address the question proposed.
>
> **
>
> Null Hypothesis (H₀): There are no differences in cranial capacities between cranial reconstructions of the same individuals.
> Alternative Hypothesis (H₁): There are differences in cranial capacities between cranial reconstructions of the same individuals.

Now that an appropriate question has been developed, along with a set of testable hypotheses, you can run the statistical analysis.

- To run a Wilcoxon signed-rank test in SPSS, utilize the following tutorial.
- Refer to Chapter 13 for getting started and understanding SPSS.
- Check all assumptions prior to running the test.

Wilcoxon Signed-Rank Test SPSS Tutorial

1. The Wilcoxon signed-rank test is handled very similar in SPSS to Mann–Whitney U. The data are loaded slightly differently; the two variables contain cranial capacities for two different cranial reconstructions of brain size (the original cranial capacity we used in the prior example and a second estimation on the same crania) and there is no grouping variable.

Homocc	Alternativecc
638.00	620.00
655.00	670.00
690.00	665.00
727.00	750.00
909.00	927.00
1002.00	980.00
1067.00	1045.00
510.00	500.00
580.00	584.00
595.00	612.00
631.50	637.00
662.50	650.00
668.50	655.00
750.00	780.00

2. Click on the **Analyze** menu and select **Nonparametric**, then **Legacy Dialogs**, and then click on **2 Related Samples**.

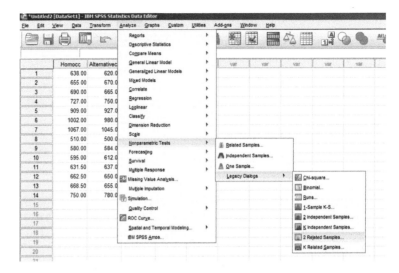

3. Click on **Homocc** and the arrow to the left of the **Test Pairs** box to move **Homocc** to the Variable 1 slot.

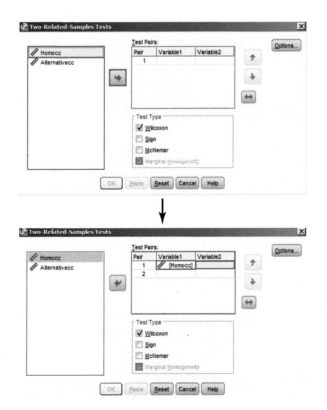

4. Click on **Alternativecc** and the same arrow to the left of the **Test Pairs** box to move **Alternativecc** to Variable 2.

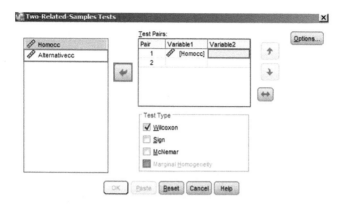

5. Click **OK**.

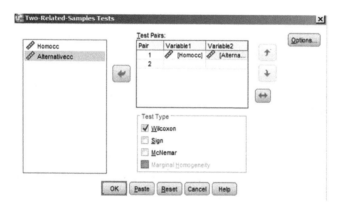

6. A separate document will appear. This is referred to as the output.

Wilcoxon Signed Ranks Test

Ranks

		N	Mean Rank	Sum of Ranks
Alternativecc - Homocc	Negative Ranks	7[a]	7.79	54.50
	Positive Ranks	7[b]	7.21	50.50
	Ties	0[c]		
	Total	14		

a. Alternativecc < Homocc
b. Alternativecc > Homocc
c. Alternativecc = Homocc

Test Statistics[a]

	Alternativecc - Homocc
Z	-.126[b]
Asymp. Sig. (2-tailed)	.900

a. Wilcoxon Signed Ranks Test
b. Based on positive ranks.

The z score is -0.126 and has a corresponding p-value of 0.9, which exceeds our p-value$_{cutoff}$ of 0.05 and indicates we should fail to reject the null hypothesis. Thus, there are no significant differences in cranial capacities between the two different cranial reconstructions.

Concluding Statement

The two approaches to reconstructing cranial capacity did not produce significantly different estimates ($Z = -0.126$, p-value $= 0.9$).

Note: If you want to save the SPSS file with the inserted data as well as the SPSS output with the results of the statistical analysis performed, then you must save each document separately (see Chapter 13).

7.9 Mann–Whitney *U* Numbers Tutorial

Osmoregulation is a complex and energy consuming process involving the active transport of ions across membranes. Fish living in saline environments work hard to maintain their internal environment, which is usually less salty than their surroundings. For example, fish living in the ocean pump salts out of their bodies, and may use upward of 50% of their energy budget to maintain homeostasis (Boeuf and Payan, 2001). One adaptation of fishes in saltwater environments is to increase the number of mitochondrial rich cells in their gills, which allows for increased transport of salts out of the body. Researchers were interested in comparing the adaptations of a hypersaline population of *Limia perugiae*, a small livebearer species that normally occupies freshwater habitats. To investigate whether the hypersaline population had more mitochondrial rich cells in their gills, they measured mitochondrial-dependent oxygen consumption in gill tissues from the hypersaline population and a freshwater population of the same species.

*This example was taken from the research conducted by Dr. Pablo Weaver.

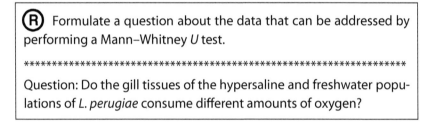

Question: Do the gill tissues of the hypersaline and freshwater populations of *L. perugiae* consume different amounts of oxygen?

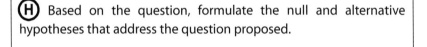

Null Hypothesis (H_0): Oxygen consumption is the same between populations.
Alternative Hypothesis (H_1): Oxygen consumption differs between populations.

Now that an appropriate testable question has been developed, along with a set of testable hypotheses, you can run the statistical analysis.

- To run a Mann–Whitney U in Numbers, utilize the following tutorial.
- More information on programming in Numbers is found in Chapter 14.
- Check all assumptions prior to running the test.

Mann–Whitney *U* Test Numbers Tutorial

1. Start by creating a table similar to the one below. You need a **Salinity** column with a nominal value assigned to each sample (1 = hypersaline and 2 = freshwater) and an **Oxygen Consumption** column.

Oxygen Consumption in Limia

Hypersaline = 1
Freshwater = 2

Salinity	Oxygen Consumption	Rank
1	-37.75	
1	-40.00833333	
1	-41.45	
1	-38.9	
1	-52.77777778	
1	-36.53333333	
1	-38.47647059	
1	-45.3	
1	-72.15	
2	-25.12727273	
2	-25.81875	
2	-19.28461538	
2	-26.73333333	
2	-43.90625	
2	-25.68125	
2	-23.20384615	
2	-23.26111111	
2	-28.44375	
2	-33.54285714	
2	-40.08125	
2	-34.86666667	

2. On the right side of the screen, select the **Sort & Filter** function. Select **Sort Entire Table**, and select **Add a Column**, and **Oxygen Consumption**, then **Ascending (1, 2, 3…)**.

Mann–Whitney U and Wilcoxon Signed-Rank | 331

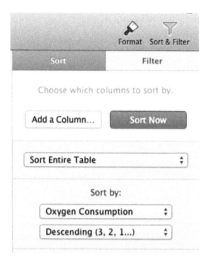

3. Enter the rank of each value in the rank column. In this dataset, all the values have the different values.

Oxygen Consumption in Limia

	Salinity	Oxygen Consumption	Rank
	2	-19.28461538	1
	2	-23.20384615	2
	2	-23.26111111	3
Hypersaline = 1	2	-25.12727273	4
Freshwater = 2	2	-25.68125	5
	2	-25.81875	6
	2	-26.73333333	7
	2	-28.44375	8
	2	-33.54285714	9
	2	-34.86666667	10
	1	-36.53333333	11
	1	-37.75	12
	1	-38.47647059	13
	1	-38.9	14
	1	-40.00833333	15
	2	-40.08125	16
	1	-41.45	17
	2	-43.90625	18
	1	-45.3	19
	1	-52.77777778	20
	1	-72.15	21

Note: If there were two numbers with the same value, you will average the rank of those numbers (see the Numbers Wilcoxon signed-rank tutorial for an example).

4. Next sort by **Salinity** to separate the hypersaline from freshwater samples and split the data into two columns.

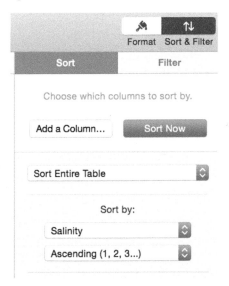

Oxygen Consumption in Limia

	Salinity	Oxygen Consumption	Rank	Salinity	Oxygen Consumption	Rank
	1	-36.53333333	11	2	-19.28461538	1
	1	-37.75	12	2	-23.20384615	2
Hypersaline = 1	1	-38.47647059	13	2	-23.26111111	3
Freshwater = 2	1	-38.9	14	2	-25.12727273	4
	1	-40.00833333	15	2	-25.68125	5
	1	-41.45	17	2	-25.81875	6
	1	-45.3	19	2	-26.73333333	7
	1	-52.77777778	20	2	-28.44375	8
	1	-72.15	21	2	-33.54285714	9
				2	-34.86666667	10
				2	-40.08125	16
				2	-43.90625	18
	N1= 9	R1=		N2= 12	R2=	

Note: In the last image, we have added the values for **N1** (the sample size of hypersaline) and **N2** (the sample size of freshwater). **N1** = 9 and **N2** = 12.

5. Next, calculate the sum of ranks for the hypersaline sample, the **R1** value. Type **=sum** and highlight the column of ranks for the hypersaline fish.

Oxygen Consumption in Limia

		Salinity	Oxygen Consumption	Rank		Salinity	Oxygen Consumption	Rank
		1	-36.53333333	11		2	-19.28461538	1
		1	-37.75	12		2	-23.20384615	2
Hypersaline = 1		1	-38.47647059	13		2	-23.26111111	3
Freshwater = 2		1	-38.9	14		2	-25.12727273	4
		1	-40.00833333	15		2	-25.68125	5
		1	-41.45	17		2	-25.81875	6
		1	-45.3	19		2	-26.73333333	7
		1	-52.77777778	20		2	-28.44375	8
		1	-72.15	21		2	-33.54285714	9
						2	-34.86666667	10
						2	-40.08125	16
						2	-43.90625	18
	N1=		9	=fx SUM ▼ E2:E10 ▼				R2=

6. And, calculate the sum of ranks for the freshwater sample, the **R2** value. Type **=sum** and highlight the column of ranks for the freshwater fish.

Salinity	Oxygen Consumption	Rank
2	-19.28461538	1
2	-23.20384615	2
2	-23.26111111	3
2	-25.12727273	4
2	-25.68125	5
2	-25.81875	6
2	-26.73333333	7
2	-28.44375	8
2	-33.54285714	9
2	-34.86666667	10
2	-40.08125	16
2	-43.90625	18
12	=fx SUM ▼ I2:I13 ▼	

7. Finally, set up the following table:

N1=	
N2=	
R1=	
R2=	
U1=	
U2=	
uu=	
ou=	
Z=	

8. Fill in the numbers we have calculated so far.

N1=	9
N2=	12
R1=	142
R2=	89
U1=	
U2=	
uu=	
ou=	
Z=	

9. Calculate **U1** using the following equation:

$$U(1) = N1 * N2 + \left(\frac{N1 * (N1 + 1)}{2}\right) - R1$$

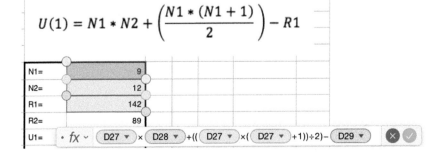

10. Calculate **U2** using the following equation:

$$U(2) = N1 * N2 + \left(\frac{N2 * (N2 + 1)}{2}\right) - R2$$

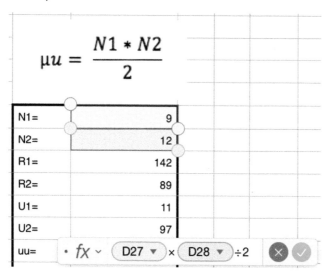

11. Calculate **μu**.

$$\mu u = \frac{N1 * N2}{2}$$

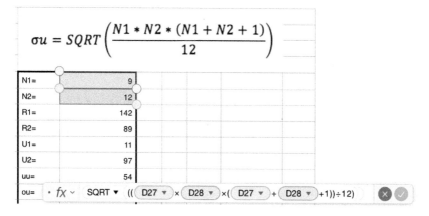

12. Calculate **σU**.

$$\sigma u = SQRT\left(\frac{N1 * N2 * (N1 + N2 + 1)}{12}\right)$$

13. Finally, calculate the z score.

$$Z\ Score = \frac{U - \mu u}{\sigma u}$$

N1=	9
N2=	12
R1=	142
R2=	89
U1=	11
U2=	97
uu=	54
ou=	14.0712472794703
Z=	fx ˅ (D31 ▼) − (D33 ▼) ÷ (D34 ▼) │

14. Here is a summary table of the results:

N1=	9
N2=	12
R1=	142
R2=	89
U1=	11
U2=	97
uu=	54
ou=	14.0712472794703
Z=	-3.05587693443042

When examining the z table, the area between -1.96 and 1.96 under the z distribution curve indicates no statistical significance. However, if the z score is <-1.96 or >1.96, then we can reject the null hypothesis. In this case, $-3.056 < -1.96$; therefore, we can reject the null hypothesis ($p < 0.05$).

Concluding Statement

Fish in hypersaline populations used significantly more oxygen than fish in freshwater populations ($U = 11$, p-value < 0.05).

7.10 Wilcoxon Signed-Rank Numbers Tutorial

One of the biggest challenges that science students face is communicating their science to a variety of audiences. In particular, writing assignments are often seen as a road block to success. Rosenfield et al. (1995) and Chesebro et al. (1992) showed that students who demonstrate high communication apprehension in group work, as well as when speaking to strangers and acquaintances, do not fare well academically. In an effort to improve academic success, communication confidence, and writing ability of Biology majors, Biology faculty teamed up with the Communications Department to develop a course in science communication. The goal of the course was to increase communication competency through a targeted intervention of speech and debate training on science-related topics. Improvement in communication was measured by assessing student writing samples pre and post the speech course intervention.

*This example was taken from the research conducted by Dr. Rob Ruiz, Dr. Kathleen Weaver, and Dr. Jerome Garcia. A selection of the total student data were used in this tutorial.

> **(R)** Formulate a question about the data that can be addressed by performing a Wilcoxon signed-rank analysis.
>
> **
>
> Question: How will the speech and debate training within a science communication course affect writing ability of biology students?

> **(H)** Based on the question, formulate the null and alternative hypotheses that address the question proposed.
>
> ***
>
> Null Hypothesis (H_0): The speech and debate training within a science communication course will not significantly change the writing scores of biology students.
>
> Alternative Hypothesis (H_1): The speech and debate training within a science communication course will significantly change the writing scores of biology students.

Now that an appropriate testable question has been developed, along with a set of testable hypotheses, you can run the statistical analysis.

- To run a Wilcoxon signed-rank in Numbers, utilize the following tutorial.
- More information on programming in Numbers is found in Chapter 14.
- Check all assumptions prior to running the test.

Wilcoxon Signed-Rank Numbers Tutorial

1. Create a table similar to the one below.

Student Number	Pre-course Mean Score	Post-course Mean Score	Difference	Absolute Value of Differences	Rank	(+) Rank	(-) Rank
1	1.75	1					
2	1.75	1					
3	1.75	2.75					
4	2.75	2.25					
5	2.5	2					
6	2.25	2.75					
7	2.5	1.75					
8	2.5	1.25					
9	1.5	3					
10	1.5	2					
11	2	1.25					
12	1.5	2.5					
13	2	2.75					
14	1.75	3					
15	2	2.5					
16	1	1.75					
17	1.25	2.25					
18	3	2					
19	1.75	2.5					
20	2	1.25					
21	2.75	2.5					
22	1.25	1.5					
23	2.25	2.25					
24	1.75	2					
25	2.5	3.75					

2. Calculate the difference between the pre-course and post-course scores. If there is no difference, label the cell **Ignore**. To calculate the difference, type in =cell2-cell1 and press enter/return.

Student Number	Pre-course Mean Score	Post-course Mean Score	Difference	Absolute Value of Differences	Rank
1	1.75	1	= fx (C2 ▼)−(B2 ▼)		
2	1.75	1			
3	1.75	2.75			

Note: Mathematically, it does not matter the order in which the difference is calculated (pre minus post or post minus pre), as the smaller value is used from the two calculated sums (sum of positive ranks and sum of negative ranks) in determination of the T value in step #9.

3. From the center of the cell, pull down and apply the equation to the rest of the cells.

Student Number	Pre-course Mean Score	Post-course Mean Score	Difference
1	1.75	1	-0.75
2	1.75	1	
3	1.75	2.75	
4	2.75	2.25	
5	2.5	2	
6	2.25	2.75	
7	2.5	1.75	
8	2.5	1.25	
9	1.5	3	
10	1.5	2	
11	2	1.25	
12	1.5	2.5	
13	2	2.75	
14	1.75	3	
15	2	2.5	
16	1	1.75	
17	1.25	2.25	
18	3	2	
19	1.75	2.5	
20	2	1.25	
21	2.75	2.5	
22	1.25	1.5	
23	2.25	2.25	
24	1.75	2	
25	2.5	3.75	

4. The table will look like the following:

Pre-course Mean Score	Post-course Mean Score	Difference	A... Di...
1.75	1	-0.75	
1.75	1	-0.75	
1.75	2.75	1	
2.75	2.25	-0.5	
2.5	2	-0.5	
2.25	2.75	0.5	
2.5	1.75	-0.75	
2.5	1.25	-1.25	
1.5	3	1.5	
1.5	2	0.5	
2	1.25	-0.75	
1.5	2.5	1	
2	2.75	0.75	
1.75	3	1.25	
2	2.5	0.5	
1	1.75	0.75	
1.25	2.25	1	
3	2	-1	
1.75	2.5	0.75	
2	1.25	-0.75	
2.75	2.5	-0.25	
1.25	1.5	0.25	
2.25	2.25	0	
1.75	2	0.25	
2.5	3.75	1.25	

5. Calculate the absolute value of the difference value. Type in =**abs**, select the first difference cell and press enter/return.

in	Difference	Absolute Value of Differences	Rank	(+) Ranl
1	-0.75	• fx ˅ ABS ▼ D2 ▼		
1	-0.75			
75	1			

6. Again, from the center of the cell, pull down and apply the equation to the rest of the cells.

Difference	Absolute Value of Differences
-0.75	0.75
-0.75	
1	
-0.5	
-0.5	
0.5	
-0.75	
-1.25	
1.5	
0.5	
-0.75	
1	
0.75	
1.25	
0.5	
0.75	
1	
-1	
0.75	
-0.75	
-0.25	
0.25	
0	
0.25	
1.25	

7. The table will look like the following:

Student Number	Pre-course Mean Score	Post-course Mean Score	Difference	Absolute Value of Differences
1	1.75	1	-0.75	0.75
2	1.75	1	-0.75	0.75
3	1.75	2.75	1	1
4	2.75	2.25	-0.5	0.5
5	2.5	2	-0.5	0.5
6	2.25	2.75	0.5	0.5
7	2.5	1.75	-0.75	0.75
8	2.5	1.25	-1.25	1.25
9	1.5	3	1.5	1.5
10	1.5	2	0.5	0.5
11	2	1.25	-0.75	0.75
12	1.5	2.5	1	1
13	2	2.75	0.75	0.75
14	1.75	3	1.25	1.25
15	2	2.5	0.5	0.5
16	1	1.75	0.75	0.75
17	1.25	2.25	1	1
18	3	2	-1	1
19	1.75	2.5	0.75	0.75
20	2	1.25	-0.75	0.75
21	2.75	2.5	-0.25	0.25
22	1.25	1.5	0.25	0.25
23	2.25	2.25	0	0
24	1.75	2	0.25	0.25
25	2.5	3.75	1.25	1.25

8. To rank the values, sort the differences in descending values. On the right side of the screen, select the **Sort & Filter** function. Select **Add a Column** and select **Absolute Values of Differences**.

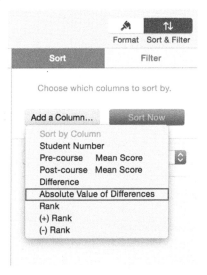

9. Select **Sort Entire Table** and **Descending (1, 2, 3…)** and **Sort Now**.

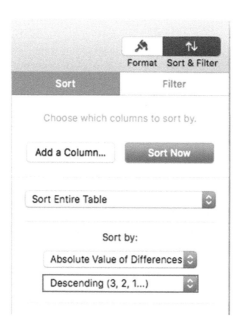

10. Then, apply the **=RANK** function. Select E3:E26 and select Preserve Row under Start and End. Then select largest is high under the largest is high option.

Student Number	Pre-course Mean Score	Post-course Mean Score	Difference	Absolute Value of Differences	Rank	(+) Rank	(-) Rank
23	2.25	2.25	ignore	ignore			
9	1.5	3	1.5	1.5	= RANK ▼ (E3 ▼), (E$3:E$26 ▼), (largest is high ▼)
8	2.5	1.25	-1.25	1.25			
14	1.75	3	1.25	1.25	Start		End
25	2.5	3.75	1.25	1.25	☑ Preserve Row		☑ Preserve Row
3	1.75	2.75	1	1	☐ Preserve Column		☐ Preserve Column

11. From the center of the cell where you just calculated the rank, pull down and apply the equation to the rest of the cells.

Difference	Absolute Value of Differences	Rank
0	ignore	
1.5	1.5	24
-1.25	1.25	
1.25	1.25	
1.25	1.25	
1	1	
1	1	
1	1	
-1	1	
-0.75	0.75	
-0.75	0.75	
-0.75	0.75	
-0.75	0.75	
0.75	0.75	
0.75	0.75	
0.75	0.75	
-0.75	0.75	
-0.5	0.5	
-0.5	0.5	
0.5	0.5	
0.5	0.5	
0.5	0.5	
-0.25	0.25	
0.25	0.25	
0.25	0.25	

Note: In our example, several of the numbers have the same value. For instance, there are three numbers with a value of 0.25. In this case, you would take the average of 1, 2, and 3 (2) to obtain the rank. The next value would start at 4, which follows 3; however, we again see there are five numbers with the same value (0.5). Take the average of 4, 5, 6, 7, and 8 (6). Repeat for the remaining numbers.

Difference	Absolute Value of Differences	Rank
0	ignore	
1.5	1.5	24
-1.25	1.25	22
1.25	1.25	22
1.25	1.25	22
1	1	18.5
1	1	18.5
1	1	18.5
-1	1	18.5
-0.75	0.75	12.5
-0.75	0.75	12.5
-0.75	0.75	12.5
-0.75	0.75	12.5
0.75	0.75	12.5
0.75	0.75	12.5
0.75	0.75	12.5
-0.75	0.75	12.5
-0.5	0.5	6
-0.5	0.5	6
0.5	0.5	6
0.5	0.5	6
0.5	0.5	6
-0.25	0.25	2
0.25	0.25	2
0.25	0.25	2

12. Fill in the rest of the table. Referring to the difference column, positive numbers will go into the (+) **Rank** column and negative numbers will go into the (−) **Rank** column.

Difference	Absolute Value of Differences	Rank	(+) Rank	(-) Rank
0	ignore			
1.5	1.5	24	24	
-1.25	1.25	22		22
1.25	1.25	22	22	
1.25	1.25	22	22	
1	1	18.5	18.5	
1	1	18.5	18.5	
1	1	18.5	18.5	
-1	1	18.5		18.5
-0.75	0.75	12.5		12.5
-0.75	0.75	12.5		12.5
-0.75	0.75	12.5		12.5
-0.75	0.75	12.5		12.5
0.75	0.75	12.5	12.5	
0.75	0.75	12.5	12.5	
0.75	0.75	12.5	12.5	
-0.75	0.75	12.5		12.5
-0.5	0.5	6		6
-0.5	0.5	6		6
0.5	0.5	6	6	
0.5	0.5	6	6	
0.5	0.5	6	6	
-0.25	0.25	2		2
0.25	0.25	2	2	
0.25	0.25	2	2	

13. Calculate the sum of ranks for both the (+) and (−) rank columns. Type **=sum** select all values in the **(+) Rank** column and press enter/return. Repeat this step for **(−) Rank**.

Rank	(+) Rank	(−) Rank
24	24	
22		22
22	22	
22	22	
18.5	18.5	
18.5	18.5	
18.5	18.5	
18.5		18.5
12.5		12.5
12.5		12.5
12.5		12.5
12.5		12.5
12.5	12.5	
12.5	12.5	
12.5	12.5	
12.5		12.5
6		6
6		6
6	6	
6	6	
6	6	
2		2
2	2	
2	2	
Sum of Ranks	fx ⌄ SUM ▾ G3:G26 ▾	

Absolute Value of Differences	Rank	(+) Rank	(-) Rank
1.5	24	24	
1.25	22		22
1.25	22	22	
1.25	22	22	
1	18.5	18.5	
1	18.5	18.5	
1	18.5	18.5	
1	18.5		18.5
0.75	12.5		12.5
0.75	12.5		12.5
0.75	12.5		12.5
0.75	12.5		12.5
0.75	12.5	12.5	
0.75	12.5	12.5	
0.75	12.5	12.5	
0.75	12.5		12.5
0.5	6		6
0.5	6		6
0.5	6	6	
0.5	6	6	
0.5	6	6	
0.25	2		2
0.25	2	2	
0.25	2	2	

· fx ∨ SUM ▼ (H3:H26 ▼)

	(+) Rank	(-) Rank
Sum of Ranks	183	117

14. Insert the following table:

(+) Rank	183
(-) Rank	117
T=	
N=	
z=	

15. *T* is the smaller of the two Rank values, in this case 117. *N* is the sample size of the dataset (without the **Ignore** cells) = 24.

(+) Rank	183
(-) Rank	117
T=	117
N=	24
Z=	

16. Next you will calculate *z* using the following equation.

$$z = \frac{T - \frac{n(n+1)}{4}}{\sqrt{\frac{n(n+1)(2n+1)}{24}}}$$

17. Here is the final output table:

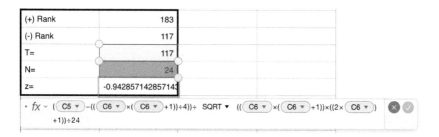

18. When examining the *z* table, the area between −1.96 and 1.96 under the *z* distribution curve indicates no statistical significance. However, if the *z* score is <−1.96 or >1.96, then we can reject the null hypothesis. In this case, our number is between −1.96 and 1.96; therefore, we cannot reject the null hypothesis.

(+) Rank	183
(-) Rank	117
T=	117
N=	24
Z=	-0.942857142857143

Concluding Statement

There is no significant difference between the pre- and post-course writing scores ($z = -0.943$, p-value > 0.05).

7.11 Mann–Whitney *U*/Wilcoxon Signed-Rank R Tutorial

The island of Hispaniola harbors an impressive diversity of freshwater fish species that reflect its unique array of aquatic habitats, from cold mountain streams to warm, coastal lagoons. Live-bearing fishes of the family Poeciliidae comprise much of the diversity of fishes on the island, with unique species colonizing and partitioning nearly every available aquatic habitat. Two closely related species show a particularly interesting distribution pattern, with *Poecilia hispaniolana* occupying the western side of the island and *Poecilia dominicensis* occupying the eastern side of the island, with few instances of sympatry (occupying the same environment) toward the middle of the island. Researchers were interested in explaining the driving factors for this biogeographic pattern and whether it could be explained by differences in the habitat preferences of the two fish species. One parameter that was measured was temperature, with each species potentially exhibiting a specific thermal preference.

*This example was taken from the research conducted by Dr. Pablo Weaver.

Ⓡ Formulate a question about the data that can be addressed by performing a Mann–Whitney *U* test.

****

Question: Do *P. hispaniolana* and *P. dominicensis* show preferences to different water temperatures?

Ⓗ Based on the question, formulate the null and alternative hypotheses that address the question proposed.

****

Null Hypothesis (H_0): The habitats of *P. hispaniolana* and *P. dominicensis* do not differ in temperature.

Alternative Hypothesis (H_1): There is a difference in the temperatures preference between *P. hispaniolana* and *P. dominicensis*.

Now that an appropriate testable question has been developed along with a set of testable hypotheses, you can run the statistical analysis.

- This tutorial focuses on running a Mann–Whitney U and Wilcoxon signed-rank test in R.
- Refer to Chapter 15 for R-specific terminology and instructions on how to invoke and construct code.
- Check all assumptions prior to running the test.

Mann–Whitney U Test Numbers Tutorial

1. Create two vectors that contain the water temperatures (in degrees Celsius) sampled from the habitat of each species of fish. Name the vectors by the species name, or something similarly appropriate. Press enter/return after constructing each vector.

```
> hispaniolana<-c(17.6, 17.8, 18.5, 18.5, 18.6, 19.5, 20.5, 20.6, 21.2, 23.6, 24.1, 24.3, 24.8, 24.8, 24.9, 25.3, 25.3, 25.4, 25.6, 25.7, 25.8)

> dominicensis<-c(21.1, 21.1, 21.6, 23.2, 24.1, 24.3, 24.6, 24.7, 24.9, 25.3, 25.4, 26.4, 26.5, 27.1, 27.2, 27.4, 27.5, 28.4, 29.1, 29.8, 30)
```

2. Type the vector names into the `wilcox.test()` function (*Note*: The Mann–Whitney U is also called the Mann–Whitney–Wilcoxon or Wilcoxon rank sum test, which is why we use the `wilcoxon.test()` function). Press enter/return.

```
> wilcox.test(hispaniolana, dominicensis)
```

Note: The `wilcox.test()` function can also be used to perform a Wilcoxon signed-rank test by simply adding the argument `paired = TRUE`, which would look like the following: `wilcox.test(hispaniolana, dominicensis, paired = TRUE)`. The output is interpreted similarly to the output for the Mann–Whitney U.

3. The following screen will appear. This is known as the output.

```
Wilcoxon rank sum test with continuity correction

data:  hispaniolana and dominicensis
W = 103, p-value = 0.003234
alternative hypothesis: true location shift is not equal to 0

Warning message:
In wilcox.test.default(hispaniolana, dominicensis) :
  cannot compute exact p-value with ties
```

In the case of the water temperature in the habitats of two fish species, the p-value is significant (p-value = 0.003234) indicating that we should reject the null hypothesis that the two species of fish prefer the same temperature in their habitat. Therefore, the fish species prefer two different habitats. A quick glance of the data shows us the water temperatures are higher for *P. dominicensis*.

Concluding Statement

The water temperatures (degrees Celsius) significantly differ ($W = 103$, p-value $= 0.003234$) between the habitats for *P. hispaniolana* and *P. dominicensis* on the island of Hispaniola. An examination of the data show that *P. dominicensis* prefers water with a warmer temperature.

8

Kruskal–Wallis

> **Learning Outcomes:**
>
> By the end of this chapter, you should be able to:
>
> 1. Understand when a Kruskal–Wallis test is applicable and why.
> 2. Use statistical programs to perform a Kruskal–Wallis and determine the significance (*p*-value) for your analysis.
> 3. Evaluate the medians or mean ranks of three or more groups or samples and construct a logical conclusion for each dataset.
> 4. Use the skills generated to perform, analyze, and evaluate your own dataset from independent research.

8.1 Kruskal–Wallis Background

In the last chapter, we covered the Mann–Whitney U test and the Wilcoxon signed-rank test, which are two-sample, nonparametric tests. Another commonly used nonparametric test is the two- or more sample Kruskal–Wallis test. The Kruskal–Wallis is considered the nonparametric analogue to the parametric, one-way ANOVA. The Kruskal–Wallis test is also considered to be similar to the Mann–Whitney U test, as it performs a comparison or analysis of independent samples.

The Kruskal–Wallis is most applicable when comparing two or more samples; the data within each of the multiple samples do not need to follow a normal distribution. As with one-way ANOVA, the test aims to determine if there is an overall difference among the various sampling groups; also similar to one-way ANOVA, the Kruskal–Wallis lacks the ability to specify where the difference lies between groups.

For example, suppose a pharmacist was testing the efficacy of three novel drugs: Drug A, Drug B, and Drug C. After running a Kruskal–Wallis test, the results indicate that there is a significant difference in patient drug response ($p < 0.05$); however, the results would not indicate which drug was the most or least effective. Was the difference between Drug B and Drug C? Or was the difference between Drug A and Drug C? In this case, we cannot determine significance between each group separately, only that there is a significant difference somewhere between drugs.

A graph can depict where the differences lie, as can interpretations of the medians or mean ranks. A test such as Dunnett's C post hoc test can also be applied to determine which of the drugs were the most/least effective. To present your results from a Kruskal–Wallis test, you want to report the chi-square value (X^2), degrees of freedom (df), p-value, and median or mean rank for each group.

8.2 Case Study 1

Snails are known to be the intermediate host for a number of parasites that later infect mammalian hosts, including humans and domestic livestock. For example, freshwater areas in Africa are at high risk for containing parasitic blood-flukes belonging to the genus *Schistosomes*. *Schistosomes* invade the digestive tract of freshwater snails, and continue their life cycle until they are excreted into the water in the form of larvae. As humans come into contact with contaminated water, they run the risk of becoming the secondary host for *Schistosomes*, leading to the disease Schistosomiasis.

In the following hypothetical example, Dr. Perkins, a parasitologist, wants to examine the parasite load of intermediate hosts from Lake Malawi, the southernmost East African rift lake. She collected 50 snails from the lake and identified three species of *Bulinus*

(*Bulinus forskalii*, *Bulinus beccarii*, and *Bulinus cernicus*) from her collection. She performed a dissection on each of the individuals and kept a tally of the number of parasites observed in the gastrointestinal tract of each individual. Dr. Perkins proposed the following question and hypotheses.

> (?) Question: "Is there a difference in the number of parasites in the three different species of *Bulinus*?"

> (H) Null Hypothesis (H_0): There is no difference in the number of parasites in the three different species of *Bulinus*.
> Alternative Hypothesis (H_1): There is a difference in the number of parasites in the three different species of *Bulinus*.

Experimental Data

Reported parasite load (number of parasites) within individuals of the three species of snails, *B. forskalii*, *B. beccarii*, and *B. cernicus* can be observed in Table 8.1. Given the experimental data, we can conclude that the median and mean values are not equal for each of the three groups and that a series of outliers are present within the dataset. From these observations, we can determine that a Kruskal–Wallis is best suited for this example.

Experimental Results

Figure 8.1 displays the SPSS output for the experimental results.

Graph

Because it is a nonparametric test, we display results from a Kruskal–Wallis test using the medians, see Figure 8.2.

Table 8.1 Kruskal–Wallis experimental data showing the number of parasites within individual snails from one of the three species of snails, *Bulinus forskalii*, *Bulinus beccarii*, and *Bulinus cernicus*. The median and mean are calculated and provided at the bottom.

	B. forskalii (n = 18)	B. beccarii (n = 13)	B. cernicus (n = 19)
	9	0	1
	12	4	0
	2	0	3
	6	1	2
	1	7	4
	4	8	0
	6	2	2
	3	5	3
	5	6	1
	7	1	5
	2	0	2
	10	5	0
	3	3	1
	2		2
	8		1
	2		1
	8		7
	3		6
			1
Median	4.50	3.00	2.00
Mean	5.17	3.23	2.21

Kruskal-Wallis Test

Ranks

	Species	N	Mean Rank
Parasites	1	18	33.44
	2	13	23.96
	3	19	19.03
	Total	50	

Test Statistics[a,b]

	Parasites
Chi-Square	9.392
df	2
Asymp. Sig.	.009

a. Kruskal Wallis Test
b. Grouping Variable: Species

Figure 8.1 Kruskal–Wallis SPSS output.

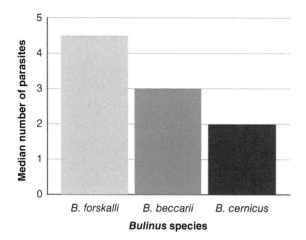

Figure 8.2 Bar graph illustrating the median number of parasites observed among the three snail species, Bulinus forskalii, Bulinus beccarii, and Bulinus cernicus.

Concluding Statement

After reviewing the experimental results, Dr. Perkins was able to determine that there was a significant difference in the number of parasites among the three different species of *Bulinus* ($X^2 = 9.392$, df = 2, $p = 0.009$) snails. The reported medians for each independent group are as follows: Species 1 (*B. forskalii*, $n_1 = 18$) with a median of 4.5; Species 2 (*B. beccarii*, $n_2 = 13$) with a median of 3; and Species 3 (*B. cernicus*, $n_3 = 19$) with a median of 2. The Kruskal–Wallis test does not allow for a direct comparison between pairs of species. A post hoc analysis, such as Dunnett's C, will need to be run in order to determine the specific pairwise differences.

Nuts and Bolts

The Kruskal–Wallis test is most applicable when certain conditions and assumptions are fulfilled.

Assumptions

1. **Data type** – The dependent variable is ordinal or interval/ratio in nature. Similar to the Mann–Whitney *U* test, the Kruskal–Wallis

test must report the dependent variable in terms of a ranking or continuous data, as in IQ or exam scores. The independent variable should be categorical with two or more groups.
2. **Distribution of data** – The Kruskal–Wallis is a nonparametric statistical test; therefore, the data follow a non-normal distribution. As we discussed in Chapter 4, the evaluation of medians allows us to assess the overall data, despite the presence of outliers skewing the data. The distribution of the groups is important. If the shape of the distribution is the same across the groups, then the medians can be used in interpretations. If the shape of the distribution is different, then the mean ranks are interpreted.
3. **Sampling groups and observations** – The Kruskal–Wallis test is commonly used when you have two or more sampling groups, unlike the Mann–Whitney U test which has only two groups. In addition, the Kruskal–Wallis test assumes that the observations are independent and gathered from different individuals or subjects; one individual cannot belong to more than one group. In the case where one individual contributed to more than one observation, the observations are not considered to be independent.
4. **Random sampling**– The observations were collected at random.

8.3 Case Study 2

Researchers from a local sleep institute are interested in studying the effects of alcohol consumption on the quality of sleep. They recruited 20 healthy people between the ages of 25 and 35 years to participate in a study. For an average sized person, one standard alcoholic drink is equivalent to 5 ounces of wine, 1.5 ounces of an 80 proof distilled drink, or 0.16 g/kg of body weight (Williams and Salamy, 1972). On the initial night of the study, each participant's body weight was measured. Then, they were randomized into four groups and asked to consume different amounts of alcohol based on their group and body weight. The following are the randomized control/treatment groups:

Control

Group 1: No alcohol consumed (0.00 g of alcohol/kg body weight)

Treatment Groups

Group 2: One standard drink (0.16 g of alcohol/kg body weight)
Group 3: Two standard drinks (0.32 g of alcohol/kg body weight)
Group 4: Three standard drinks (0.48 g of alcohol/kg body weight)

All the participants were given an equal amount of liquid to consume, with varying amounts of alcoholic content, depending on their assigned group. The participants were blinded to their group assignment (they were not told to which group they were assigned), and the taste of alcohol was masked. Alcohol was administered 60 minutes prior to "bedtime" for all participants, and they were allowed to sleep for an 8-hour period. The next day, they were asked to rank their quality of sleep on a scale from 1 to 5:

1 = Could not sleep, 2 = Had somewhat difficulty in sleeping, 3 = Decent night's sleep, 4 = Slept comfortably, 5 = Slept great

> **(R)** Formulate a question about the data that can be addressed by performing a Kruskal–Wallis test.
>
> **
>
> Question: Does the amount of alcohol consumed before bedtime affect the quality of sleep?

> **(H)** Formulate the null and alternative hypotheses that address the question proposed.
>
> **
>
> Null Hypothesis (H_0): There is no difference in the overall quality of sleep following the consumption of differing levels of alcohol in a standard drink.
>
> Alternative Hypothesis (H_1): There is a difference in the overall quality of sleep following the consumption of differing levels of alcohol in a standard drink.

Experimental Data

For the summary of collected sleep satisfaction scores, see Table 8.2. Even though the data are not highly skewed, because the dataset is

Table 8.2 Kruskal–Wallis experimental data showing the summary of collected sleep satisfaction scores. The median and mean are calculated and provided at the bottom.

	Control	One Drink	Two Drinks	Three Drinks
	5	3	2	2
	3	4	2	1
	4	3	3	2
	4	4	1	1
	2	4	1	1
Median	4	4	2	1
Mean	3.6	3.6	1.8	1.4

particularly small and is composed of ranked responses, it is appropriate to use a Kruskal–Wallis test for the analysis of these data.

Experimental Results

Figure 8.3 illustrates the SPSS output for the experimental results.

Kruskal-Wallis Test

Ranks

Concentration		N	Mean Rank
Score	1	5	14.90
	2	5	15.20
	3	5	6.90
	4	5	5.00
	Total	20	

Test Statistics[a,b]

	Score
Chi-Square	12.766
df	3
Asymp. Sig.	.005

a. Kruskal Wallis Test
b. Grouping Variable: Concentration

Figure 8.3 Kruskal–Wallis SPSS output.

Graph

Refer to Figure 8.4 for a graphic illustration of the results.

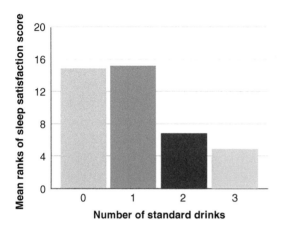

Figure 8.4 Bar graph illustrating the mean ranks of sleep satisfaction score for the four treatment groups.

Concluding Statement

After running the Kruskal–Wallis analysis, the output indicated a significant difference in the quality of sleep between the four different treatment groups ($X^2 = 12.766$, df $= 3$, $p = 0.005$). Thus, we reject the null hypothesis and the alternative hypothesis is supported. The ranks chart reported the mean ranks for each independent group; the mean rank of 14.90 for the control (0 g/kg, $n_1 = 5$) group, 15.20 for one standard drink (0.16 g/kg, $n_2 = 5$), 6.90 for two standard drinks (0.32 g/kg, $n_3 = 5$), and 5.00 for three standard drinks (0.48 g/kg, $n_4 = 5$). Recall that the output suggests there is a difference between groups, but does not specify which groups are significant *vis-à-vis* other groups. To determine individual effects for each group relative to others, a post hoc test needs to be utilized.

Tutorials

8.4 Kruskal–Wallis Excel Tutorial

Snails within the genus *Oreohelix* are known for having a high degree of plasticity in their shell and reproductive morphologies, which makes documenting diversity challenging. However, species identification is especially necessary for groups living in threatened habitats. One example is the dry forests of Central Washington which are prone to large forest fires that have destroyed snail populations throughout the area. In an effort to describe *Oreohelix* snail diversity in the region, a study aimed to add a radula (tooth) count dataset to the existing shell, reproductive, and molecular data and formally describe the species within the region.

*This example was taken from the research conducted by Vanessa Morales, Robert Candelaria, and Dr. Kathleen Weaver.

(R) Formulate a question about the data that can be addressed by performing a Kruskal–Wallis analysis.

****

Question: Does radula tooth count differ by species within the genus *Oreohelix*?

(H) Based on the question, formulate the null and alternative hypotheses that address the question proposed.

****

Null Hypothesis (H_0): Radula tooth count is the same across the *Oreohelix* species examined.

Alternative Hypothesis (H_1): Radula tooth count is different across the *Oreohelix* species examined.

Now that an appropriate testable question has been developed, along with a set of testable hypotheses, you can run the statistical analysis.

- Utilize the following tutorial to run a Kruskal–Wallis test in SPSS.
- Refer to Chapter 13 for getting started and understanding SPSS.
- Check all assumptions prior to running the test.

Kruskal–Wallis Excel Tutorial

1. Arrange the data in Excel to appear similar to the following.

	A	B
1	Species	Tooth Count
2	O. junii	59
3	O. junii	75
4	O. junii	55
5	O. junii	57
6	O. junii	51
7	O. junii	45
8	O. chelanensis	49
9	O. chelanensis	51
10	O. chelanensis	24
11	O. chelanensis	43
12	O. chelanensis	46
13	O. tinensis	42
14	O. tinensis	47
15	O. tinensis	47
16	O. tinensis	51
17	O. tinensis	53
18	O. tinensis	47
19	O. crumensis	49
20	O. crumensis	59
21	O. crumensis	47

2. Insert a new column between column A and column B and label this column as **Group**. Assign each categorical variable (e.g., species) a number.

	A	B	C
1	Species	Group	Tooth Count
2	O. junii	1	59
3	O. junii	1	75
4	O. junii	1	55
5	O. junii	1	57
6	O. junii	1	51
7	O. junii	1	45
8	O. chelanensis	2	49
9	O. chelanensis	2	51
10	O. chelanensis	2	24
11	O. chelanensis	2	43
12	O. chelanensis	2	46
13	O. tinensis	3	42
14	O. tinensis	3	47
15	O. tinensis	3	47
16	O. tinensis	3	51
17	O. tinensis	3	53
18	O. tinensis	3	47
19	O. crumensis	4	49
20	O. crumensis	4	59
21	O. crumensis	4	47

3. Rearrange data in ascending order based on tooth count. Then label column D as **Rank**.

	A	B	C	D
1	Species	Group	Tooth Count	Rank
2	O. chelanensis	2	24	
3	O. tinensis	3	42	
4	O. chelanensis	2	43	
5	O. junii	1	45	
6	O. chelanensis	2	46	
7	O. tinensis	3	47	
8	O. tinensis	3	47	
9	O. crumensis	4	47	
10	O. tinensis	3	47	
11	O. chelanensis	2	49	
12	O. crumensis	4	49	
13	O. junii	1	51	
14	O. chelanensis	2	51	
15	O. tinensis	3	51	
16	O. tinensis	3	53	
17	O. junii	1	55	
18	O. junii	1	57	
19	O. junii	1	59	
20	O. crumensis	4	59	
21	O. junii	1	75	

4. Next assign each value a **rank**. Remember to take the mean rank when two or more values (**Tooth Count**) are the same. For example, $(6+7+8+9)/4 = 7.5$ in the example below.

	A	B	C	D
1	Species	Group	Tooth Count	Rank
2	O. chelanensis	2	24	1
3	O. tinensis	3	42	2
4	O. chelanensis	2	43	3
5	O. junii	1	45	4
6	O. chelanensis	2	46	5
7	O. tinensis	3	47	7.5
8	O. tinensis	3	47	7.5
9	O. crumensis	4	47	7.5
10	O. tinensis	3	47	7.5
11	O. chelanensis	2	49	10.5
12	O. crumensis	4	49	10.5
13	O. junii	1	51	13
14	O. chelanensis	2	51	13
15	O. tinensis	3	51	13
16	O. tinensis	3	53	15
17	O. junii	1	55	16
18	O. junii	1	57	17
19	O. junii	1	59	18.5
20	O. crumensis	4	59	18.5
21	O. junii	1	75	20

5. Separate the data based on the categorical variable (e.g., Group 1, Group 2, Group 3, and Group 4). The data should be organized and appear similar to the following.

Group	Tooth Count	Rank	Group	Tooth Count	Rank	Group	Tooth Count	Rank	Group	Tooth Count	Rank
1	45	4	2	24	1	3	42	2	4	47	7.5
1	51	13	2	43	3	3	47	7.5	4	49	10.5
1	55	16	2	46	5	3	47	7.5	4	59	18.5
1	57	17	2	49	10.5	3	47	7.5			
1	59	18.5	2	51	13	3	51	13			
1	75	20				3	53	15			

6. Determine the sum of the ranks for each categorical variable. To calculate the sum, in an empty box type in = **sum**. Double-click the summation function that appears.

	F	G	H	I	J
1	Group	Tooth Count	Rank		Gro
2	1	45	4		2
3	1	51	13		2
4	1	55	16		2
5	1	57	17		2
6	1	59	18.5		2
7	1	75	20		
8					
9			=sum		
10			SUM		
11			SUMIF		
12			SUMIFS		
13			SUMPRODUCT		
14			SUMSQ		
15			SUMX2MY2		
16			SUMX2PY2		
			SUMXMY2		

7. Highlight the numerical rank values for the first categorical variable. Press enter/return.

	F	G	H	I
1	Group	Tooth Count	Rank	
2	1	45	4	
3	1	51	13	
4	1	55	16	
5	1	57	17	
6	1	59	18.5	
7	1	75	20	
8				
9			=SUM(H2:H7|	

8. The resulting value is the sum of the ranks for the first categorical variable. Label this value as **R1**.

	F	G	H
1	Group	Tooth Count	Rank
2	1	45	4
3	1	51	13
4	1	55	16
5	1	57	17
6	1	59	18.5
7	1	75	20
8			
9		R1 =	88.5

9. Calculate the sum of the ranks for each remaining categorical variable. Label the result values as **R2**, **R3**, and **R4**, respectively.

Group	Tooth Count	Rank	Group	Tooth Count	Rank	Group	Tooth Count	Rank	Group	Tooth Count	Rank
1	45	4	2	24	1	3	42	2	4	47	7.5
1	51	13	2	43	3	3	47	7.5	4	49	10.5
1	55	16	2	46	5	3	47	7.5	4	59	18.5
1	57	17	2	49	10.5	3	47	7.5			
1	59	18.5	2	51	13	3	51	13			
1	75	20				3	53	15			
	R1 =	88.5		R2 =	32.5		R3 =	52.5		R4 =	36.5

10. Determine the sample size for each categorical variable. Label as **N1**, **N2**, **N3**, and **N4**, respectively.

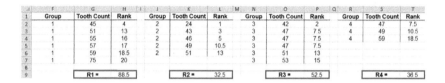

11. Determine the total sample size by calculating the **summation** of N1, N2, N3, and N4. Press enter/return.

n = 20

12. Calculate the **H Statistic** value by utilizing the following equation:

$$H\ Statistic = \left(\frac{12}{n(n+1)}\right)\left(\frac{R1^2}{N1} + \frac{R2^2}{N2} + \frac{R3^2}{N3} + \frac{R4^2}{N4}\right) - 3(n+1)$$

13. After you insert the values from your data that correspond to the variables in the equation, you can solve for **H**.

$$\text{H Statistic} = 6.145238095$$

14. Next, calculate the **degrees of freedom (df)**.

$$df = \#of\ categories - 1$$

In this case, we calculate our df = 4 −1 = 3.

15. Using the degrees of freedom (3), refer to a chi-square table to determine the significance (*p*-value).

Note: If the chi-square value from the table is greater than or equal to the H statistic, then there is a significant difference among the groups, and the *p*-value is less than 0.05. If the chi-square value is less than the H statistic, then there is no statistical difference, and the *p*-value is greater than 0.05.

Percentage Points of the Chi-Square Distribution

Degrees of Freedom	Probability of a larger value of x^2								
	0.99	0.95	0.90	0.75	0.50	0.25	0.10	0.05	0.01
1	0.000	0.004	0.016	0.102	0.455	1.32	2.71	3.84	6.63
2	0.020	0.103	0.211	0.575	1.386	2.77	4.61	5.99	9.21
3	0.115	0.352	0.584	1.212	2.366	4.11	6.25	7.81	11.34
4	0.297	0.711	1.064	1.923	3.357	5.39	7.78	9.49	13.28
5	0.554	1.145	1.610	2.675	4.351	6.63	9.24	11.07	15.09
6	0.872	1.635	2.204	3.455	5.348	7.84	10.64	12.59	16.81

Referring to the chart, the H statistic (6.145) is less than 7.81, which means the *p*-value is greater than 0.05.

Concluding Statement

The research findings indicated no statistical difference in radula tooth count among the sampled species of *Oreohelix* ($X^2 = 7.81$, df = 3, $p > 0.05$). Therefore, we failed to reject the null hypothesis.

Note: Excel does not provide an automated feature for calculating post hoc tests. Other software, such as SPSS and R, have automated functions that are easy to use. Thus, it is suggested another program be used for post hoc tests other than Excel.

8.5 Kruskal–Wallis SPSS Tutorial

Many parents complain that video game violence has increased tremendously within the past 10 years. Bungie Studios has developed three new video games they consider to be only mildly violent. However, parents of children who normally play these types of video games tend to have very different opinions.

For this study, 15 parents ranging from 30 to 40 years of age were asked to watch their 5–10-year-old children test one video game. This experiment was replicated two more times for a total of 3 video games, 45 parents and 45 children. After playing each game for 30 minutes, parents were asked to rate the video game using a scale from 1 to 20 (1 = no violence, 20 = extremely violent).*

*This example was generated for teaching purposes. These are not published data.

> **(R)** Formulate a question about the data that can be addressed by performing a Kruskal–Wallis analysis.
>
> **
>
> Question: Do the median violence scores differ based on the video game played?

> **(H)** Based on the question, formulate the null and alternative hypotheses that address the question proposed.
>
> **
>
> Null Hypothesis (H_0): Median violence scores will not change based on the video game played.
> Alternative Hypothesis (H_1): Median violence scores will change based on the video game played.

Now that an appropriate testable question has been developed along with a set of testable hypotheses, you can run the statistical analysis.

- Utilize the following tutorial to run a Kruskal–Wallis test in SPSS.
- Refer to Chapter 13 for getting started and understanding SPSS.
- Check all assumptions prior to running the test.

Kruskal–Wallis SPSS Tutorial

1. Input the following dataset (formatted in Excel) into SPSS so that it is formatted like the SPSS example to the right of the Excel data. In the variable **Game** in SPSS, the videos were converted from letters to numbers: A = 1, B = 2, C = 3. **Data from Excel Data in SPSS (Truncated).**

Data in Excel

	A	B	C	D
1	Video Game A	B	C	
2		3	18	8
3		2	17	8
4		3	16	8
5		4	10	8
6		5	10	8
7		4	11	9
8		3	12	10
9		4	15	7
10		3	15	8
11		2	14	9
12		2	13	9
13		2	14	7
14		5	15	7
15		5	15	10
16		3	14	9

Data in SPSS

	Game	ParentalRating
1	1.00	3.00
2	1.00	2.00
3	1.00	3.00
4	1.00	4.00
5	1.00	5.00
6	1.00	4.00
7	1.00	3.00
8	1.00	4.00
9	1.00	3.00
10	1.00	2.00
11	1.00	2.00
12	1.00	2.00
13	1.00	5.00
14	1.00	5.00
15	1.00	3.00
16	2.00	18.00
17	2.00	17.00
18	2.00	16.00
19	2.00	10.00
20	2.00	10.00
21	2.00	11.00
22	2.00	12.00
23	2.00	15.00

2. Click on the **Analyze** menu, then select **Nonparametric Tests, Legacy Dialogs,** and finally **K Independent Samples.**

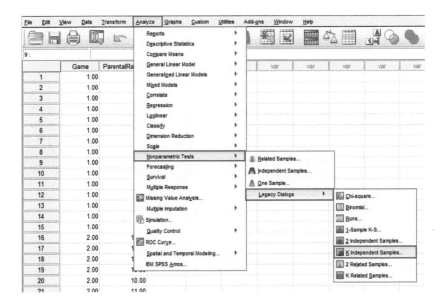

3. The following **Tests for Several Independent Samples** box should pop up. Click on **ParentalRating** and then click on the arrow next to the **Test Variable List** box.

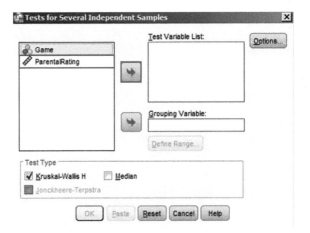

4. Click on **Game** and the arrow next to **Grouping Variable.**

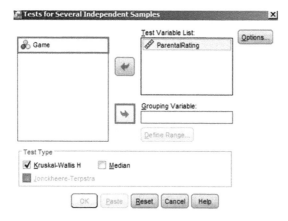

5. Click on **Define Range**.

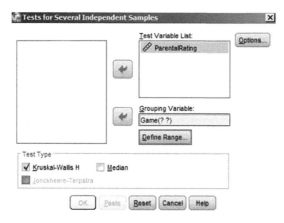

6. Fill in the range of values for the **Game** variable. Click **Continue**.

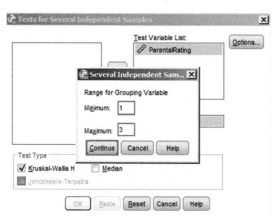

7. Make sure the **Kruskal–Wallis H** box is checked and then click **OK**.

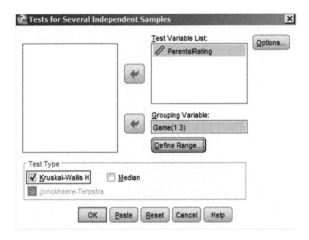

8. A separate document will appear. This is referred to as the output.

Kruskal-Wallis Test

Ranks

	Game	N	Mean Rank
ParentalRating	1.00	15	8.00
	2.00	15	37.87
	3.00	15	23.13
	Total	45	

Test Statistics[a,b]

	ParentalRating
Chi-Square	39.071
df	2
Asymp. Sig.	.000

a. Kruskal Wallis Test
b. Grouping Variable: Game

The p-value (<0.0001) associated with the X^2 value (39.071) is significant, indicating we can reject the null hypothesis that there are no differences among the video games. The mean ranks appear as if all three games are significantly different. However, to ascertain where the differences lie among the groups, we should and will run a post hoc test.

9. Click on the **Analyze** menu, then select **Compare Means** and **One-Way ANOVA**.

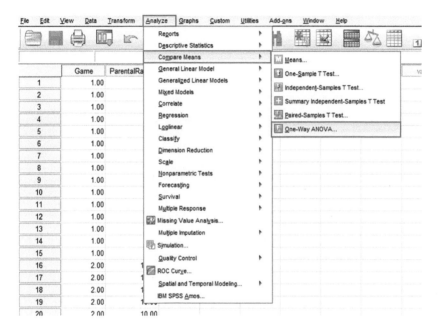

10. Similar to steps 4 and 5 above, assign **ParentalRating** and **Game** as the **Dependent** and **Factor** variables, respectively, by highlighting each variable and then clicking the arrow next to the box to which they belong. Click on **Post Hoc**.

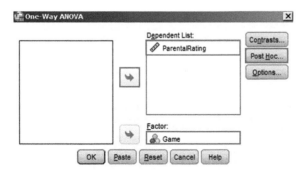

11. Click the box to check **Dunnett's C**. Click **Continue**.

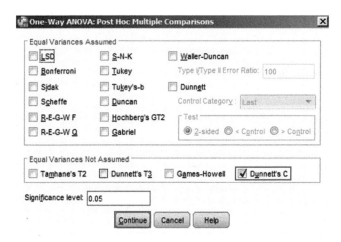

12. Click **OK**.

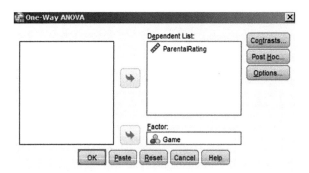

13. A separate output document will appear.

Post Hoc Tests

Multiple Comparisons

Dependent Variable: ParentalRating
Dunnett C

(I) Game	(J) Game	Mean Difference (I-J)	Std. Error	95% Confidence Interval	
				Lower Bound	Upper Bound
1.00	2.00	-10.60000*	.67706	-12.3721	-8.8279
	3.00	-5.00000*	.38214	-6.0002	-3.9998
2.00	1.00	10.60000*	.67706	8.8279	12.3721
	3.00	5.60000*	.66285	3.8651	7.3349
3.00	1.00	5.00000*	.38214	3.9998	6.0002
	2.00	-5.60000*	.66285	-7.3349	-3.8651

*. The mean difference is significant at the 0.05 level.

All values are significant, indicating the three games received significantly different ratings from the parents. Using the negative sign to indicate a group has a lower rating than the group it is being compared to, we see the confidence intervals demonstrate that game A (group 1) is the least violent (has the smallest values), while game B (group 2) is the most violent (largest values).

Concluding Statement

In this sample of video games, there are significant differences in parental ratings of violent content across the three games ($X^2 = 39.071$, df = 2, $p < 0.0001$) with game A being the least violent and game B being the most violent.

Note: If you want to save the SPSS file with the inserted data as well as the SPSS output with the results of the statistical analysis performed, then you must save each document separately (see Chapter 13).

8.6 Kruskal–Wallis Numbers Tutorial

The combination of human disturbance and drought is known to negatively impact arid and semi-arid habitats. In particular, natural seed banks and seed germination are susceptible to continued disturbance, including friction with soil particles, damage from wildfires, or grazing by wildlife and/or domestic animals. Fabião et al. (2013) were interested in increasing the nursery production for *Retama sphaerocarpa*, an ornamental plant from the Mediterranean region with potential importance to ecological restoration of Mediterranean ecosystems as well as bioenergy production. *R. sphaerocarpa* seeds, like other arid plants, are hardy and can survive in nature for years without dessication. In order to maximize germination of *R. sphaerocarpa* seeds within controlled nursery conditions, the researchers tested several methods to increase the permeability of the seed coat, including hot water, sulfuric acid, or no treatment.

Ⓡ Formulate a question about the data that can be addressed by performing a Kruskal–Wallis statistical analysis.

****

Question: Does seed treatment that changes the permeability of the seed coat affect the germination success of *R. sphaerocarpa*?

Ⓗ Based on the question, formulate the null and alternative hypotheses that address the question proposed.

****

Null Hypothesis (H_0): Germination success will be the same across seed treatment type.
Alternative Hypothesis (H_1): Germination success will be different across seed treatment type.

Now that an appropriate testable question has been developed, along with a set of testable hypotheses, you can run the statistical analysis.

- To run Kruskal–Wallis test in Numbers, utilize the following tutorial.
- More information on programming in Numbers is found in Chapter 14.
- Check all assumptions prior to running the test.

Kruskal–Wallis Numbers Tutorial

1. Here is the dataset for the Kruskal–Wallis tutorial.

Germination Percentage

Treatment	Control	Sulphuric Acid	Hot Water
	11	79	29
	11	79	28
	11	77	27
	11	79	28
	12	84	27
	13	77	30
	11	79	32
	10	78	25
	11	78	27
	11	79	28
	12	82	25
	14	81	30
	10	79	31
	9	77	32
	10	79	33
	12	85	28
	13	76	31
	11	82	27
	13	79	28

2. First organize the data into two columns, **Treatment** type (control, sulfuric acid, or hot water) and **Germination Percentage**. Add a column for **Rank**.

Treatment	Germination Percentage	Rank	
Control	9		
Control	10		
Control	10		
Control	10		
Control	11		
Control	11		
Control	11		
Control	11		
Control	11		
Control	11		
Control	11		
Control	11		
Control	12		
Control	12		
Control	12		
Control	13		
Control	13		
Control	13		
Control	14		
Hot Water	25		
Hot Water	25		
Hot Water	27		
Hot Water	27		
Hot Water	27		
Hot Water	27		
Hot Water	28		
Hot Water	28		
Hot Water	28		
Hot Water	28		
Hot Water	28		
Hot Water	29		
Hot Water	30		
Hot Water	30		
Hot Water	31		
Hot Water	31		
Hot Water	32		
Hot Water	32		
Hot Water	33		
Sulphuric Acid	76		
Sulphuric Acid	77		
Sulphuric Acid	77		
Sulphuric Acid	77		
Sulphuric Acid	78		
Sulphuric Acid	78		
Sulphuric Acid	79		
Sulphuric Acid	79		
Sulphuric Acid	79		
Sulphuric Acid	79		
Sulphuric Acid	79		
Sulphuric Acid	79		
Sulphuric Acid	79		
Sulphuric Acid	79		
Sulphuric Acid	81		
Sulphuric Acid	82		
Sulphuric Acid	82		
Sulphuric Acid	84		
Sulphuric Acid	85		

3. Sort the data by **Germination Percentage**.

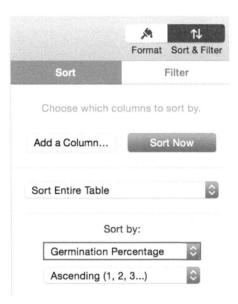

4. Then rank the data. Note that in our example, several of the numbers have the same value. For instance, there are 3 numbers with a value of 10. In this case, you would take the average of 2, 3, and 4 (= 3) to obtain the rank.

Treatment	Germination Percentage	Rank
Control	9	1
Control	10	3
Control	10	3
Control	10	3
Control	11	8.5
Control	11	8.5
Control	11	8.5
Control	11	8.5
Control	11	8.5
Control	11	8.5
Control	11	8.5
Control	11	8.5
Control	12	14
Control	12	14
Control	12	14
Control	13	17
Control	13	17
Control	13	17
Control	14	19
Hot Water	25	20.5
Hot Water	25	20.5
Hot Water	27	23.5
Hot Water	27	23.5
Hot Water	27	23.5
Hot Water	27	23.5
Hot Water	28	28
Hot Water	28	28
Hot Water	28	28
Hot Water	28	28
Hot Water	28	28
Hot Water	29	31
Hot Water	30	32.5
Hot Water	30	32.5
Hot Water	31	34.5
Hot Water	31	34.5
Hot Water	32	36.5
Hot Water	32	36.5
Hot Water	33	38
Sulphuric Acid	76	39
Sulphuric Acid	77	40
Sulphuric Acid	77	41
Sulphuric Acid	77	42
Sulphuric Acid	78	43
Sulphuric Acid	78	44
Sulphuric Acid	79	45
Sulphuric Acid	79	46
Sulphuric Acid	79	47
Sulphuric Acid	79	48
Sulphuric Acid	79	49
Sulphuric Acid	79	50
Sulphuric Acid	79	51
Sulphuric Acid	79	52
Sulphuric Acid	81	53
Sulphuric Acid	82	54
Sulphuric Acid	82	55
Sulphuric Acid	84	56
Sulphuric Acid	85	57

5. Next, resort the data by **Treatment**.

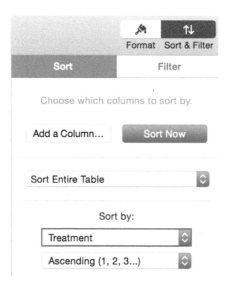

6. Organize the data into three sets of columns (one for each treatment). For all three treatments, determine the sample size (N). In this example, **N1**, **N2**, and **N3** all equal 19. In the following steps, we will calculate **R1**, **R2**, and **R3**.

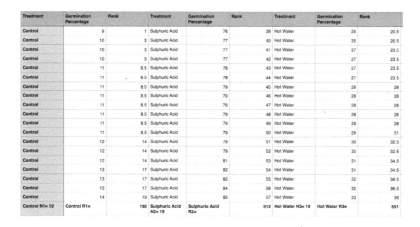

7. Before performing your calculations, create a list off to the side. In this example, N1 = N2 = N3 = 19.

All the variables you will need in order to calculate the H statistic can be pulled from this list.

N1=	
N2=	
N3=	
R1=	
R2=	
R3=	
n=	
H Statistic=	

8. To calculate **R1** (Control), take the sum of the ranks within control. Type =**sum(range of cells)**.

A	B	C	D	E
Treatment	Germination Percentage	Rank	Treatment	Germination Percentage
Control		9	1 Sulphuric Acid	
Control		10	3 Sulphuric Acid	
Control		10	3 Sulphuric Acid	
Control		10	3 Sulphuric Acid	
Control		11	8.5 Sulphuric Acid	
Control		11	8.5 Sulphuric Acid	
Control		11	8.5 Sulphuric Acid	
Control		11	8.5 Sulphuric Acid	
Control		11	8.5 Sulphuric Acid	
Control		11	8.5 Sulphuric Acid	
Control		11	8.5 Sulphuric Acid	
Control		11	8.5 Sulphuric Acid	
Control		12	14 Sulphuric Acid	
Control		12	14 Sulphuric Acid	
Control		12	14 Sulphuric Acid	
Control		13	17 Sulphuric Acid	
Control		13	17 Sulphuric Acid	
Control		13	17 Sulphuric Acid	
Control		14	19 Sulphuric Acid	
Control N1= 19	Control R1=	• *fx* ˅ SUM ▼ C2:C20 ▼ | ✕ ✓		

9. To calculate **R2** (Sulfuric Acid), take the sum of the ranks within sulfuric acid. Type =**sum(range of cells)**.

D	E	F	G	H
Treatment	Germination Percentage	Rank	Treatment	Germination Percentage
Sulphuric Acid	76		39 Hot Water	
Sulphuric Acid	77		40 Hot Water	
Sulphuric Acid	77		41 Hot Water	
Sulphuric Acid	77		42 Hot Water	
Sulphuric Acid	78		43 Hot Water	
Sulphuric Acid	78		44 Hot Water	
Sulphuric Acid	79		45 Hot Water	
Sulphuric Acid	79		46 Hot Water	
Sulphuric Acid	79		47 Hot Water	
Sulphuric Acid	79		48 Hot Water	
Sulphuric Acid	79		49 Hot Water	
Sulphuric Acid	79		50 Hot Water	
Sulphuric Acid	79		51 Hot Water	
Sulphuric Acid	79		52 Hot Water	
Sulphuric Acid	81		53 Hot Water	
Sulphuric Acid	82		54 Hot Water	
Sulphuric Acid	82		55 Hot Water	
Sulphuric Acid	84		56 Hot Water	
Sulphuric Acid	85		57 Hot Water	
Sulphuric Acid N2= 19	Sulphuric Acid R2=	· fx ∨ SUM ▼ F2:F20 ▼ | ✕ ✓		

10. To calculate **R3** (Hot Water), take the sum of the ranks within hot water. Type =**sum(range of cells)**.

Treatment	Germination Percentage	Rank
Hot Water	25	20.5
Hot Water	25	20.5
Hot Water	27	23.5
Hot Water	27	23.5
Hot Water	27	23.5
Hot Water	27	23.5
Hot Water	28	28
Hot Water	28	28
Hot Water	28	28
Hot Water	28	28
Hot Water	28	28
Hot Water	29	31
Hot Water	30	32.5
Hot Water	30	32.5
Hot Water	31	34.5
Hot Water	31	34.5
Hot Water	32	36.5
Hot Water	32	36.5
Hot Water	33	38
Hot Water N3= 19	Hot Water R3=	

11. Here are the calculations so far.

Control N1=	19
Sulphuric Acid N2=	19
Hot Water N3=	19
Control R1=	190
Sulphuric Acid R2=	912
Hot Water R3=	551
n=	
H statistic =	

12. Next calculate n, which is the sum of N1, N2, and N3.

Control N1=		19
Sulphuric Acid N2=		19
Hot Water N3=		19
Control R1=		190
Sulphuric Acid R2=		912
Hot Water R3=		551
n=	• fx ⌄ SUM ▼ L7:L9 ▼	
H statistic =		

13. Finally, calculate the H statistic using the following equation.

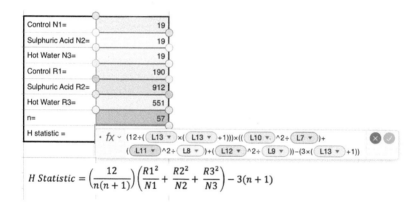

$$H\ Statistic = \left(\frac{12}{n(n+1)}\right)\left(\frac{R1^2}{N1} + \frac{R2^2}{N2} + \frac{R3^2}{N3}\right) - 3(n+1)$$

14. Here are the final calculations.

Control N1=	19
Sulphuric Acid N2=	19
Hot Water N3=	19
Control R1=	190
Sulphuric Acid R2=	912
Hot Water R3=	551
n=	57
H statistic =	49.7931034482758

15. Calculate the degrees of freedom (df).

$$df = \#of\ categories - 1$$

$$df = 3 - 1 = 2$$

16. Using the degrees of freedom, refer to an H statistic significance chart to determine the *p*-value.

degrees of freedom	0.995	0.99	0.975	0.95	0.90	0.10	0.05	0.025	0.01	0.005
1	--	--	0.001	0.004	0.016	2.706	3.841	5.024	6.635	7.879
2	0.01	0.02	0.051	0.103	0.211	4.605	5.991	7.378	9.21	10.597

From the chart, we see that the H statistic (49.793) is greater than 10.597 which means that the *p*-value is less than 0.005.

Concluding Statement

Within controlled nursery conditions, the researchers found that the three methods (hot water, sulfuric acid, or no treatment) affect the permeability of the seed coat and germination success differently ($H = 49.793$, $df = 2$, $p < 0.005$); therefore, the null hypothesis can be rejected.

Note: Numbers does not provide an automated feature for calculating post hoc tests. Other software, such as SPSS and R, have automated functions that are easy to use. Thus, it is suggested another program be used for post hoc tests other than Numbers.

8.7 Kruskal–Wallis R Tutorial

Putty-nosed monkeys have been shown to use several vocalizations, including distinct alarm calls, "pyows" and "hacks," which scientists think are used in combination to signify different predators or other threats. In an effort to decode the alarm patterns of putty-nosed monkeys, Arnold and Zuberbuhler (2006) used a loud speaker to play combinations of pyow–hack (P–H) calls and other vocalizations and measured group movements (meters moved from original location) after the calls.

> **R** Formulate a question about the data that can be addressed by performing a Kruskal–Wallis test.

****

Question: Do putty-nosed monkeys respond differently to vocalizations played (pyow-hack, other, vs. none)?

> **H** Based on the question, formulate the null and alternative hypotheses that address the question proposed.

****

Null Hypothesis (H_0): The response of putty-nosed monkeys will not differ based on type of call.
Alternative Hypothesis (H_1): Putty-nosed monkeys will respond differently to the types of calls played by the researchers.

Now that an appropriate testable question has been developed along with a set of testable hypotheses, you can run the statistical analysis.

- This tutorial focuses on running a Kruskal–Wallis test in R.
- Refer to Chapter 15 for R specific terminology and instructions on how to invoke and construct code.
- Check all assumptions prior to running the test.

Kruskal–Wallis R Tutorial

1. We will be creating a data frame of the two vectors for calls (entitled **calls**) and meters moved from location (entitled **distance**) using the data and script below. The lines of programming will merge the two vectors into a single data frame, called **monkeys**, using the `data.frame()` function. To do this, first we must create a vector for distance like we would when we normally create a vector. Do not press enter/return.

```
> distance<-c(110, 110, 95, 93, 118, 125, 136, 145, 165, 92, 80, 65, 91,
162, 153, 186, 65, 100, 94, 308, 92, 65, 250, 154, 96, 35, 190, 76, 126,
43, 10, 48, 25, 21, 49, 65, 62, 21, 16, 0, 52, 106, 26, 16, 61, 42,
11, 52, 23, 20, 48, 64, 62, 20, 18, 62, 49, 18, 0, 23, 14, 0, 45, 32,
0, 5, 2, 200, 45, 35, 10, 18, 180, 0, 0, 3, 68, 0, 22, 16, 0, 44, 31,
0, 4, 75, 0)
```

2. Next, we will add in the programming to create a vector of calls. To avoid typing the call classification for every meter count, we can assign values 1–3 to the call types, whereby **1** = pyows and hacks, **2** = other calls, and **3** = no calls.

The values in the **distance** vector are ordered by call so that we can use a short cut to program the calls in R. Each call type has 29 associated values in the **distance** vector. Instruct R to repeat the 3 call values (using the **rep()** function) 29 times each (using the **c()** function), as there are 29 observations for each call classification. Do not press enter/return.

```
> distance<-c(110, 110, 95, 93, 118, 125, 136, 145, 165, 92, 80, 65, 91,
162, 153, 186, 65, 100, 94, 308, 92, 65, 250, 154, 96, 35, 190, 76, 126,
43, 10, 48, 25, 21, 49, 65, 62, 21, 16, 0, 52, 106, 26, 16, 61, 42,
11, 52, 23, 20, 48, 64, 62, 20, 18, 62, 49, 18, 0, 23, 14, 0, 45, 32,
0, 5, 2, 200, 45, 35, 10, 18, 180, 0, 0, 3, 68, 0, 22, 16, 0, 44, 31,
0, 4, 75, 0); calls=rep(rep(1:3), c(29,29,29))
```

- Creates a vector of calls 1-3. The ":" means "through"
- Instructs R to repeat each call (1-3) 29 times.

3. Next, we need to tell R the two vectors are in a single data frame. We can name the data frame at the same time (we will call it **monkeys**). Press enter/return.

```
> distance<-c(110, 110, 95, 93, 118, 125, 136, 145, 165, 92, 80, 65, 91,
162, 153, 186, 65, 100, 94, 308, 92, 65, 250, 154, 96, 35, 190, 76, 126,
43, 10, 48, 25, 21, 49, 65, 62, 21, 16, 0, 52, 106, 26, 16, 61, 42,
11, 52, 23, 20, 48, 64, 62, 20, 18, 62, 49, 18, 0, 23, 14, 0, 45, 32,
0, 5, 2, 200, 45, 35, 10, 18, 180, 0, 0, 3, 68, 0, 22, 16, 0, 44, 31,
0, 4, 75, 0); calls=rep(rep(1:3), c(29,29,29)); monkeys=data.frame
(distance, calls)
```

Creates a data frame using the **distance** and **calls** vectors

Check how the data frame was constructed by calling up the data frame (**monkeys**). Press enter/return. An excerpt from the data frame is included below.

```
> monkeys
   distance calls
1       110    1
2       110    1
3        95    1
4        93    1
5       118    1
6       125    1
7       136    1
8       145    1
9       165    1
10       92    1
11       80    1
12       65    1
13       91    1
14      162    1
15      153    1
16      186    1
17       65    1
18      100    1
19       94    1
20      308    1
21       92    1
22       65    1
23      250    1
24      154    1
25       96    1
26       35    1
27      190    1
28       76    1
29      126    1
30       43    2
```

4. Now that the data frame is loaded, the individual vectors (**calls**, **distance**) from the data frame (**monkeys**) can be called upon for testing in the `kruskal.test()` function using the format of `data.frame$vector`. Press enter/return.

```
> kruskal.test(monkeys$distance~monkeys$calls)
```

5. The following screen will appear. This is known as the output.

```
        Kruskal-Wallis rank sum test

data:  monkeys$distance by monkeys$calls
Kruskal-Wallis chi-squared = 48.011, df = 2, p-value = 3.755e-11
```

The output shows a *p*-value less than 0.05 with a corresponding chi-square value ($X^2 = 48.011$, df = 2), indicating the results are significant and the null hypothesis should be rejected. Thus, there are differences in the distances the monkeys traveled in response to different call types.

6. A multiple comparison test (post hoc) can be applied to test where the differences are among the calls. Load the **pgirmess** package using the **library()** function. Press enter/return. If you do not have the package installed, you will get a message that reads "Error in library(pgirmess): there is no package called 'pgirmess'." Refer to Chapter 15 for instructions on how to install packages.

```
> library(pgirmess)
```

7. Apply the **kruskalmc()** function to the data, similar to step 4, and using the **data.frame$vector** format. Press enter/return.

```
> kruskalmc(monkeys$distance, monkeys$calls)
```

8. The following screen will appear. This is known as the output.

```
> kruskalmc(monkeys$distance, monkeys$calls)
Multiple comparison test after Kruskal-Wallis
p.value: 0.05
Comparisons
     obs.dif  critical.dif  difference
1-2  33.37931 15.87987      TRUE
1-3  44.00000 15.87987      TRUE
2-3  10.62069 15.87987      FALSE
```

The **TRUE** and **FALSE** statements refer to whether the observed differences exceed the critical difference at which the differences between the groups become significant. In this case, the distance moved from pyows and hacks (group 1) is significantly different than other calls (group 2) and no calls (group 3). Thus, the putty-nosed monkeys travel greater distances in response to pyows and hacks than other calls and no calls.

Concluding Statement

In this sample of putty-nosed monkeys, there are significant differences ($X^2 = 48.011$, df $= 2$, p-value < 0.05) in the distance they travel in response to different types of calls, traveling further in response to pyows and hacks than to other calls or no calls.

9

Chi-Square Test

> **Learning Outcomes**
>
> By the end of this chapter, you should be able to:
>
> 1. Determine the observed and expected chi-square values as well as the degrees of freedom associated with each scenario.
> 2. Use statistical programs to perform a chi-square test and determine significance.
> 3. Evaluate the relationship between the observed and expected and construct a logical conclusion for each scenario.
> 4. Use the skills acquired to perform, analyze, and evaluate your own dataset from independent research.

9.1 Chi-Square Background

The last test we will cover involving two or more samples is known as the Pearson's chi-square test (X^2 test). The chi-square test (also known as chi-squared test) examines the difference between expected and observed distributions. Specifically, we will look at a goodness-of-fit test, comparing the expected frequency (which is the value that we expect to see based on the literature background material or a hypothesis generated as part of an experiment) to the observed frequency (which is the value actually observed as part of an experiment or study). The goodness-of-fit test compares the distribution of your data to a specified distribution. The default in some programs is a uniform distribution (all expected values are the same), but

An Introduction to Statistical Analysis in Research: With Applications in the Biological and Life Sciences,
First Edition. Kathleen F. Weaver, Vanessa C. Morales, Sarah L. Dunn, Kanya Godde and Pablo F. Weaver.
© 2018 John Wiley & Sons, Inc. Published 2018 by John Wiley & Sons, Inc.
Companion Website: www.wiley/com/go/weaver/statistical_analysis_in_research

researchers can define and test other distributions. Often researchers use the chi-square test in genetics for tests of Hardy–Weinberg equilibrium and for comparing expected and observed offspring phenotypes. The chi-square test is used on categorical variables.

9.2 Case Study 1

The tiger leech, *Haemadipsa picta*, is known to rely on sensory mechanisms to locate prey. Field observations have recorded cases where leeches moved toward warm or moving objects, in addition to detecting sound and vibrations from incoming prey. Kmiecik and colleagues (2008) conducted a comparative study to further examine leech behavior when introduced to a moving or warm object. The following question was proposed: "Does movement or heat play a larger role in leech attraction?"

Approximately 75 leeches were collected from the Belian Trail and surrounding forest area of the Maliau Basin Conservation, Sabah, Malaysia. From the collected groups, 64 were randomly selected to participate in a movement and heat experiment ($N = 32$ per experiment). For both experiments, conditions were created to mimic those of their natural habitat.

For the experiment testing attraction to a moving object, two pendulums were placed at opposite ends of the 12 cm plastic tray. One static pendulum served as the control, while the other pendulum served as the treatment and was put into motion perpendicular to the tray. The area occupied by each pendulum accounted for 12% of the test area, while the "no response" area accounted for 76% of the test plot (see Figure 9.1). Each trial consisted of one leech being placed on the tray and recording its movement. The following behaviors were recorded:

1. Movement toward the moving pendulum counted as a reaction to the treatment.
2. Movement toward the static pendulum counted as a reaction to the control.
3. All other movements made outside their path counted as no response.

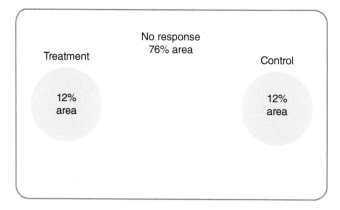

Figure 9.1 Illustration depicts the experimental setup (treatment, no response, and control) utilized in both case studies assessing leech attraction.

The findings from the following examples reflect the results from the original study by Kmiecik and colleagues (2008). Figure 9.1 is a representation of the experimental setup utilized to carry out the movement and heat experiment. Note that the area around each pendulum (treatment and control) accounts for 12% of the study plot, respectively, while the area of "No Response" accounts for 76% of the study plot. Depending on the experiment, treatment may consist of the moving pendulum (part 1) or heated cup (part 2), while the control may consist of the static pendulum (part 1) or room temperature cup (part 2).

> **? Question:** What is the probability that the leeches will not respond to the treatment or the control? What is the probability that the leeches will respond to the control or the treatment?
>
> ***
>
> The leeches have a 76% probability of not responding to the treatment or the control. The leeches have a 12% probability of responding to the control and a 12% probability of responding to the treatment.

Generalized Hypotheses

The hypotheses for chi-square are similar in structure to the tests we have looked at in previous chapters. The null hypothesis (H_0)

Table 9.1 Chi-square test experimental data showing the expected behavior distribution for 32 total trials.

Treatment (12%)	Control (12%)	No Response (76%)
4	4	24

states that there is no significant difference between the expected and observed data. The alternative hypothesis (H_1) states that there is a significant difference between the expected and observed data.

Hypotheses simplified:

$H_0 : X_{obs} = X_{exp}$

$H_1 : X_{obs} > X_{exp}$ OR $X_{obs} < X_{exp}$

The following are hypotheses generated for the previous example:

Null Hypothesis (H_0): Leech movement will not show a preference for any location; therefore, subject distribution will match area distribution. Leeches will have 76% chance of being in the no response area and 12% chance each of being in the response area (either the static (control) or the moving pendulum (treatment)).

The prediction is for no significant difference between the observed and expected data. Numbers were calculated based on the expected distribution of leeches on the study plot (76% No Response, 12% Treatment, 12% Control (Kmiecik et al., 2008)). Table 9.1 reports the expected behavioral distribution for 32 total trials.

Alternative Hypothesis (H_1): Leeches will show a preference for the response area (either the static (control) or the moving pendulum (treatment)). Thus, there is a significant difference between the observed and expected data. The following observations were made while conducting the movement experiment. Table 9.2 reports the observed behavioral distribution for 32 total trials:

Table 9.2 Chi-square test experimental data showing the observed behavior distribution for 32 total trials.

Treatment	Control	No Response
2	2	28

Assumptions

The chi-square test does not make any assumptions about the distribution of the data. The following assumptions must be satisfied in order to run a chi-square:

1. **Data type** – Two or more categorical (ordinal or nominal) variables.
2. **Independence** – The samples are independent.

Nuts and Bolts

To determine whether there is a difference between the observed and expected data, we must first calculate the **observed chi-square (X^2_{obs}) value**, using the following equation:

$$X^2_{obs} = \sum \frac{(\text{Observed} - \text{Expected})^2}{\text{Expected}}$$

Note: This formula measures the difference between the observed and expected values. The larger the difference between the two values, the larger the chi-statistic (X^2_{obs}) will be and the more likely the observed patterns will have a statistically significant difference from the expected patterns. In the minimum limit, if all observed values equal the expected values, then the chi-squared statistic simply equals 0.

The researchers expected to observe random behavior (4 responding to the treatment, 4 responding to the control, and 24 having no response); however, there was a difference in the observed behavior (2 responding to the treatment, 2 responding to the control, and 28 having no response). How different was the observed response from expected? The following observed chi-square (X^2_{obs}) value was calculated to find out:

Observed = 2 (Treatment), 2 (Control), 28 (No response)
Expected = 4 (Treatment), 4 (Control), 24 (No response)

$$X^2_{obs} = \sum \frac{(2-4)^2}{4} + \frac{(2-4)^2}{4} + \frac{(28-24)^2}{24} = 2.67$$

In order to evaluate the hypotheses, the observed X^2 value (X^2_{obs}) is compared to the expected X^2 value (X^2_{exp}) from the chi-square table (this is different from the observed and expected distribution in the cross-tabulation table). First, the degrees of freedom (df) need to be determined. Degrees of freedom can be calculated by summing the total number of groups, then subtracting by 1.

Degrees of freedom (df): Total number of groups − 1

In the case of the movement experiment, there were three expected behavioral response categories (treatment, control, and no response). Therefore, the degrees of freedom can be calculated as:

$$df = 3 - 1 = 2$$

The degrees of freedom and the significance level $\alpha = 0.05$ will be used in combination to locate the X^2_{exp} value in the chi-square distribution table. (See table 9.3, which is a simplified distribution table.)

The X^2_{obs} value (calculated) is compared to the X^2_{exp} value (distribution table). If the X^2_{obs} value is greater than or equal to the X^2_{exp} value, then we can reject the null. However, if the X^2_{obs} value is less than the X^2_{exp} value, then we cannot reject the null. This comparison will determine whether or not the experimental observed data are significantly different from the hypothesized expected data. For a visual simplification, see the following:

Table 9.3 Percentage points of the chi-square distribution.

Degrees of Freedom	Probability of a larger value of X^2								
	0.99	0.95	0.90	0.75	0.50	0.25	0.10	0.05	0.01
1	0.000	0.004	0.016	0.102	0.455	1.32	2.71	3.84	6.63
2	0.020	0.103	0.211	0.575	1.386	2.77	4.61	5.99	9.21
3	0.115	0.352	0.584	1.212	2.366	4.11	6.25	7.81	11.34
4	0.297	0.711	1.064	1.923	3.357	5.39	7.78	9.49	13.28
5	0.554	1.145	1.610	2.675	4.351	6.63	9.24	11.07	15.09
6	0.872	1.635	2.204	3.455	5.348	7.84	10.64	12.59	16.81

$X_{obs}^2 \geq X_{exp}^2$ (reject H_0, support H_1 thus indicating a significant difference)

$X_{obs}^2 < X_{exp}^2$ (support H_0, reject H_1 thus indicating no significant difference)

The following table is a summation of the data gathered.

Chi-Square Analysis

Critical p-value	0.05 (Set by us)
Degrees of freedom	2 (Calculated by us)
X_{obs}^2	2.67 (Calculated by us)
X_{exp}^2	? (From chi-square test)

The following steps demonstrate how to read the chi-square distribution table in order to determine the X_{exp}^2 value.

1. Move down the left side of the table to locate the degrees of freedom corresponding with the experiment. (Remember, the calculated df = 2). Move to the right of the row, following the degrees of freedom "2."

Degrees of Freedom	Probability of a larger value of x^2								
	0.99	0.95	0.90	0.75	0.50	0.25	0.10	0.05	0.01
1	0.000	0.004	0.016	0.102	0.455	1.32	2.71	3.84	6.63
2	0.020	0.103	0.211	0.575	1.386	2.77	4.61	5.99	9.21
3	0.115	0.352	0.584	1.212	2.366	4.11	6.25	7.81	11.34
4	0.297	0.711	1.064	1.923	3.357	5.39	7.78	9.49	13.28
5	0.554	1.145	1.610	2.675	4.351	6.63	9.24	11.07	15.09
6	0.872	1.635	2.204	3.455	5.348	7.84	10.64	12.59	16.81

2. Locate the critical value, 0.05, in the distribution table.

Degrees of Freedom	Probability of a larger value of x^2								
	0.99	0.95	0.90	0.75	0.50	0.25	0.10	0.05	0.01
1	0.000	0.004	0.016	0.102	0.455	1.32	2.71	3.84	6.63
2	0.020	0.103	0.211	0.575	1.386	2.77	4.61	5.99	9.21
3	0.115	0.352	0.584	1.212	2.366	4.11	6.25	7.81	11.34
4	0.297	0.711	1.064	1.923	3.357	5.39	7.78	9.49	13.28
5	0.554	1.145	1.610	2.675	4.351	6.63	9.24	11.07	15.09
6	0.872	1.635	2.204	3.455	5.348	7.84	10.64	12.59	16.81

3. Move down the 0.05 column until you reach the exact value where the two arrows meet as seen below.

Degrees of Freedom	Probability of a larger value of x²								
	0.99	0.95	0.90	0.75	0.50	0.25	0.10	0.05	0.01
1	0.000	0.004	0.016	0.102	0.455	1.32	2.71	3.84	6.63
2	0.020	0.103	0.211	0.575	1.386	2.77	4.61	5.99	9.21
3	0.115	0.352	0.584	1.212	2.366	4.11	6.25	7.81	11.34
4	0.297	0.711	1.064	1.923	3.357	5.39	7.78	9.49	13.28
5	0.554	1.145	1.610	2.675	4.351	6.63	9.24	11.07	15.09
6	0.872	1.635	2.204	3.455	5.348	7.84	10.64	12.59	16.81

From the graph, we determined the X^2_{exp} value to be 5.99. We can now fill in our chi-square analysis table. Table 9.4 is a summation of the data.

Table 9.4 Chi-square analysis table summarizing the data gathered by means of calculating the chi-square statistic or determined from the chi-square distribution table.

Critical p-value	0.05 (Set by us)
Degrees of freedom	2 (Calculated by us)
X^2_{obs}	2.67 (Calculated by us)
X^2_{exp}	**5.99** (From chi-square)

Given that the hypothesized expected data (X^2_{exp}), 5.99, is greater than the experimental observed data (X^2_{obs}), 2.67, the null hypothesis fails to be rejected.

$$X^2_{obs}(2.67) < X^2_{exp}(5.99)$$

Therefore, the subjects did not favor the stimulus and leech behavior follows a random pattern.

Concluding Statement

Given that the hypothesized expected data (X^2_{exp}) is greater than the experimental observed data (X^2_{obs}), the null hypothesis fails to be rejected. The leeches were behaviorally distributed similar to the specified distribution that reflects the area of the regions of the tray.

9.3 Case Study 2

Let us walk through an example from beginning to end.

For an experiment testing the attractiveness to a warm object, one opaque polymer cup (control) was placed at the end of a 12 cm plastic tray and left at room temperate; a second opaque polymer cup (treatment) was heated by placing a 110-watt bulb inside the cup for 90 seconds and placed on the same tray as the control, but at the opposite end. Each trial consisted of one leech placed on the tray and recording its movement. The following behaviors were recorded:

1. Movement toward the heated cup and repeatedly touching their heads against the cup counted as a reaction to the treatment.
2. Movement toward the cup held at room temperature and repeatedly touching their heads against that cup served as a reaction to the control.
3. All other movements made outside their paths counted as no response.

(H) Formulate the null and alternative hypotheses that address the question originally proposed.

****

Null Hypothesis (H_0): Leeches will be distributed according to the spacial design of the experiment (76%, 12%, 12%) and not show a preference for any location; therefore, subject distribution will match area distribution.

Leeches will have 76% chance of being in the no response area and 12% chance of being in either the response area (either the control or treatment). Again, the percentages are based on the area of each treatment in relation to the overall area of the study plot. *(There is no significant difference between the observed and expected data.)*

Alternative Hypothesis (H_1): Leeches will show a preference for the response area (either the control or the treatment). *(There is a significant difference between the observed and expected data.)*

Table 9.5 Expected behavioral distribution for 32 total trials.

Treatment (12%)	Control (12%)	No Response (76%)
4	4	24

Table 9.6 Observed behavioral distribution for 32 total trials.

Treatment	Control	No Response
18	4	10

Table 9.5 reports the expected behavioral distribution for 32 total trials. Table 9.6 reports the observed behavioral distribution for 32 total trials. Calculate the observed chi-square (X^2_{obs}) value using the following equation:

$$X^2_{obs} = \sum \frac{(\text{Observed} - \text{Expected})^2}{\text{Expected}}$$

The researchers expected to observe random behavior (**4** responding to the treatment, **4** responding to the control, and **24** having no response); however, there was a difference in the observed behavior (**18** responding to the treatment, **4** responding to the control, and **10** having no response). How different was the observed response from expected? Again, calculate the observed chi-square (X^2_{obs}) value to find out:

Observed = 18 (Treatment), 4 (Control), 10 (No response)
Expected = 4 (Treatment), 4 (Control), 24 (No response)

$$X^2_{obs} = \sum \frac{(18-4)^2}{4} + \frac{(4-4)^2}{4} + \frac{(10-24)^2}{24} = 57.17$$

The heat experiment also had three expected behavioral responses (treatment, control, and no response). The degrees of freedom can be calculated as:

$$df = 3 - 1 = 2$$

Table 9.7 is a summation of the data gathered.

Chi-Square Test | 403

Table 9.7 Summation of the data gathered.

Critical p-value	0.05 (Set by us)
Degrees of freedom	2 (Calculated by us)
X^2_{obs}	57.17 (Calculated by us)
X^2_{exp}	? (From chi-square)

Chi-Square Analysis

Once again, use the chi-square distribution table to determine the X^2_{exp} value and whether or not the null hypothesis can be supported:

1. Move down the left side of the table to locate the degrees of freedom corresponding to the experiment. (Remember, the calculated df = 2). Move to the right of the row, following the degrees of freedom "**2**."

Degrees of Freedom	Probability of a larger value of x^2								
	0.99	0.95	0.90	0.75	0.50	0.25	0.10	0.05	0.01
1	0.000	0.004	0.016	0.102	0.455	1.32	2.71	3.84	6.63
2	0.020	0.103	0.211	0.575	1.386	2.77	4.61	5.99	9.21
3	0.115	0.352	0.584	1.212	2.366	4.11	6.25	7.81	11.34
4	0.297	0.711	1.064	1.923	3.357	5.39	7.78	9.49	13.28
5	0.554	1.145	1.610	2.675	4.351	6.63	9.24	11.07	15.09
6	0.872	1.635	2.204	3.455	5.348	7.84	10.64	12.59	16.81

2. Locate our critical value, 0.05, in the distribution table.

Degrees of Freedom	Probability of a larger value of x^2								
	0.99	0.95	0.90	0.75	0.50	0.25	0.10	0.05	0.01
1	0.000	0.004	0.016	0.102	0.455	1.32	2.71	3.84	6.63
2	0.020	0.103	0.211	0.575	1.386	2.77	4.61	5.99	9.21
3	0.115	0.352	0.584	1.212	2.366	4.11	6.25	7.81	11.34
4	0.297	0.711	1.064	1.923	3.357	5.39	7.78	9.49	13.28
5	0.554	1.145	1.610	2.675	4.351	6.63	9.24	11.07	15.09
6	0.872	1.635	2.204	3.455	5.348	7.84	10.64	12.59	16.81

3. Move down the 0.05 column until you reach the exact value where the two arrows meet as seen below.

Degrees of Freedom	Probability of a larger value of x^2								
	0.99	0.95	0.90	0.75	0.50	0.25	0.10	0.05	0.01
1	0.000	0.004	0.016	0.102	0.455	1.32	2.71	3.84	6.63
2	0.020	0.103	0.211	0.575	1.386	2.77	4.61	5.99	9.21
3	0.115	0.352	0.584	1.212	2.366	4.11	6.25	7.81	11.34
4	0.297	0.711	1.064	1.923	3.357	5.39	7.78	9.49	13.28
5	0.554	1.145	1.610	2.675	4.351	6.63	9.24	11.07	15.09
6	0.872	1.635	2.204	3.455	5.348	7.84	10.64	12.59	16.81

From the graph, the X^2_{exp} value was determined to be **5.99**. We can now fill in Table 9.7

Recall, if X^2_{obs} is greater than or equal to the X^2_{exp}, then the null hypothesis must be rejected, thus supporting the alternative hypothesis that leech behavior is not based on the distribution of the experimental area and subjects favor the response area.

If the *p*-value is significant, refer to a cross-tabulation table. The cross-tabulation table provides the observed counts and the expected counts, which can be utilized to determine where the difference lies: the treatment, the control, or a combination of the two. A replica is shown below for teaching purposes, the actual cross-tabulation table will appear differently depending on the statistical software you use.

Chi-square cross-tabulation table:

	Treatment (12%)	Control (12%)	No Response (76%)
Expected	4	4	24
Observed	18	4	10

When examining the table above, compare the observed and the expected values. If one or more categories (e.g., treatment, control) show a large deviation in the observed than the expected count, then that is the source of the nonrandom conclusion. In this example, the leeches were observed to be drawn to the treatment (heat) **18** times

Table 9.8 Chi-square cross-tabulation table.

Critical p-value	0.05 (Set by us)
Degrees of freedom	2 (Calculated by us)
X^2_{obs}	57.17 (Calculated by us)
X^2_{exp}	**5.99** (From chi-square)

rather than the expected **4** times, whereas the number of times the leeches were drawn to the control did not deviate from the expected. Thus, the cross-tabulation table tells us that differences between the expected and observed values can be attributed to the treatment (heat) rather than the control response.

Concluding Statement

The heat experiment resulted in an observed value greater than the hypothesized expected value. As a result, the null hypothesis is rejected, indicating a significant difference between the expected and observed data. Furthermore, the cross-tabulation identified the treatment as the source of significance. Thus, we can conclude that the heat stimulus is the source of the preferential pattern and that leeches displayed an increased attraction to the heat stimulus. In answer to our original question, "Does movement or heat play a larger role in leech attraction?" we can say with 95% confidence that heat plays a role in leech attraction while movement has no affect.

Tutorials

9.4 Chi-Square Excel Tutorial

Example Dataset

Due to an increase in endangered species, mating habits of mammals are being more widely studied. One pattern to consider with mating habits is female preference, or sexual selection. Although it is well known that most species choose their mate based on preference, some species (because of their dispersed distribution across the landscape) exhibit a uniform distribution among mates, which allows every male to have an equal chance to reproduce.

With this in mind, a group of biologists decided to study the mating habits of polar bears to determine whether or not mating was preferential or uniform. In this study, 5 male and 100 female polar bears were randomly selected. If mating was uniform, the expected distribution of females to males should be 20:1. For the study, all polar bears were secluded in a separate habitat for 30 days, and their mating habits were recorded.

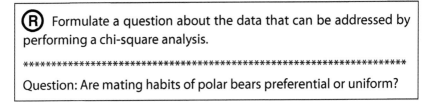

Question: Are mating habits of polar bears preferential or uniform?

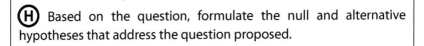

Null Hypothesis (H_0): Mating of polar bears is uniform so all males have an equal chance of mating with females.
Alternative Hypothesis (H_1): Mating of polar bears is not uniform, but rather preferential.

Now that an appropriate question has been developed, along with a set of testable hypotheses, you can run the statistical analysis.

- This tutorial focuses on running a chi-square test in Excel.
- Refer to Chapter 12 for tips and tools when using Excel.
- Check all assumptions prior to running the test.

Chi-Square Excel Tutorial

Like any other statistic, one of the purposes of performing a chi-square analysis is to determine a *p*-value. In Excel, there is one way to calculate the observed chi-square value; however, there are two ways to determine the *p*-value. The first method of computation (the longer version) requires you to calculate the observed X^2 value in

Excel and reference the chi-square distribution table to determine the *p*-value. Although this method is more time consuming, it is necessary to learn, especially for students majoring in Biology who would take courses such as Genetics and Evolution. The second method of computation (the shorter version) does not require you to calculate the observed X^2 value in order to determine the *p*-value. Instead, based on the observed and expected frequency, the *p*-value can be directly calculated using the chi-test function in Excel.

The first part of this tutorial will address the first method of computation and will walk you through calculating the observed X^2 value and how to determine its corresponding *p*-value; whereas, the second part will address the second method of computation and will walk you through how to use the chi-test function to determine the *p*-value.

Observed Chi-Square Value Computation

1. Before inputting the data onto the Excel spreadsheet, create a table similar to the following.

Category						O = Observed
Polar Bears	Observed	Expected	O-E	(O-E)^2	(O-E)^2/E	E = Expected

2. Under the **Category** column is the label **Polar Bears**. Utilizing the dataset provided in the overview, input labels Male 1 through Male 5 as shown below.

Category						O = Observed
Polar Bears	Observed	Expected	O-E	(O-E)^2	(O-E)^2/E	E = Expected
Male 1						
Male 2						
Male 3						
Male 4						
Male 5						

3. Under the **Observed** column, input the observed frequencies. In the case of the polar bear example, mating frequency was being studied. So the amount of females that mated with each male would be inserted under this column as shown below.

Category						O = Observed
Polar Bears	Observed	Expected	O-E	(O-E)^2	(O-E)^2/E	E = Expected
Male 1	22					
Male 2	12					
Male 3	17					
Male 4	24					
Male 5	25					

4. Under the **Expected** column, input the expected frequencies. Because there were 100 females and 5 males, if the distribution of females is uniform (equal probability), then each male would have mated with 20 females. This is the expected frequency.

Category						O = Observed
Polar Bears	Observed	Expected	O-E	(O-E)^2	(O-E)^2/E	E = Expected
Male 1	22	20				
Male 2	12	20				
Male 3	17	20				
Male 4	24	20				
Male 5	25	20				

5. Calculate the difference between the observed frequency and the expected frequency. Under the **O-E** column, type in an equals sign (=) then click on the first value under the **Observed** frequency column.

Category						O = Observed
Polar Bears	Observed	Expected	O-E	(O-E)^2	(O-E)^2/E	E = Expected
Male 1	22	20	=D7			
Male 2	12	20				
Male 3	17	20				
Male 4	24	20				
Male 5	25	20				

Note: The sequence seen in the box (right arrow) corresponds to the selected data value (left arrow).

Chi-Square Test

6. Type in a minus sign (−) then click on the first value under the **Expected** frequency column. Then press enter/return.

Category						O = Observed
Polar Bears	Observed	Expected	O-E	(O-E)^2	(O-E)^2/E	E = Expected
Male 1	22	20	=D7-E7			
Male 2	12	20				
Male 3	17	20				
Male 4	24	20				
Male 5	25	20				

7. Calculate the difference between the observed and expected frequencies for the remaining categories, Males 2 through 5. To apply the command to the rest of the column, click the lower right corner of the cell. A black cross will appear; this is the fill handle. Drag it all the way to the bottom of the column.

Category						O = Observed
Polar Bears	Observed	Expected	O-E	(O-E)^2	(O-E)^2/E	E = Expected
Male 1	22	20	2			
Male 2	12	20				
Male 3	17	20				
Male 4	24	20				
Male 5	25	20				

8. The resulting values should appear.

Category						O = Observed
Polar Bears	Observed	Expected	O-E	(O-E)^2	(O-E)^2/E	E = Expected
Male 1	22	20	2			
Male 2	12	20	-8			
Male 3	17	20	-3			
Male 4	24	20	4			
Male 5	25	20	5			

9. To calculate the square of the difference of the frequencies, click on the first empty cell under the **(O-E) ^2** column so that it appears outlined as below.

Category						O = Observed
Polar Bears	Observed	Expected	O-E	(O-E)^2	(O-E)^2/E	E = Expected
Male 1	22	20	2			
Male 2	12	20	-8			
Male 3	17	20	-3			
Male 4	24	20	4			
Male 5	25	20	5			

10. Under the Formulas tab in the toolbar, select **Math & Trig**. Scroll down the formula options and double click on the **Power** function.

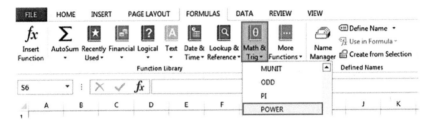

11. The following box will appear on the right of the screen. Click on the icon corresponding to **number**.

12. Click on the first value listed under the O-E column. Click on the icon to the right (as depicted in step 11) when finished.

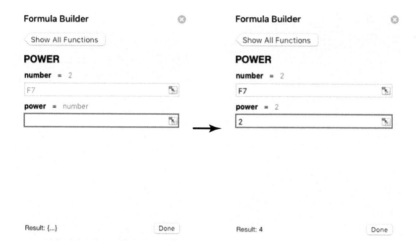

13. In the white box corresponding to the power, type in 2.

14. Click **Done**.

15. The squared value of the frequency difference should appear in the cell originally selected. To apply the command to the rest of the column, use the fill handle to drag it to the bottom of the column.

Category						O = Observed
Polar Bears	Observed	Expected	O-E	(O-E)^2	(O-E)^2/E	E = Expected
Male 1	22	20	2	4		
Male 2	12	20	-8			
Male 3	17	20	-3			
Male 4	24	20	4			
Male 5	25	20	5			

16. Next, calculate the value of the squared frequency difference divided by the expected frequency. To do so, click on the first empty cell below the **(O-E)^2/E** column. Type in an equals sign (=) and select the first value under the **(O-E)^2** column to represent the first part of the equation.

Category						O = Observed
Polar Bears	Observed	Expected	O-E	(O-E)^2	(O-E)^2/E	E = Expected
Male 1	22	20	2	4	=G7	
Male 2	12	20	-8	64		
Male 3	17	20	-3	9		
Male 4	24	20	4	16		
Male 5	25	20	5	25		

17. Type in a forward slash (/) and select the first value under the **Expected** column. Then press enter/return.

Category						O = Observed
Polar Bears	Observed	Expected	O-E	(O-E)^2	(O-E)^2/E	E = Expected
Male 1	22	20	2	4	=G7/E7	
Male 2	12	20	-8	64		
Male 3	17	20	-3	9		
Male 4	24	20	4	16		
Male 5	25	20	5	25		

18. The desired quotient will appear. To apply the command to the rest of the column, drag the fill handle to the bottom of the column.

Category						O = Observed
Polar Bears	Observed	Expected	O-E	(O-E)^2	(O-E)^2/E	E = Expected
Male 1	22	20	2	4	0.2	
Male 2	12	20	-8	64		
Male 3	17	20	-3	9		
Male 4	24	20	4	16		
Male 5	25	20	5	25		

19. The complete table should look similar to the one below.

Category						O = Observed
Polar Bears	Observed	Expected	O-E	(O-E)^2	(O-E)^2/E	E = Expected
Male 1	22	20	2	4	0.2	
Male 2	12	20	-8	64	3.2	
Male 3	17	20	-3	9	0.45	
Male 4	24	20	4	16	0.8	
Male 5	25	20	5	25	1.25	

20. To determine the chi-square value, find the sum of the quotients. First, below the table, type in **SUM =** .

Category						O = Observed
Polar Bears	Observed	Expected	O-E	(O-E)^2	(O-E)^2/E	E = Expected
Male 1	22	20	2	4	0.2	
Male 2	12	20	-8	64	3.2	
Male 3	17	20	-3	9	0.5	
Male 4	24	20	4	16	0.8	
Male 5	25	20	5	25	1.3	
				SUM =		

21. In the cell to the right, type in an equals sign (=) and the word **SUM**. From the functions that appear, double click on the **SUM** function icon.

Category						O = Observed
Polar Bears	Observed	Expected	O-E	(O-E)^2	(O-E)^2/E	E = Expected
Male 1	22	20	2	4	0.2	
Male 2	12	20	-8	64	3.2	
Male 3	17	20	-3	9	0.5	
Male 4	24	20	4	16	0.8	
Male 5	25	20	5	25	1.3	
				SUM =	=SUM	

Most Recently Used
SUM
Functions
SUM
SUMIF
SUMIFS
SUMPRODUCT
SUMSQ
SUMX2MY2
SUMX2PY2
SUMXMY2

22. Select all values within the **(O-E)^2/E** column. Then press enter/return.

Category						O = Observed
Polar Bears	Observed	Expected	O-E	(O-E)^2	(O-E)^2/E	E = Expected
Male 1	22	20	2	4	0.2	
Male 2	12	20	-8	64	3.2	
Male 3	17	20	-3	9	0.5	
Male 4	24	20	4	16	0.8	
Male 5	25	20	5	25	1.3	
				SUM =	=SUM(H7:H11)	

23. The resulting sum is the chi-square value.

Category						O = Observed
Polar Bears	Observed	Expected	O-E	(O-E)^2	(O-E)^2/E	E = Expected
Male 1	22	20	2	4	0.2	
Male 2	12	20	-8	64	3.2	
Male 3	17	20	-3	9	0.5	
Male 4	24	20	4	16	0.8	
Male 5	25	20	5	25	1.3	
				SUM =	5.9	

24. Label this new value by typing **Chi-square Value** in the empty cell to the right of the resulting number.

SUM =		5.9 Chi-square Value

25. Calculate the degrees of freedom by utilizing the following equation:

    Degrees of freedom (df): Total number of groups − 1

    Because there were five males being observed and each male is considered as a separate group, the df value is as follows:

    Degrees of freedom (df): $5 - 1 = 4$

26. At this point, you can either reference a chi-square distribution chart to determine the *p*-value given the calculated chi-square value and degrees of freedom, or you can proceed to the second part of this tutorial, ***p*-Value Computation**, to learn how to use Excel to determine the *p*-value for you.

p-Value Computation

Calculating the observed chi-square value is not a necessary step in determining the *p*-value generated by a set of data. If you are already using Excel, take advantage of the chi-test function that the program includes. A benefit from using the chi-test function is that it calculates the exact *p*-value, unlike the chi-square distribution table. With the table, you are only able to determine the *p*-value range in which your data fall under. If you are concerned with having an exact *p*-value, then referencing a distribution table is not for you.

The following section of this tutorial will guide you through the short way of completing a chi-square analysis and how to directly calculate the *p*-value.

1. To begin the short version of the chi-square analysis, only the observed and expected values need to be entered into the spreadsheet.

Category						O = Observed
Polar Bears	Observed	Expected	O-E	(O-E)^2	(O-E)^2/E	E = Expected
Male 1	22	20	2	4	0.2	
Male 2	12	20	-8	64	3.2	
Male 3	17	20	-3	9	0.5	
Male 4	24	20	4	16	0.8	
Male 5	25	20	5	25	1.3	

2. Next type in "**Probability value=**" somewhere below your data. This will be the label for the value you are about to calculate.

Category						O = Observed
Polar Bears	Observed	Expected	O-E	(O-E)^2	(O-E)^2/E	E = Expected
Male 1	22	20	2	4	0.2	
Male 2	12	20	-8	64	3.2	
Male 3	17	20	-3	9	0.5	
Male 4	24	20	4	16	0.8	
Male 5	25	20	5	25	1.3	
				SUM =		5.9 Chi-square Value
				Probability value =		

3. In the empty cell to the right, type in **=chitest**. Double click on the **CHITEST** function that appears.

Category						O = Observed
Polar Bears	Observed	Expected	O-E	(O-E)^2	(O-E)^2/E	E = Expected
Male 1	22	20	2	4	0.2	
Male 2	12	20	-8	64	3.2	
Male 3	17	20	-3	9	0.5	
Male 4	24	20	4	16	0.8	
Male 5	25	20	5	25	1.3	
				SUM =		5.9 Chi-square Value
				Probability value =	=CHITEST	
					Most Recently Used	
					CHITEST	
					Compatibility Functions	
					CHITEST	

4. Enter in the array for **Observed** values by typing in the cell references D7:D11.

Category						
Polar Bears	Observed	Expected	O-E	(O-E)^2	(O-E)^2/E	O = Observed E = Expected
Male 1	22	20	2	4	0.2	
Male 2	12	20	-8	64	3.2	
Male 3	17	20	-3	9	0.5	
Male 4	24	20	4	16	0.8	
Male 5	25	20	5	25	1.3	
				SUM =	5.9 Chi-square Value	
				Probability value =	=CHITEST(D7:D11)	

5. Next type in a comma (,) and add the array for the **Expected** data E7:E11.

Category						
Polar Bears	Observed	Expected	O-E	(O-E)^2	(O-E)^2/E	O = Observed E = Expected
Male 1	22	20	2	4	0.2	
Male 2	12	20	-8	64	3.2	
Male 3	17	20	-3	9	0.5	
Male 4	24	20	4	16	0.8	
Male 5	25	20	5	25	1.3	
				SUM =	5.9 Chi-square Value	
				Probability value =	=CHITEST(D7:D11,E7:E11)	

6. Press enter/return. The resulting number is the probability value, also known as the *p*-value.

Probability value =   0.20674184

Based on the output, the generated *p*-value was greater than 0.05, suggesting that the observed frequency of polar bear mating was not significantly different than the expected frequency.

Concluding Statement

Polar bear mating is uniform, and the null hypothesis is supported. Females do not exhibit a preference in the type of male with which they mate ($X^2 = 5.9$, df $= 4$, $p > 0.05$).

9.5 Chi-Square SPSS Tutorial

Prairies are becoming more endangered and fragmented, resulting in a loss of biodiversity in plant, insect, and animal groups. One particular species of concern is the Regal Fritillary Butterfly (*Speyeria idalia*). *S. idalia* has been known to reside in native prairies in central and eastern United States; however, a survey completed in 1995 (Debinski and Kelly, 1998) reported that *S. idalia* was only found in 11 of the 52 native prairies in the state of Iowa. In 2001, Ries and Debinski conducted a study that examined the emigration rates of *S. idalia* and their ability to leave a habitat patch. Emigration of *S. idalia* involved (1) approaching the edge of a habitat patch and (2) committing to crossing the edge of the habitat patch.

Study areas consisted of 30 plots (40 × 40 m), 2–3 plots within each of the 11 native Iowa prairies. While three of the interior study edges remained within the prairie, the fourth edge was considered the exterior edge of both the study area and the prairie itself. The exterior edge was exposed to the following characteristics: crop, tree line, road, or field. If *S. idalia* was observed crossing the exterior edge leading to one of the aforementioned boundaries, then it was marked as an exit occurrence. On the other hand, if *S. idalia* was observed crossing one of the three edges leading to the interior of the prairie, then there was no exit occurrence recorded. Ries and Debinski attempted to determine whether emigration rates were influenced by the type and quality of neighboring habitat.

*The dataset in this tutorial was designed for teaching purposes and is not the original dataset from the Ries and Debinski (2001) study.

(R) Formulate a question about the data that can be addressed by performing a chi-square analysis.

****

Question: Is butterfly emigration rate influenced by the type and quality of neighboring habitat or is it uniform?

> (H) Based on the question, formulate the null and alternative hypotheses that address the question proposed.
>
> ***
>
> Null Hypothesis (H_0): Butterfly behavior will be uniform; therefore, subjects will have a 25% chance of choosing one of the four exit boundaries.
>
> Alternative Hypothesis (H_1): Butterfly behavior will not be uniform; therefore, subjects will not follow random distribution and favor a neighboring habitat.

Now that an appropriate testable question has been developed, along with a set of testable hypotheses, you can run the statistical analysis.

- To run a chi-square test in SPSS, utilize the following tutorial.
- Refer to Chapter 13 for getting started and understanding SPSS.
- Check all assumptions prior to running the test.

Chi-Square SPSS Tutorial

1. Your data should now look similar to the following.

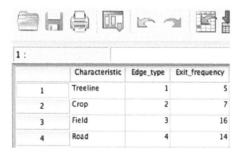

2. Now that your data are organized, we must set the observed frequency (**Exit_frequency**) as a weighted variable. Click on the Data tab in the toolbar located along the top of the page. Scroll to the very bottom and select **Weight Cases**.

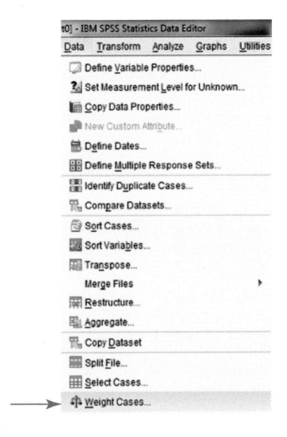

3. The following screen will appear.

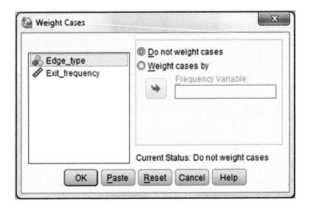

4. Select the radio button that is labeled as **Weight cases by**.

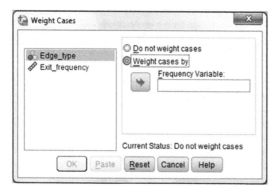

5. Click on the **Exit_frequency**, or observed frequency, and click the corresponding arrow to move the variable to the **Frequency Variable** selection. Then, select **OK**.

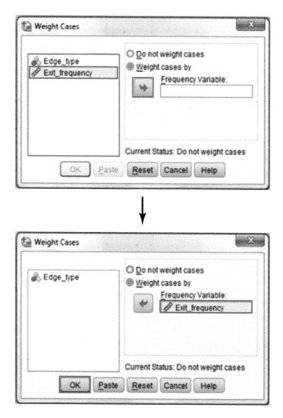

6. The following screen will appear confirming that you have made the selected changes. This screen can be minimized to view the SPSS spreadsheet with your data.

7. Confirm at the lower, right hand corner of the screen that **Weight On** is present. This ensures that you have weighted the frequency cases.

8. After you have weighted the cases, click on the **Analyze** tab in the toolbar located along the top of the page. Scroll down to the **Nonparametric Tests** option, select **Legacy Dialogs** and select **Chi-square**.

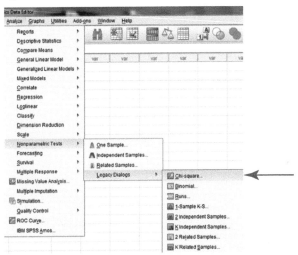

9. The following screen will appear. Click on the test variable, in this case **Edge_type**, and click the corresponding arrow to move the variable to the **Test Variable List**.

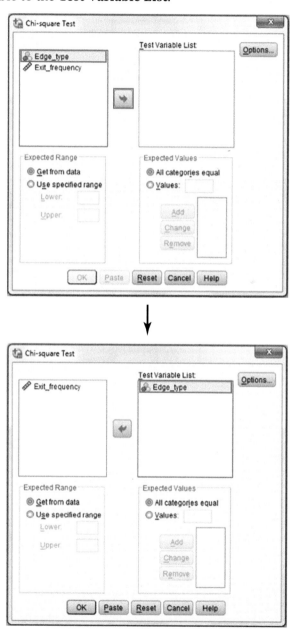

10. Notice the **Expected Values** selection. In this example, all expected values or frequencies are presumed equal. In other

words, *S. idalia* had an equal opportunity to select from the four different exit boundaries (Tree line, Crop, Field, and Road). The default for SPSS is **All categories equal**. Make sure this radio button is selected as in the picture.

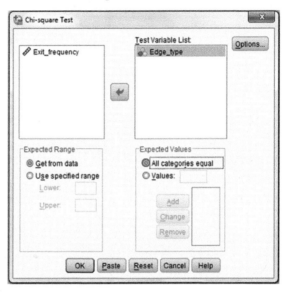

Note: In a situation where the expected values are not equal, you must manually type in the expected values (single value at a time) and select **Add**.

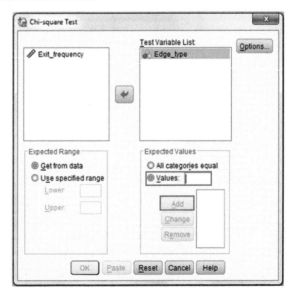

11. Click **OK**.

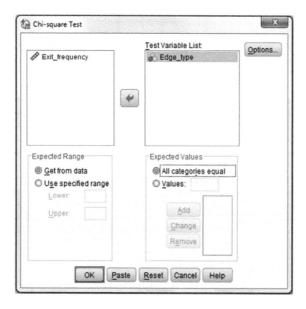

12. A separate document will appear. This is referred to as the output.

Chi-Square Test
Frequencies

Edge_type

	Observed N	Expected N	Residual
1	5	10.5	-5.5
2	7	10.5	-3.5
3	16	10.5	5.5
4	14	10.5	3.5
Total	42		

Test Statistics

	Edge_type
Chi-Square	8.095[a]
df	3
Asymp. Sig.	.044

a. 0 cells (.0%) have expected frequencies less than 5. The minimum expected cell frequency is 10.5.

According to the output, the p-value resulting from the chi-square analysis was statistically significant ($X^2 = 8.095$, df = 3, $p < 0.05$). This suggests that *S. idalia* did not equally select between tree line (1), crop (2), field (3), and road (4). Looking at the observed and expected counts, from the cross-tabulation table in the output, *S. idalia* emigrates from field and road habitats (16, 14) at a higher rate than tree line and crop habitats (5, 7).

Concluding Statement

There is a significant difference between the expected and observed behavior of *S. idalia* ($X^2 = 8.095$, df = 3, $p < 0.05$). The results suggest that tree line and crop edge areas serve as suitable barriers by preventing free movement of butterflies from prairie habitat to neighboring habitat.

Note: If you want to save the SPSS file with the inserted data as well as the SPSS output with the results of the statistical analysis performed, then you must save each document separately (see Chapter 13).

9.6 Chi-Square Numbers Tutorial

In Section 9.3, we walked through how to calculate the X^2 and find the corresponding p-value using a chi-square distribution table. In this tutorial, we will show you an alternative method for finding the p-value using Numbers.

Recap from case study 2: For an experiment testing the attractiveness to a warm object, one opaque polymer cup (control) was placed at the end of a 12 cm plastic tray and left at room temperate; a second opaque polymer cup (treatment) was heated by placing a 110-watt bulb inside the cup for 90 seconds and placed on the same tray as the control, but at the opposite end. Each trial consisted of one leech placed on the tray and recording its movement.

*Example is based on research by Kmiecik and colleagues (2008).

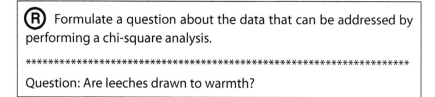

Question: Are leeches drawn to warmth?

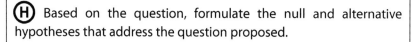

Null Hypothesis (H₀): Leeches will be distributed according to the spacial design of the experiment (76%, 12%, 12%) and will not show a preference for any location; therefore, subject distribution will match area distribution.

Leeches will have 76% chance of being in the no response area and 12% chance of being in the response area (either the control or treatment). Again, the percentages are based on the area of each treatment in relation to the overall area of the study plot. *(There is no significant difference between the observed and expected data.)*
Alternative Hypothesis (H₁): Leeches will show a preference for the response area (either the control or the treatment). *(There is a significant difference between the observed and expected data.)*

Now that an appropriate testable question has been developed, along with a set of testable hypotheses, you can run the statistical analysis.

- To run a chi-square test in Numbers, utilize the following tutorial.
- More information on programming in Numbers is found in Chapter 14.
- Check all assumptions prior to running the test.

Chi-Square Numbers Tutorial

The following tutorial will demonstrate how to determine the probability value using the chi-test function.

1. Start by creating a table similar to the one below.

Leech Example - Part 2 Warmth

	Expected	Observed
Treatment (12%)	4	18
Control (12%)	4	4
No Response (76%)	24	10
chi-square p-value		

2. Type **=chitest**

Leech Example - Part 2 Warmth

	Expected	Observed
Treatment (12%)	4	18
Control (12%)	4	4
No Response (76%)	24	10
chi-square p-value	fx ~ CHITEST ▾ (actual-values), (expected-values)	

3. Highlight the **observed** and **expected** numbers.

Leech Example - Part 2 Warmth

	Expected	Observed
Treatment (12%)	4	18
Control (12%)	4	4
No Response (76%)	24	10
chi-square p-value	fx ~ CHITEST ▾ (C2:C4 ▾), (B2:B4 ▾)	

4. The resulting chi-square p-value is < 0.05.

chi-square p-value	0.0000000000000385847827158881

Concluding statement

There is a significant difference between the expected and observed leech behavior ($p < 0.05$). The results suggest that leeches are drawn to warmth.

Note: In Numbers, unlike Excel, you are unable to calculate the observed chi-square value. This is due to how Numbers reports the quotient value ((O-E)^/E)). Numbers calculates the whole part of the quotient and ignores the fractional component. The rounding will cause miscalculation of the observed chi-square value; therefore, if utilizing Numbers to run a chi-square analysis, you can only compute the probability value.

9.7   Chi-Square R Tutorial

Forensic anthropologists and bioarchaeologists need to estimate sex from human skeletal remains. While the pelvis and long bones (specifically, arms and legs) work best, the skull can also provide information regarding sex.

To establish how well a particular trait of the skull can discriminate between males and females, tests need to be conducted by multiple observers on different populations that contain documentation of sex. In an effort to do so, data were collected from the Hamann-Todd Skeletal Collection by scoring the mastoid process on a 5-point scale (1–5) to establish the degree of femininity or masculinity of the trait. In terms of volume, smaller mastoids are considered to be indicative of a female and larger mastoids indicative of males. A lower score corresponds to a lower volume, and thus, a feminine appearance. Conversely, a higher score reflects a larger volume and a more masculine appearance. Females typically score 1–3 with occasional 4s and 5s.

A variety of skeletal collections provide this information, drawing on the recorded demographic information (sex, age, etc.) that was known about the individual prior to their death. The recorded scores are then compared to the documented sex of the individuals.

*This example was taken from the research study, Godde (2014).

> **(R)** Formulate a question about the data that can be addressed by performing a chi-square analysis.
>
> **
>
> Question: Do the scores 1–3 of the mastoid process in females follow a specific distribution?

> **(H)** Based on the question, formulate the null and alternative hypotheses that address the question proposed.
>
> **
>
> Null Hypothesis (H_0): Distribution of mastoid process scores follow a specified distribution in females.
> Alternative Hypothesis (H_1): Distribution of mastoid process scores deviate from the distribution in females.

Now that an appropriate testable question has been developed along with a set of testable hypotheses, you can run the statistical analysis.

- This tutorial focuses on running a chi-square test in R.
- Refer to Chapter 15 for R-specific terminology and instructions on how to invoke and construct code.
- Check all assumptions prior to running the test.

Chi-Square R Tutorial

1. Input the data at the command prompt by creating a vector entitled **score** to store the number of observed instances the particular mastoid process scores (1–3) were observed in females.

```
> score<-c(27,18,9)
```

This script informed R that there were 27 females whose mastoid was scored as 1, 18 mastoids scored as a 2, and 9 mastoids as a 3.

2. When running a chi-square test, R will assume that it is processing a vector of observed counts and run a chi-square

goodness-of-fit test. Thus, it is important to report the counts in the vector, rather than the score given to each individual skull. Unless we specify otherwise, R automatically assigns equal probability to each category (a uniform distribution). In other words, there is an equal probability for females to score a 1, 2, or 3.

3. Use the function **chisq.test()** to invoke the test on the **score** vector.

```
> chisq.test(score)
```

4. The following document will appear. This is referred to as the output.

```
> chisq.test(score)

    Chi-squared test for given probabilities

data:  score
X-squared = 9, df = 2, p-value = 0.01111
```

According to the output, the chi-square value and corresponding p-value ($X^2 = 9$, p-value $= 0.0111$) indicate the null hypothesis should be rejected and the scores are significantly different than expected from a goodness-of-fit test (which assumes an equal probability of belonging to scores 1–3).

5. As we only want to know the observed and expected counts for each score category (1–3), we will store the **chisq.test()** function in R so that we can display these counts individually and avoid producing unnecessary results. To do this, we will use the "←" symbol to store the chi-square results to **chi**. Type in the following script and press enter/return.

```
> chi<-chisq.test(score)
```

6. We can display the observed and expected values to compare the differences and determine why the p-value was significant. Using the methodology to refer to a particular vector within a data frame (**data.frame$vector**), we will call up the observed count vector (**chi$observed**) from the **chi** data frame at the command

prompt, as well as the expected count vector (**chi$expected**) from the **chi** data frame by typing in the following two scripts and pressing enter/return after each one.

> chi$observed

> chi$expected

7. The results are expressed in counts and should look like the output below.

```
> chi$observed
[1] 27 18  9
> chi$expected
[1] 18 18 18
```

As you can see, the **chi$observed** vector is composed of the counts we input in step 1. The **chi$expected** vector is the uniform distribution (equal probability of belonging in each category) the chi-square procedure assumed using a sample size of the sum of our counts (27 + 18 + 9 = 54. 54 divided by 3 is 18). We can see the significant *p*-value is due to scores 1 and 3 where the observed counts (27 and 9) differ greatly from the expected counts (both 18). Thus, in females, scores of 1 were reported more frequently than expected, while scores of 3 were reported less frequently.

8. As stated in step 7, the default in R is for the goodness-of-fit test in the chi-square function to assume a uniform distribution. You can also specify a particular distribution by storing a second vector (called **p** below) of the probability each female will be scored as 1, 2, or 3.

> p<-c(50,40,10)

In this case, we are going to suggest that 50% of the sample should have scored a 1, 40% a 2, and 10% a 3. The null hypothesis then becomes that the mastoid scores recorded do not differ from a distribution where 50% of the sample is scored as a 1, 40% as a 2,

and 10% as a 3. Type in the programming above to store a distribution other than the default and press enter/return.
9. To specify your own distribution, we will use the arguments `p =` and `rescale.p =` within the `chisq.test()` function. The argument `p =` specifies the vector of probabilities you wish to test, while setting `rescale.p =` to `TRUE` tells R that the probabilities need to be rescaled to decimal values. Type in the script below and press enter/return.

```
> chisq.test(score, p=p, rescale.p=TRUE)
```

10. Interpret the results from the output in a manner similar to step 4.

```
> chisq.test(score, p=p, rescale.p=TRUE)

    Chi-squared test for given probabilities

data:  score
X-squared = 3, df = 2, p-value = 0.2231
```

The *p*-value (0.2231) corresponding to the chi-square value (3) is not significant, indicating the null hypothesis was supported and the reported scores fit the specified distribution (about 50% of the mastoids were scored as 1, 40% as a 2, and 10% as a 3).

In the results from steps 4–7, the *p*-value is significant (<0.05), indicating the null hypothesis should be rejected. The tests thus indicate the observed scores do not follow the uniform distribution of equal probability across each score. An examination of the cross-tabulation table shows that females tend to have more mastoid process scores of 1 (more feminine appearance) and fewer mastoid scores of 3 (more neutral appearance) than expected. Similarly, if we test the distribution we specified in steps 8–10, we find the *p*-value is not significant (>0.05), and thus the null hypothesis is supported that the reported scores follow the distribution we specified across scores 1–3 (50%, 40%, 10%).

Concluding Statement

Steps 4–7: In this sample, the null hypothesis is rejected, indicating statistically significant differences ($X^2 = 9$, df = 2, $p < 0.05$) exist between the observed data and the specified distribution of females having equal probability to be classified in each mastoid process score 1–3. The cross-tabulation table demonstrates that females were scored as a 1 (feminine) more than expected, while they were scored as a 3 (neutral) less than expected.

Steps 8–10: In this sample, the null hypothesis is supported ($X^2 = 3$, df = 2, $p > 0.05$) that female mastoid process scores will be distributed where approximately 50% will be scored as a 1, 40% will be scored as a 2, and 10% will be scored as a 3.

10

Pearson's and Spearman's Correlation

> **Learning Outcomes:**
>
> 1. Evaluate the relationship between two variables.
> 2. Use statistical programs to run a correlation analysis to determine a significant relationship (p-value) and the strength of the relationship (r, ρ).
> 3. Use the skills acquired to analyze and evaluate your own dataset from independent research.

10.1 Correlation Background

To determine if one variable varies with another, we use **correlation** to represent the relationship between the two variables. Assessing the correlation between two variables may help to establish the strength of that relationship as well as determine if the relationship is positive, negative, or nonexistent. Correlation is used to determine the strength of a linear relationship between two variables. Graphing the data points on a scatter plot will give the researcher a better idea of the type of relationship that exists between the two variables (Figure 10.1).

10.2 Example

While visiting the Galapagos Islands, Dr. Jackson, an ornithologist, was interested in the inheritance of beak size in *Geospiza fortis*,

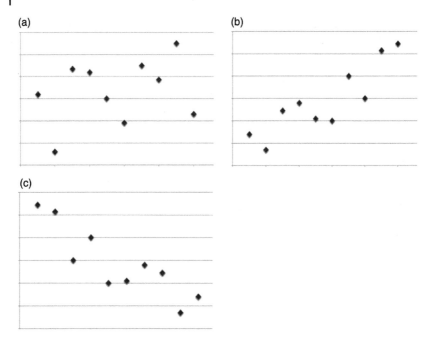

Figure 10.1 Scatter plot of data points depicting (a) no relationship, (b) a positive relationship, and (c) a negative relationship.

Darwin's medium ground finch. Dr. Jackson may decide to run a correlation analysis to determine whether there is a relationship between the variable of parent beak size and the variable of offspring beak size. He proposes the following question: "Is there a relationship between parental and offspring beak size in Darwin's medium ground finches?"

> **(H)** Based on the question, formulate the null and alternative hypotheses that address the question proposed.
>
> ***
>
> Null Hypothesis (H_0): There is no statistical relationship between parental and offspring beak size.
>
> Alternative Hypothesis (H_1): There is a statistical relationship between parental and offspring beak size.

Concluding Statements

Figure 10.2a shows that as the parent's beak size increases, the offspring's beak size also increases, illustrating a direct relationship or positive correlation. Thus, it is possible that offspring are inheriting beak size traits from the parents.

Figure 10.2b demonstrates that as adult beak size increases, the offspring beak size decreases, which represents negative correlation.

Figure 10.2c shows that as parent beak size increases, there is no change in offspring beak size. Therefore, there is no relationship (no correlation) between the beak size of the parents and the beak size of their offspring. Perhaps environmental factors play a larger role in the beak of the offspring.

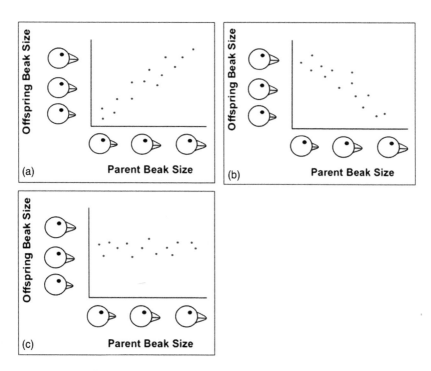

Figure 10.2 Different relationships between parent and offspring beak size. (a) shows a positive relationship, (b) shows a negative relationship, and (c) shows no relationship between the two variables.

Nuts and Bolts

A correlation analysis provides a quantifiable value and direction for the relationship between the two variables, but the output generated cannot determine cause and effect. The quantifiable values are the **probability value** (p) indicating the significance of the relationship and the **correlation coefficients** (r, ρ) indicating the strength of the relationship.

The probability value is analyzed with respect to the critical value (0.05 or another of your selection) to determine significance. In this situation, when you are assessing the relationship between two variables, the probability value indicates whether there is an association between the variables. In other words:

$p \leq 0.05$ (statistically significant relationship)
$p > 0.05$ (NOT statistically significant relationship)

Once you determine if there is a relationship between variables based on the p-value, determining the strength of the relationship is the next step. The strength of the relationship is represented with the r (Pearson's) or ρ (Spearman's) values. When the r- or ρ-value is closer to -1 or $+1$, a perfect (or near perfect) relationship may exist. A value closer to zero (reflecting a horizontal line) indicates no relationship. Examples of the correlational coefficient in the output could be $r = 0.46$ or $r = 0.78$. In either situation, the relationship may be significant ($p \leq 0.05$), although with $r = 0.78$, the value is closer to 1; therefore, the correlation may be described as a strong relationship. When $r = 0.46$, the value is closer to 0; therefore, the correlation may be described as a weak relationship as the spread of the data is larger and would indicate that the data do not fit the line well. Some disciplines use 0.7 and -0.7 as the cutoff between strong and weak relationships when interpreting r. Considering this cutoff, $r = 0.78$ would be a strong correlation and $r = 0.46$ would be a weak correlation. Spearman's ρ is interpreted similarly. To visualize the strength of the correlation, or spread of the data, see the scatter plots in Figure 10.3.

Hypotheses

A correlation has a set of hypotheses that addresses whether there is a relationship between the variables. Below, ρ represents the

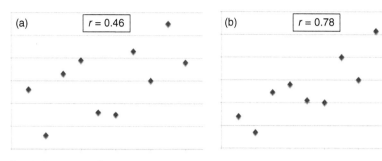

(a) r = 0.46

Note the large spread of the data. Would you expect this to have a weak or strong relationship?

(b) r = 0.78

Note the smaller spread of the data. Would you expect this to have a weak or strong relationship?

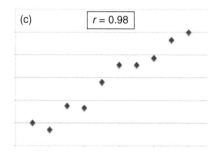

(c) r = 0.98

Note the tight spread of the data. Would you expect this to have a weak or strong relationship?

Figure 10.3 Representation of the strength of correlation based on the spread of data on a scatterplot, with higher *r* values indicating stronger correlation.

population correlation and is used for both Pearson's and Spearman's coefficient.

Hypotheses simplified:

H_0: There is no correlation between the two variables ($\rho = 0$)
H_1: The two variables are correlated ($\rho \neq 0$)

Pearson's Correlation Versus Spearman's Correlation

The two commonly used correlation analyses are Pearson's correlation (parametric) and Spearman's rank-order correlation (nonparametric). The Pearson and Spearman analyses provide the researcher with a *p*-value (i.e., significance level) and an *r*-or *ρ*-value (i.e., strength of the relationship). Both correlations will determine the direction of correlation (positive, negative, or nonexistent). When deciding on whether to use the parametric or nonparametric analysis, the decision is based on the assumptions similar to those made in previous chapters. Pearson's and Spearman's correlations are most appropriate when certain conditions and assumptions are fulfilled.

Assumptions

Let us discuss the assumptions of the correlation analysis in more depth. The following assumptions must be satisfied in order to run **Pearson's correlation:**

1. **Data type**–Both variables are continuous.
2. **Distribution of data**–Pearson's correlation examines the linear relationship between two variables from a **normal distribution** with homogeneity of variance (homoscedasticity) and no significant outliers.
3. **Random sampling**–The observations were collected at random.

Unlike Pearson's correlation, **Spearman's rank-order** does not make any assumptions about the distribution of the data. The following assumptions must be satisfied in order to run Spearman's correlation:

1. **Data type**–The variables can be ordinal, interval, or ratio (see Chapter 1).
2. **Distribution of data**–The data must be **monotonic**. This means that either both values from the two variables increase/decrease together, or as one variable increases the other decreases.
3. **Random sampling**–Spearman's assumes that the observations were **randomly** collected from a population.

Homoscedasticity (also known as homogeneity of variance) is the assumption that the independent variable's data points have relatively the same amount of variance (refer to Figure 10.4a, and 10.4b) and **heteroscedasticity** refers to an uneven variance between data points (refer to Figure 10.4c). To determine if the data meet linear distribution assumptions, the relationship between x and y must be linear (this can easily be examined by graphing the data on an *XY* scatter plot), the data points must be in the form of an elliptical or circular shape (*XY* scatter plot has an elliptical or circular shape, see Figure 10.4a and 10.4b), and the data must be free of any outliers (refer to Figure 10.4d). Nonlinear data (e.g., data reflecting an exponential or inverse exponential graph; refer to Figure 10.4e), or data containing outliers, are primary examples of data that violate the normality assumption. If this is the case, transforming the data is the best alternative to still run Pearson's correlation.

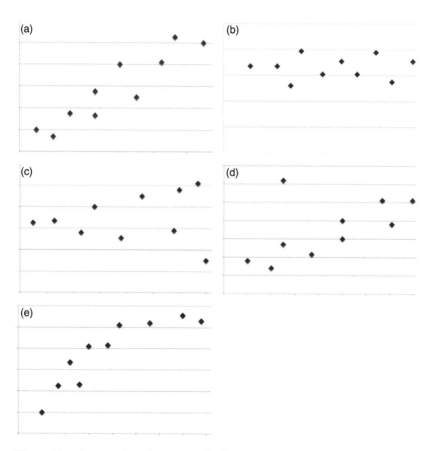

Figure 10.4 Scatter plots illustrating the features used to determine normality of a dataset: (a) homoscedastic data that display both a linear and elliptical shape satisfies the normality assumption, (b) homoscedastic data that display an elliptical shape satisfies the normality assumption, (c) heteroscedastic data that is funnel shaped, rather than elliptical or circular violates the normality assumption, (d) the presence of outliers violates the normality assumption, and (e) data that are non-linear also violate the normality assumption.

Table 10.1 Comparison of Pearson's correlation and Spearman's correlation with respect to the statistics, distribution, and appropriate variables to be analyzed with each test.

	Pearson's Correlation (parametric)	Spearman's Correlation (nonparametric)
Statistics	Tests for a linear relationship	Tests for a monotonic relationship
Distribution	Normal, linear, homoscedastic	No assumptions
Variables	Continuous data only	Ordinal, interval, or ratio data

Keeping in mind the assumptions and the way in which the data are measured (e.g., continuous or ordinal) will help with the selection of the appropriate test. Table 10.1 compares Pearson's and Spearman's tests. *Note*: Although, correlation cannot tell the researcher cause and effect between two variables, but it can give a quantifiable value to the relationship. After a relationship has been established (whether positive or negative), we may also want to know if one of the variables has any predictive capability (see Chapter 11).

Let us refer back to the finch example: Suppose our Pearson's correlation showed that the parent genetically influenced offspring's beak size. If we want to predict the actual size (millimeters) of the offspring's beak, can we predict offspring's beak size (millimeters) based on parent's beak size (millimeters)? A linear regression is needed to determine this predictability.

10.3 Case Study – Pearson's Correlation

It is a common perception that the more a student studies for an exam, the better he or she will perform. Suppose a Kinesiology professor was interested in testing this hypothesis on her nutrition class. On the first exam, she asked her class to write the number of hours they spent studying the night before the exam. She collected the data and proposed the following question: "Does student performance on a test correlate to how many hours they studied the previous night?"

Note: The following dataset is a hypothetical situation and created solely for teaching purposes.

> **(H)** Based on the question, formulate the null and alternative hypotheses that address the question proposed.
>
> ***
>
> Null Hypothesis (H_0): There is no relationship between the number of hours studied and the grade earned on the exam.
>
> Alternative Hypothesis (H_1): There is a significant relationship between the number of hours studied and the grade earned on the exam.

Refer to Table 10.2 for a summary of the experimental data.

Experimental Data

Table 10.2 Pearson's correlation experimental data showing study hours and exam scores for 28 students.

Study Hours	Exam Score
5.0	65
4.0	55
3.0	45
5.0	71
7.0	83
8.5	87
8.0	83
8.0	97
4.0	66
5.0	69
2.0	57
8.5	94
6.0	59
7.5	85
7.0	79
8.0	81
6.0	67

(*Continued*)

Table 10.2 (Continued)

Study Hours	Exam Score
6.0	70
5.5	65
7.0	83
6.0	69
8.5	92
8.0	82
8.0	62
6.0	57
7.5	83
5.0	93
8.0	89

Experimental Results

Pearson's correlation yielded a p-value < 0.0001 as indicated in Figure 10.5. A resulting p-value less than the critical value (0.05) indicates a significant correlation. Therefore, a relationship exists between the number of hours studied (the night prior to the exam) and the exam score ($p < 0.0001$), and we can reject the null hypothesis. The r-value (0.745) is positive and strong, which indicates that

		Correlations	
		NumberofHours Studiedfor Nutrition	Nutrition ExamScore
NumberofHours StudiedforNutrition	Pearson Correlation	1	.745**
	Sig. (2-tailed)		.000
	N	28	28
NutritionExamScore	Pearson Correlation	.745**	1
	Sig. (2-tailed)	.000	
	N	28	28

**. Correlations is significant at the 0.01 level (2-tailed).

Figure 10.5 Pearson's correlation SPSS output.

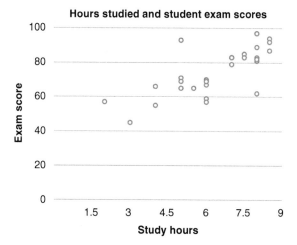

Figure 10.6 Scatter plot illustrating number of hours studied and student exam scores for 28 students.

there is a direct relationship or positive correlation between the number of hours studied and exam score.

Concluding Statement

There is a positive correlation ($r = 0.745$, $p \leq 0.0001$) between the number of hours studied and the grade earned on the nutrition exam. Thus, the positive correlation suggests there is a relationship between the amount of studying and exam score. However, this correlation does not imply a *causal* relationship between studying and exam grades. Refer to Figure 10.6 for a graphic illustration of the results.

10.4 Case Study – Spearman's Correlation

The previous example looked at a relationship between the hours studied and how well a student performed on an exam. Considering that both the variables are continuous, and the data are normally distributed, Pearson's correlation is the recommended statistic to appropriately evaluate the data. Now, suppose that the Kinesiology professor wanted to correlate the hours studied with how prepared the students felt prior to the exam. Because we are trying to measure how prepared students felt going into the exam (based on a Likert scale, 1 = not prepared to 5 = very prepared), the variable is considered ordinal or ranked. These types of variables are best analyzed using Spearman's rank-order correlation.

We might predict that the more hours invested in studying would ultimately lead to a stronger feeling of preparedness. The professor proposed the following question: "Does a significant relationship exist between the number of hours studied and the level of preparedness expressed by students?"

> **(H)** Based on the question, formulate the null and alternative hypotheses that address the question proposed.
>
> **
>
> Null Hypothesis (H_0): There is no relationship between the number of hours studied and the level of preparedness expressed by students.
> Alternative Hypothesis (H_1): There is a relationship between the number of hours studied and the level of preparedness expressed by students.

Experimental Data

The following data are hypothetical and will allow us to examine the relationship between student study hours and preparedness (Table 10.3).

Experimental Results

After performing Spearman's correlation, the analysis indicates a correlation coefficient (rho, or ρ) of 0.336 and a *p*-value of 0.080 (see Figure 10.7); therefore, we fail to reject our null hypothesis using the critical value of 0.05. Our results indicate that there is no significant relationship between the number of hours studied and how prepared the students felt for the exam. In other words, the students who dedicated more hours to studying did not feel more prepared for the exam than those who dedicated fewer hours.

Table 10.3 Spearman's correlation experimental data showing hours studied and feeling of preparedness (Likert scale) for 28 students.

Study Hours	Preparedness
5.0	3
4.0	3
3.0	2
5.0	2
7.0	4
8.5	3
8.0	2
8.0	3
4.0	4
5.0	3
2.0	2
8.5	5
6.0	4
7.5	3
7.0	3
8.0	3
6.0	2
6.0	2
5.5	4
7.0	2
6.0	2
8.5	4
8.0	3
8.0	3
6.0	2
7.5	2
5.0	2
8.0	4

Concluding Statement

There is no significant correlation ($\rho = 0.336$, $p = 0.080$) between the number of hours a student studied and how prepared they felt for the exam. Thus, the correlation suggests that there is no relationship between the time spent studying and how prepared students felt for exams. See Figure 10.8 for a graphic illustration of the results.

Correlations			Number of Hours Studied for Nutrition	Preparedness
Spearman's rho	Number of Hours Studied for Nutrition	Correlation Coefficient	1.000	.336
		Sig. (2-tailed)	.	.080
		N	28	28
	Preparedness	Correlation Coefficient	.336	1.000
		Sig. (2-tailed)	.080	.
		N	28	28

Figure 10.7 Spearman's correlation SPSS output.

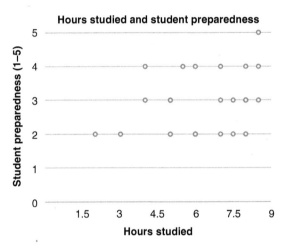

Figure 10.8 Scatter plot illustrating number of hours studied and feeling of preparedness based on a Likert scale (1–5) for 28 students.

Tutorials

10.5 Pearson's Correlation Excel and Numbers Tutorial

When human skeletal remains are found, forensic anthropologists must assist in the identification process by reconstructing who the person was when they were alive. To do this, forensic anthropologists create a biological profile, which includes estimates of specific demographics including age, sex, ancestry, and stature. Stature, or height, can be recreated using measurements of long bones (more specifically, most arm and leg bones) input into well-established discriminant functions (equations derived from long bone length and heights), where the results will provide a range of stature that should encompass the person's height at death.

A forensic anthropologist is interested in deriving discriminant functions that are population specific, or tailored to an individual population, to publish in a forensic journal and for use by forensic practitioners. She measures the long bones of 18 modern skeletons from rural America that have documentation of stature prior to/at death. Before inputting the skeletal measurements into discriminant analysis to derive a new discriminant function equation, the anthropologist verifies each of the long bone lengths she collected is correlated to stature in her sample. She first started by correlating femoral (femur) length (in centimeters) to stature (in centimeters).

*The following tutorial shows the screenshots for Excel, but you can use the same equation from this tutorial in Numbers.

**This example was taken from the research conducted by Dr. Kanya Godde.

(R) Formulate a question about the data that can be addressed by performing a correlation.

****

Question: Are stature and femoral length associated with one another in rural Americans?

(H) Based on the question, formulate the null and alternative hypotheses that address the question proposed.

****

Null Hypothesis (H_0): There is no relationship between stature and femoral length in rural Americans.

Alternative Hypothesis (H_1): There is a relationship between stature and femoral length in rural Americans.

Now that an appropriate question has been developed, along with a set of testable hypotheses, you can run the statistical analysis.

- This tutorial focuses on running Pearson's correlation in Excel.
- Refer to Chapter 12 for tips and tools when using Excel.
- Check all assumptions prior to running the test, and complete the tutorial **regardless** of whether the assumptions were met.

Pearson's Correlation Excel Tutorial

1. Input data into the Excel spreadsheet. Type the column labels into cells **A1** and **B1**. The data will appear in cells **A2:A19** and **B2:B19**.

Height (cm)	Femur maximum length (cm)
147	40.2
155	40.6
152	42.5
157	43.3
160	43.3
163	43.5
164	43.7
165	44
166	44
160	44.2
164	44.6
157	44.7
157	44.9
168	45.6
173	45.8
175	46.3
167	46.7
164	48.5

2. In cell **A20**, label the type of correlation you will be performing (**Pearson's r**). In cell **A21**, provide a label indicating the adjacent cell will contain the ***p*-value**.

	A
1	Height (cm)
2	147
3	155
4	152
5	157
6	160
7	163
8	164
9	165
10	166
11	160
12	164
13	157
14	157
15	168
16	173
17	175
18	167
19	164
20	Pearson's r
21	p-value

3. In cell **B20**, use the **=CORREL** function to test the two arrays (**A2:A19** and **B2:B19**) or set of values for the two variables. The function will automatically run a Pearson's correlation. Press enter/return.

	A	B
1	Height (cm)	Femur maximum length (cm)
2	147	40.2
3	155	40.6
4	152	42.5
5	157	43.3
6	160	43.3
7	163	43.5
8	164	43.7
9	165	44
10	166	44
11	160	44.2
12	164	44.6
13	157	44.7
14	157	44.9
15	168	45.6
16	173	45.8
17	175	46.3
18	167	46.7
19	164	48.5
20	Pearson's r	=CORREL(A2:A19, B2:B19)

4. The result is Pearson's *r* correlation.

	A	B
1	Height (cm)	Femur maximum length (cm)
2	147	40.2
3	155	40.6
4	152	42.5
5	157	43.3
6	160	43.3
7	163	43.5
8	164	43.7
9	165	44
10	166	44
11	160	44.2
12	164	44.6
13	157	44.7
14	157	44.9
15	168	45.6
16	173	45.8
17	175	46.3
18	167	46.7
19	164	48.5
20	Pearson's r	0.724045386

5. In cell B21, use the = `TDIST` function to calculate the *p*-value. The *r*-value, degrees of freedom, and number of tails will be defined in the function.

	A	B	C
1	Height (cm)	Femur maximum length (cm)	
2	147	40.2	
3	155	40.6	
4	152	42.5	
5	157	43.3	
6	160	43.3	
7	163	43.5	
8	164	43.7	
9	165	44	
10	166	44	
11	160	44.2	
12	164	44.6	
13	157	44.7	
14	157	44.9	
15	168	45.6	
16	173	45.8	
17	175	46.3	
18	167	46.7	
19	164	48.5	
20	Pearson's r	0.724045386	
21	p-value	=TDIST(SQRT(16*(B20^2)/(1-(B20^2))), 16, 2)	

Let us look at the function and programming closer, so we understand the construction of the code.

=TDIST(SQRT(16*(B20^2)/(1-(B20^2))), 16, 2)

We cannot just put the *r*-value into = `TDIST`, as this tests a *t* distribution. So, we have to convert *r* into *t* by taking the square root of the product of the square of Pearson's *r* (already in cell **B20**), multiplied by **N − 2**. This whole calculation is then divided by 1 minus the square of *r*.

=TDISTSQRT(16*(B20^2)/(1-(B20^2))), 16, 2)

To program this, we take the sample size (**N**), which in this case is 18, and subtract 2 to yield 16. Next, we multiply this against the square of *r* by using the "*" symbol. To raise a value to a specific power in Excel, we use the "^" symbol. So, if we want to square something, it would be "^2." When we put all of this programming together, we get **16*(B20^2)**.

=TDIST(SQRT(16*(B20^2)(1-(B20^2))), 16, 2)

Next, we take the square root of what we just coded (**16*(B^2)**). To do this, we will add the function =**SQRT** and a set of parentheses.

=TDISTSQRT(16*(B20^2)/(1-(B20^2))), 16, 2)

Note: The closed parenthesis goes to the end of the calculations of the =**SQRT** function.

Now, we have to divide what we have constructed so far by 1 minus the square of *r*. So, we add a "/" symbol to tell Excel to divide. Then, we add **1−** and a set of parentheses. In the set of parentheses, we place our programming to square Pearson's *r*. By doing this, we have developed the following bit of code: **/1−(B20^2)**.

=TDIST(SQRT(16*(B20^2/(1-(B20^2))), 16, 2)

We have now finished the coding for converting r to t. Next, we will add a ",", the degrees of freedom ($N - 2$), and the number of tails we want to test (let us say 2 for a two-tailed test), separated by another comma. Press enter/return.

$$=\text{TDIST}(\text{SQRT}(16*(B20^\wedge 2)/(1-(B20^\wedge 2))), 16, 2)$$

6. The result is the *p*-value.

	A	B
1	Height (cm)	Femur maximum length (cm)
2	147	40.2
3	155	40.6
4	152	42.5
5	157	43.3
6	160	43.3
7	163	43.5
8	164	43.7
9	165	44
10	166	44
11	160	44.2
12	164	44.6
13	157	44.7
14	157	44.9
15	168	45.6
16	173	45.8
17	175	46.3
18	167	46.7
19	164	48.5
20	Pearson's r	0.724045386
21	p-value	0.000680098

From the output we generated, we see a high positive correlation ($r = 0.724$) that is statistically significant (*p*-value = 0.0007), indicating height and femoral length are correlated.

Concluding Statement

In this sample of rural Americans, femoral length (cm) and stature (cm) are highly positively correlated ($r = 0.724$) and this correlation is statistically significant (*p*-value = 0.0007).

10.6 Spearman's Correlation Excel Tutorial

If you found that femoral length and stature are not normally distributed in the Pearson's correlation Excel tutorial, consider applying a nonparametric test. Let us look at the data again, but this time with the nonparametric Spearman's correlation.

1. Sort the **Height** column from lowest to highest or highest to lowest. Then, in cell **C1**, type in a label indicating the **C** column will be for the **Rank of Height**.

	A	B	C
1	Height (cm)	Femur maximum length (cm)	Rank of Height
2	147	40.2	
3	152	42.5	
4	155	40.6	
5	157	43.3	
6	157	44.7	
7	157	44.9	
8	160	43.3	
9	160	44.2	
10	163	43.5	
11	164	43.7	
12	164	44.6	
13	164	48.5	
14	165	44	
15	166	44	
16	167	46.7	
17	168	45.6	
18	173	45.8	
19	175	46.3	

2. Next, use the **=RANK** function in cell **C2** to rank the column of heights. Highlight the cells whose rank you want displayed in **C2**, followed by a comma and the array you want to rank **A2:A19**. Press enter/return.

	A	B	C
1	Height (cm)	Femur maximum length (cm)	Rank of Height
2	147	40.2	=RANK(A2,A$2:A$19)
3	152	42.5	
4	155	40.6	
5	157	43.3	
6	157	44.7	
7	157	44.9	
8	160	43.3	
9	160	44.2	
10	163	43.5	
11	164	43.7	
12	164	44.6	
13	164	48.5	
14	165	44	
15	166	44	
16	167	46.7	
17	168	45.6	
18	173	45.8	
19	175	46.3	

Note: The use of **absolute references applied** to the row number (denoted by "$") prepares us to be able to copy the formula down the entire **C** column.

3. Copy and paste this formula from **C2:C19** by using the fill handle.

	A	B	C
1	Height (cm)	Femur maximum length (cm)	Rank of Height
2	147	40.2	18
3	152	42.5	
4	155	40.6	
5	157	43.3	
6	157	44.7	
7	157	44.9	
8	160	43.3	
9	160	44.2	
10	163	43.5	
11	164	43.7	
12	164	44.6	
13	164	48.5	
14	165	44	
15	166	44	
16	167	46.7	
17	168	45.6	
18	173	45.8	
19	175	46.3	

4. The ranks will appear.

	A	B	C
1	Height (cm)	Femur maximum length (cm)	Rank of Height
2	147	40.2	18
3	152	42.5	17
4	155	40.6	16
5	157	43.3	13
6	157	44.7	13
7	157	44.9	13
8	160	43.3	11
9	160	44.2	11
10	163	43.5	10
11	164	43.7	7
12	164	44.6	7
13	164	48.5	7
14	165	44	6
15	166	44	5
16	167	46.7	4
17	168	45.6	3
18	173	45.8	2
19	175	46.3	1

5. Next, we have to deal with ties among the ranks. To do this, we will take the **mean rank**. So, if two heights are the same, we average the

rank. For instance, a height of 160 cm appears twice, and according to Excel, those heights have been ranked as 11th. Really, one is ranked as 11 and one as 12 (notice 12 is not listed in the ranks and Excel jumps to 13). If we take the mean of 11 and 12 (11+12/2 = 11.5), we find the answer is 11.5. We will change both ranks for a height of 160 cm to **11.5**.

	A	B	C
1	Height (cm)	Femur maximum length (cm)	Rank of Height
2	147	40.2	18
3	152	42.5	17
4	155	40.6	16
5	157	43.3	13
6	157	44.7	13
7	157	44.9	13
8	160	43.3	11
9	160	44.2	11
10	163	43.5	10
11	164	43.7	7
12	164	44.6	7
13	164	48.5	7
14	165	44	6
15	166	44	5
16	167	46.7	4
17	168	45.6	3
18	173	45.8	2
19	175	46.3	1

↓

	A	B	C
1	Height (cm)	Femur maximum length (cm)	Rank of Height
2	147	40.2	18
3	152	42.5	17
4	155	40.6	16
5	157	43.3	13
6	157	44.7	13
7	157	44.9	13
8	160	43.3	11.5
9	160	44.2	11.5
10	163	43.5	10
11	164	43.7	7
12	164	44.6	7
13	164	48.5	7
14	165	44	6
15	166	44	5
16	167	46.7	4
17	168	45.6	3
18	173	45.8	2
19	175	46.3	1

6. Similarly, if we have three heights of equal value, and thus equal rank, we find the average of the ranks. For example, a height of 164 cm is found three times in this sample. Excel has ranked them

all with a 7, but if we examine the rank column closer, we see that Excel skips from 7th rank to 10th rank, which means it allows for the last two of the three heights of 164 cm to be the 8th and the 9th rank. In this case, we average across ranks 7, 8, and 9 (7 + 8 + 9/3 = 8) and find the mean is 8. We replace the three ranks corresponding to 164 cm with a rank of 8.

	A	B	C
1	Height (cm)	Femur maximum length (cm)	Rank of Height
2	147	40.2	18
3	152	42.5	17
4	155	40.6	16
5	157	43.3	13
6	157	44.7	13
7	157	44.9	13
8	160	43.3	11.5
9	160	44.2	11.5
10	163	43.5	10
11	164	43.7	7
12	164	44.6	7
13	164	48.5	7
14	165	44	6
15	166	44	5
16	167	46.7	4
17	168	45.6	3
18	173	45.8	2
19	175	46.3	1

↓

	A	B	C
1	Height (cm)	Femur maximum length (cm)	Rank of Height
2	147	40.2	18
3	152	42.5	17
4	155	40.6	16
5	157	43.3	13
6	157	44.7	13
7	157	44.9	13
8	160	43.3	11.5
9	160	44.2	11.5
10	163	43.5	10
11	164	43.7	8
12	164	44.6	8
13	164	48.5	8
14	165	44	6
15	166	44	5
16	167	46.7	4
17	168	45.6	3
18	173	45.8	2
19	175	46.3	1

Continue to do this to all ties until you have finished working with the height column ranks.

7. Calculate the ranks for **femoral maximum length** using steps 1–6.

	A	B	C	D
1	Height (cm)	Femur maximum length (cm)	Rank of Height	Rank of Length
2	147	40.2	18	18
3	152	42.5	17	16
4	155	40.6	16	17
5	157	43.3	14	14.5
6	157	44.7	14	7
7	157	44.9	14	6
8	160	43.3	11.5	14.5
9	160	44.2	11.5	9
10	163	43.5	10	13
11	164	43.7	8	12
12	164	44.6	8	8
13	164	48.5	8	1
14	165	44	6	10.5
15	166	44	5	10.5
16	167	46.7	4	2
17	168	45.6	3	5
18	173	45.8	2	4
19	175	46.3	1	3

8. In cell **B20,** label the type of correlation you will be performing (Spearman's) and the symbol representing the correlation (ρ). In cell **B21**, provide a label indicating the adjacent cell will contain the *p*-value.

	A	B
1	Height (cm)	Femur maximum length (cm)
2	147	40.2
3	152	42.5
4	155	40.6
5	157	43.3
6	157	44.7
7	157	44.9
8	160	43.3
9	160	44.2
10	163	43.5
11	164	43.7
12	164	44.6
13	164	48.5
14	165	44
15	166	44
16	167	46.7
17	168	45.6
18	173	45.8
19	175	46.3
20		Spearman's ρ
21		*p*-value

9. Use the **=CORREL** function to calculate Spearman's rho on the rank arrays **C2:C19** and **D2:D19**.

	A	B	C	D
1	Height (cm)	Femur maximum length (cm)	Rank of Height	Rank of Length
2	147	40.2	18	18
3	152	42.5	17	16
4	155	40.6	16	17
5	157	43.3	14	14.5
6	157	44.7	14	7
7	157	44.9	14	6
8	160	43.3	11.5	14.5
9	160	44.2	11.5	9
10	163	43.5	10	13
11	164	43.7	8	12
12	164	44.6	8	8
13	164	48.5	8	1
14	165	44	6	10.5
15	166	44	5	10.5
16	167	46.7	4	2
17	168	45.6	3	5
18	173	45.8	2	4
19	175	46.3	1	3
20		Spearman's	=CORREL(C2:C19, D2:D19)	

10. Press enter/return. The result is Spearman's rho.

	A	B	C
1	Height (cm)	Femur maximum length (cm)	Rank of Height
2	147	40.2	18
3	152	42.5	17
4	155	40.6	16
5	157	43.3	14
6	157	44.7	14
7	157	44.9	14
8	160	43.3	11.5
9	160	44.2	11.5
10	163	43.5	10
11	164	43.7	8
12	164	44.6	8
13	164	48.5	8
14	165	44	6
15	166	44	5
16	167	46.7	4
17	168	45.6	3
18	173	45.8	2
19	175	46.3	1
20		Spearman's ρ	0.718738525

11. Calculate the *p*-value as in step 4 of Pearson's correlation tutorial above. Substitute **C20** for **B20** in both places, as **C20** is the correlation value for Spearman's that is analogous to Pearson's.

Pearson's and Spearman's Correlation | 461

	A	B	C	D
1	Height (cm)	Femur maximum length (cm)	Rank of Height	Rank of Length
2	147	40.2	18	18
3	152	42.5	17	16
4	155	40.6	16	17
5	157	43.3	14	14.5
6	157	44.7	14	7
7	157	44.9	14	6
8	160	43.3	11.5	14.5
9	160	44.2	11.5	9
10	163	43.5	10	13
11	164	43.7	8	12
12	164	44.6	8	8
13	164	48.5	8	1
14	165	44	6	10.5
15	166	44	5	10.5
16	167	46.7	4	2
17	168	45.6	3	5
18	173	45.8	2	4
19	175	46.3	1	3
20		Spearman's ρ	0.718738525	
21		p-value	=TDIST(SQRT(16*(C20^2)/(1-(C20^2))), 16, 2)	

12. Press enter/return. The result is the *p*-value.

	A	B	C
1	Height (cm)	Femur maximum length (cm)	Rank of Height
2	147	40.2	18
3	152	42.5	17
4	155	40.6	16
5	157	43.3	14
6	157	44.7	14
7	157	44.9	14
8	160	43.3	11.5
9	160	44.2	11.5
10	163	43.5	10
11	164	43.7	8
12	164	44.6	8
13	164	48.5	8
14	165	44	6
15	166	44	5
16	167	46.7	4
17	168	45.6	3
18	173	45.8	2
19	175	46.3	1
20		Spearman's ρ	0.718738525
21		p-value	0.000777451

From the output we generated, we see a positive high correlation ($\rho = 0.7187$) that is statistically significant (*p*-value = 0.0008), indicating height and femoral length are highly correlated.

Concluding Statement

In this sample of rural Americans, femoral length (centimeter) and stature (centimeter) are highly positively correlated ($r = 0.7187$) and this correlation is statistically significant (p-value $= 0.0008$).

10.7 Pearson/Spearman's Correlation SPSS Tutorial

Patients infected with *Helicobacter pylori* may develop autoantibodies that result in damage to gastric parietal cells in the stomach. Parietal cells produce intrinsic factor, which is required for vitamin B_{12} absorption (vB12). Consequently, *H. pylori* infections can often lead to vitamin B_{12} deficiency.

Researchers at the Al-Quds University observed that many people who had vitamin B_{12} deficiency were also positive for *H. pylori* infection (Hilu et al., 2015). They obtained blood serum samples from patients who were deficient in vitamin B_{12} and ran an ELISA to determine the levels of vitamin B_{12} and anti-*H. pylori* IgA.

*The data used for this tutorial were taken from the study conducted by Hilu et al. (2015). The data were simplified for display; however, the trend was kept.

> **(R)** Formulate a question about the data that can be addressed by performing a Pearson Correlation.
>
> ***
>
> Question: Is there a relationship between levels of anti-*H. pylori* IgA and vitamin B_{12}?

> **(H)** Based on the question, formulate the null and alternative hypotheses that address the question proposed.
>
> ***
>
> Null Hypothesis (H_0): There is no relationship between levels of anti-*H. pylori* IgA and vitamin B_{12}.
> Alternative Hypothesis (H_1): There is a relationship between levels of anti-*H. pylori* IgA and vitamin B_{12}.

Now that an appropriate testable question has been developed along with a set of testable hypotheses, you can run the statistical analysis.

- Utilize the following tutorial to run Pearson or Spearman's correlation in SPSS.
- Refer to Chapter 13 for getting started and understanding SPSS.
- Check all assumptions prior to running the test.

Pearson's and Spearman's Correlation SPSS Tutorial

1. Your data should look similar to the following.

	IgA	vB12
1	49.3	151.56
2	60.4	150.98
3	64.5	150.09
4	52.6	150.37
5	58.9	150.58
6	55.6	151.23
7	48.4	150.23
8	49.5	150.69
9	56.2	150.94
10	56.9	150.03
11	61.4	150.73
12	44.3	151.43
13	49.3	151.85
14	55.8	149.72
15	59.8	150.02
16	64.3	149.01
17	59.3	148.76
18	64.5	148.31
19	68.8	148.24
20	58.9	149.62
21	67.4	149.94

2. You can now run Pearson's correlation. To begin your analysis, click on the **Analyze** tab in the toolbar located along the top of the page. Scroll down to the **Correlate** option and select **Bivariate**.

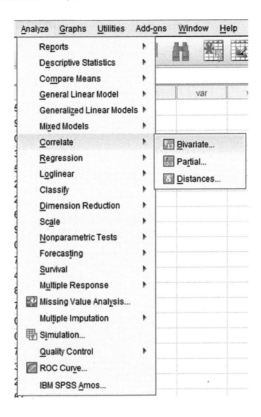

3. The following screen will appear.

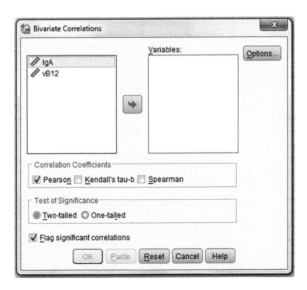

4. Click on the variable (**IgA**) that appears in the box to the left. Then click on the corresponding arrow to move the variable over to the **Variables** list.

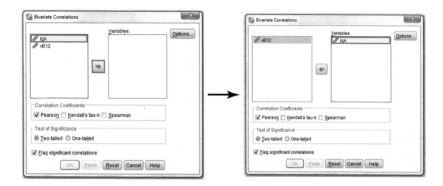

5. Next click on the variable remaining in the box to the left (**vB12**) and click the corresponding arrow to move the variable over to the **Variables** list.

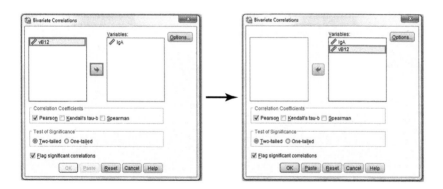

6. Check the **Pearson** box under **Correlation Coefficients** and **Two-tailed** under **Test of Significance**.

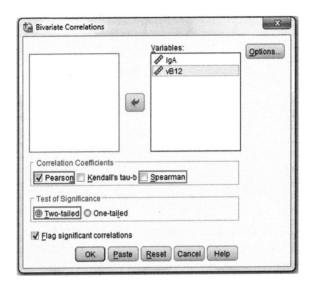

Note: To run Spearman's correlation, simply click **Spearman**.

7. Click **OK**.

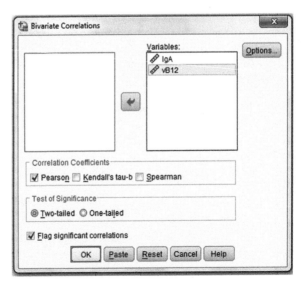

8. A separate document will appear. This is referred to as the output.

Correlations

		IgA	vB12
IgA	Pearson Correlation	1	-.690**
	Sig. (2-tailed)		.001
	N	21	21
vB12	Pearson Correlation	-.690**	1
	Sig. (2-tailed)	.001	
	N	21	21

**. Correlation is significant at the 0.01 level (2-tailed).

Pearson's correlation yielded a p-value = 0.001. A p-value less than the critical value (0.05) indicates a significant correlation. Therefore, a relationship exists between vitamin B_{12} deficiency and anti-*H. pylori* IgA ($p = 0.001$); therefore, we can reject the null hypothesis and fail to reject the alternative hypothesis. The results also indicate that $r = -0.690$ (a negative value), which demonstrates a negative relationship between the two variables. In other words, the greater the infection (*H. pylori*), the lesser vitamin B_{12} is absorbed into the body.

Concluding Statement

There is a weak negative correlation ($r = -0.690$, $p = 0.001$) between the levels of anti-*H. pylori* IgA and vitamin B_{12}.

Note: If you want to save the SPSS file with the inserted data as well as the SPSS output with the results of the statistical analysis performed, then you must save each document separately (see Chapter 13).

10.8 Pearson/Spearman's Correlation R Tutorial

Sexual selection in primates differs among groups. Primates with a more competitive social structure, such as chimpanzees who live in multi-male/multi-female groups, have a large testicular volume (measured in cubic centimeters) in relation to body mass (kg). Larger testicles in chimpanzees allow for sperm competition; chimpanzee males can produce enough sperm to compete with other males' sperm when females copulate with two males in close succession. The more sperm a male produces, the greater the likelihood his sperm will fertilize the ovum.

However, monogamous groups, like Hamadryas baboons, are usually associated with smaller testicular volume in relation to body mass, as they do not need to regularly participate in sperm competition for successful production of offspring. When comparing primate testicular volume across species, it is often done so in conjunction with body mass in order to make the testicular volumes comparable across species of different sizes.

Before utilizing the measures together, it is best to run a correlation to verify if there is a relationship between testicular volume and body size. Usually, age is also taken into account to ensure males are at the age of reproduction when testing correlation hypotheses between testicular volume and body mass.

*The data used for this tutorial were taken from the study conducted by Jolly and Phillips-Conroy (2003).

> **(R)** Formulate a question about the data that can be addressed by performing a correlation.
>
> **
>
> Question: Are testicular volume and body mass of Hamadryas baboons associated with one another?

> **(H)** Based on the question, formulate the null and alternative hypotheses that address the question proposed.
>
> **
>
> Null Hypothesis (H_0): There is no relationship between Hamadryas baboon's testicular volume and body mass.
> Alternative Hypothesis (H_1): There is a relationship between Hamadryas baboon's testicular volume and body mass.

Now that an appropriate testable question has been developed along with a set of testable hypotheses, you can run the statistical analysis.

- This tutorial focuses on running correlations in R.
- Refer to Chapter 15 for R-specific terminology and instructions on how to invoke and construct code.
- Check all assumptions prior to running the test.

Pearson's and Spearman's Correlation R Tutorial

1. At the command prompt, type the following testicular volumes (reported in cubic centimeters or cc) into a vector and store it to a name that describes its contents (e.g., **tv**). Press enter/return.

```
> tv<-c(17, 18, 19, 16, 17, 25, 24, 25, 25, 26, 24,
  25, 24, 24, 24, 25, 26, 23, 23, 23, 24, 25, 24,
  27, 28, 21, 27, 27, 26, 18)
```

2. Create a second vector for body mass (in kilogram) by typing in the body mass measurements from the Hamadryas baboons and storing it to a name such as **bodymass**. Press enter/return.

```
> bodymass<-c(11,  12, 14, 14, 17, 20, 20, 20, 21,
  22, 23, 23, 24, 24, 24, 25, 25, 25, 25, 25, 26,
  26, 26, 26, 26, 26, 26, 25, 25, 26)
```

3. The **Hmisc** package contains a function that calculates both the correlation value (r or ρ) and the p-value. If you already have it, load the **Hmisc** package by using the `library()` function. If you are unsure if you have installed the package, type in `library(Hmisc)` and press enter/return.

   If you do not have the package installed, you should see an error message similar to the one below. To install the package, follow the procedures outlined in Chapter 15.

```
> library(Hmisc)
Error in library(Hmisc): there is no package called 'Hmisc'
> |
```

If the package is installed, you will see the function execute normally, as in the image below.

```
> library(Hmisc)
Loading required package: lattice
Loading required package: survival
Loading required package: Formula
Loading required package: ggplot2
```

4. Apply the **rcorr()** function to both vectors and select the type of correlation to run with the **type=** argument. The **type=** argument allows for the practitioner to select a **pearson** or **spearman** correlation. Type in your selection of correlation procedure within quotation marks using the knowledge from examining your assumptions (in this case, the data are not normally distributed; therefore, the statistical analysis should be the nonparametric Spearman's rank-order correlation). Press enter/return.

```
> rcorr(tv, bodymass, type="spearman")
```

Note: To switch to Pearson's correlation, simply type in "**pearson**" instead of "**spearman**" in the script.

5. The output will look similar to the one below.

```
      x     y
x  1.00  0.41
y  0.41  1.00

n= 30


P
      x          y
x                0.0238
y   0.0238
```

The values on the diagonal opposite of the line of 1.0 in the upper matrix are the p-value (0.41), indicating the strength of the correlation (which is weak). The lower matrix contains the p-value (0.0238).

In this case, the p-value is 0.0238 and is significant when using $\alpha = 0.05$, causing us to reject the null hypothesis and state that in this sample of baboons, testicular volume is correlated with body mass. The correlation coefficient ρ of 0.41 is positive, indicating a positive correlation between the two variables. Therefore, as body mass increases, so does testicular volume.

Concluding Statement

In this sample of baboons, testicular volume (cubic centimeter) is significantly weakly positively correlated ($\rho = 0.41$, p-value $= 0.0238$) to body mass (kilogram).

11

Linear Regression

> **Learning Outcomes:**
>
> By the end of this chapter, you should be able to:
>
> 1. Evaluate the relationship between independent and dependent variables.
> 2. Use statistical programs to run a regression analysis and determine the significance of F and interpret the R^2 value for your analysis.
> 3. Use the skills acquired to analyze and evaluate your own dataset from independent research.

11.1 Linear Regression Background

As previously explained in Chapter 10, a correlation analysis can potentially determine whether a linear relationship exists between two variables, as well as indicate the strength of the relationship (r, ρ). It is important to remember that the output generated from the correlation analysis can give the researcher a quantifiable value describing the relationship, but it is unable to determine if X (the independent variable) can predict Y (the dependent variable). If researchers are interested in prediction, then a simple linear regression is an appropriate test to run.

A simple linear regression is part of the general linear model (GLM), which also includes an ANOVA, an analysis of covariance (ANCOVA), and t-tests, among others. Simple linear regression

An Introduction to Statistical Analysis in Research: With Applications in the Biological and Life Sciences,
First Edition. Kathleen F. Weaver, Vanessa C. Morales, Sarah L. Dunn, Kanya Godde and Pablo F. Weaver.
© 2018 John Wiley & Sons, Inc. Published 2018 by John Wiley & Sons, Inc.
Companion Website: www.wiley/com/go/weaver/statistical_analysis_in_research

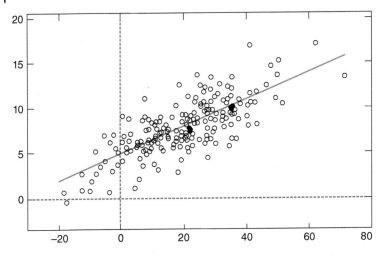

Figure 11.1 Scatter plot with regression line representing a typical regression analysis.

examines the level of change of one variable (independent or explanatory) due to another variable (dependent or response). In addition, a linear regression quantifies this change and provides a measure (R^2, or the coefficient of determination) of the variation in the dependent variable (Y) explained by the independent variable (X), see Figure 11.1 for a graphical representation of a typical regression analysis. The linear model can be utilized to predict Y based on the values of X. Under these circumstances, regression is typically applied to develop a linear equation that best describes the relationship between two variables. In other words, addressing the questions: "Is there a relationship between X and Y. If so, what is the linear equation that best describes the relationship between X and Y?"

If you have more than two independent variables, then multiple regression can be used to analyze the linear relationship among multiple variables. For example, an epidemiologist may consider running a simple linear regression on how the level of alcohol consumption determines the degree of liver dysfunction. However, because alcohol consumption is often linked with cigarette smoking, the same epidemiologist may consider looking at an additional independent variable, cigarette smoking, to see how alcohol affects the liver when cigarette smoking is taken into account. Throughout this chapter, we will be referring to a simple linear regression when referencing linear regression or regression analysis.

The equation for a line is depicted in the following equation:

$$Y = mX + b$$

This relationship is illustrated in the following graph:

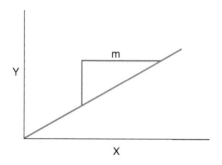

Terms and variables in the equation are:

X = Independent (explanatory) variable
Y = Dependent (response) variable
m = Slope (change in y/change in x)
b = y-intercept (the point where the line crosses the y-axis)

The analogous regression equation is

$$Y = b_0 + b_1 X_1$$

Where

X_1 = Value of first observation of independent variable
Y = Dependent variable
b_0 = y-intercept
b_1 = Value of first observation of slope

The regression equation describes the regression line that is fit through the data points. It also allows you to predict values of Y from values of X. Similar to depicting the data points of a correlation analysis in the form of an XY scatter plot, a dataset can also be expressed as an XY scatter plot for linear regression, along with the best-fit line plotted through the points. In other words, the regression line reflects the best linear model associated with the data. It is important to be able to draw conclusions from a graph, specifically from the orientation of the slope (whether it is positive or negative). The orientation of the slope gives insight into the type of relationship x and y have with one another. We will be discussing this in more detail with respect to the R^2 value. Figure 11.2 depicts different slope orientations while examining the relationship between x and y.

Regression Nuts and Bolts

A regression analysis determines whether there is an existing relationship between the independent and dependent variables, and if there is, a regression can also describe the quantifiable change in the dependent variable due to the change in the independent variable.

The p-value associated with the F-value in the ANOVA table reflects if the independent variable in the simple linear regression model has a relationship with the dependent variable. If $p \leq 0.05$, then the relationship between the independent and dependent variable is significant and there is an association between the two variables. If $p > 0.05$, then there is no significant relationship between the independent and dependent variable.

$p \leq 0.05$ (**statistically significant**)

$p > 0.05$ (**NOT statistically significant**)

There are three coefficients that are evaluated by a p-value: (1) the ANOVA F-value (as described above), (2) the t-value for slope, and (3) the t-value for the intercept. The t-value in the coefficients table is from a t-test that estimates the significance of the regression coefficients in the model (i.e., the slope and y-intercept). In simple linear regression, the F value will be identical to the slope t-value. In other types of regression, these values will differ and interpretations will become more complex.

Another important factor to consider is the R^2 value. The R^2 value is defined as the square of the correlation coefficient and is an indication of how well the linear model describes the data. Specifically, R^2 explains how much of the variation in the Y variable can be explained

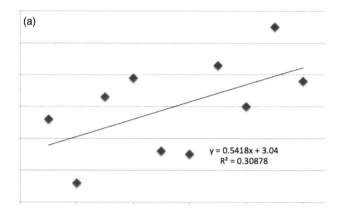

Positive slope (m = 0.5418) indicates that as x increases, y also increases. An R^2 = 0.31 indicates a weak relationship between x and y in some fields. R^2 can be interpreted as saying that X explains 31% of the variation in the Y variable. How can the above information serve in a concluding statement?

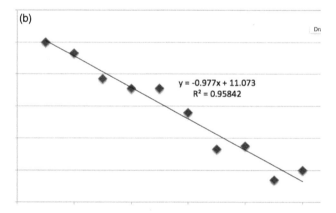

Negative slope (m = -0.977) indicates that as x increases, y decreases. An R^2 = 0.96 indicates a strong relationship between x and y. R^2 can be interpreted as saying that X explains 96% of the variation in the Y variable. How can the above information serve in a concluding statement?

Figure 11.2 Graphs depicting the spread around the trend line. Orientation of the slope determines the type of relationship between x and y and R^2 describes the strength of the relationship.

by the variation in the X variable, which is related to the ability of the independent variable to predict the dependent variable.

The closer R^2 is to 1, the better the fit and the closer the data points are to the regression line. Alternatively, the closer R^2 is to 0, the worse the fit and the further away the data points are to the regression line. Any outliers (or data points that do not follow the trend) will possibly result in a lower R^2 value. For example, an $R^2 = 0.59$ indicates that the data points are not closely located along the regression line. Another way to look at R^2 is to turn the R^2 value into a percentage. If $R^2 = 0.59$, this means that 59% of the variation in the dependent variable (y) is explained by the independent variable (x). The remaining 41% is due to unconsidered factors influencing the data. The magnitude of R^2 is interpreted differently by various disciplines. Thus, check with your advisor for what is considered weak and strong in your field.

Generalized Hypotheses

Based on the linear model, $Y = b_0 + b_1 X_1$, a simple linear regression analysis attempts to predict the dependent variable given one independent variable. If the dependent variable can be predicted from the independent variable, then the slope and intercept of the linear model increases or decreases based on the X-value. With this in mind, the following general hypotheses can be formulated for a regression analysis.

Association of the independent variable and the dependent variable (from ANOVA table, assessed with an F-value):

H_0: The independent variable (X) has no association with the dependent variable (Y).
H_1: The independent variable (X) has an association with the dependent variable (Y).

Slope (from coefficients table, assessed with a t-value):

H_0: The slope of the linear model is zero; therefore, the independent variable (X) has no relationship with the dependent variable (Y).
H_1: The slope of the linear model is not zero; therefore, there is a relationship between the independent and dependent variable.

***y*-intercept (from coefficients table, assessed with a *t*-value):**

H_0: The *y*-intercept is equal to zero.
H_1: The *y*-intercept is not equal to zero.

A good model rejects the null hypothesis in all three categories: (1) association of independent and dependent variables, (2) slope, and (3) *y*-intercept. The two coefficients generated by a simple linear regression (slope and intercept) can be input into the regression equation $Y = b_0 + b_1 X_1$ to predict values of *Y* from *X* and to describe the regression line. Before a linear regression can be applied, the following assumptions must first be satisfied.

Assumptions

1. **Data type** – A simple linear regression assumes that both the independent and dependent variable are **continuous**.
2. **Distribution** – The regression analysis assumes a **normal distribution of the residuals (errors)**. The residuals can be tested using the skewness and kurtosis tests described in Chapter 2. Also, there are no significant **outliers**.
3. **Independent samples** – Observations should be **independent**.
4. **Homogeneous variance** – In addition to meeting the normality assumption, a regression analysis assumes a **homogenous variance**. When graphed, the distribution of the residuals should not display heteroscedasticity (sideways cone shape). Instead, they should be homoscedastic (points relatively evenly distributed to each other across a graph because they have approximately the same variance). This can be confirmed through graphing the residuals in a residual plot (refer to the box on the following page).
5. There is a **linear relationship** between the independent and dependent variables.

As a reminder, running an analysis without meeting the assumption(s) may mean that the results are not valid. In the case that one or more of these assumptions are not fully satisfied, then either transforming one of the variables or invoking the nonparametric version of the regression would be best.

> To test for **homoscedasticity**, or homogeneity of variance, the residuals (the difference between the observed value and the value predicted by the linear regression equation) must be plotted against the fitted values (the values predicted by the linear regression equation) where the residuals are plotted on the Y axis and the fitted values on the X axis.
>
> In the graph below we see that the variance is relatively homogeneously distributed.
>
>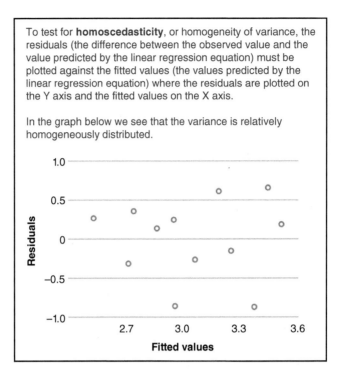

11.2 Case Study

An undergraduate group conducted a study examining the number of Streptomycin resistant strains of bacteria in a number of water sites found at varying distances from a Colorado cattle farm. A total of 12 different sites were examined, starting from Greeley, CO (the site of cattle farming) to Nederland, CO (73 km from cattle farming). Cattle farms are known to use large amounts of antibiotics and the concern is that an increase in antibiotic usage may also increase antibiotic resistant bacteria found in the watershed downstream from the farm. The undergraduate group was interested to know if bacterial resistance could be predicted based on distance from Greeley, CO.

> Ⓡ Formulate a question about the data that can be addressed by performing a linear regression.
>
> **
>
> Question: Does distance from Greeley, CO predict the number of resistant bacteria?

> **(H)** Based on the question, formulate the null and alternative hypotheses that address the question proposed.
>
> ***
>
> Independent Variable Association with Dependent Variable:
>
> Null Hypothesis (H_0): The distance from Greeley, CO is not associated with the number of resistant bacteria.
> Alternative Hypothesis (H_1): The distance from Greeley, CO is associated with the number of resistant bacteria.
>
> Slope Hypotheses:
>
> Null Hypothesis (H_0): The slope of the regression coefficient for the distance from Greeley, CO is zero.
> Alternative Hypothesis (H_1): The slope of the regression coefficient for the distance from Greeley, CO is not zero.
>
> y-intercept Hypotheses:
>
> Null Hypothesis (H_0): The y-intercept will be zero.
> Alternative Hypothesis (H_1): The y-intercept is not zero.

The students grew bacteria on Streptomycin plates and then counted the number (No.) of colonies on each plate. Table 11.1 shows the average value of six replicates for each site.

Experimental Data

Table 11.1 Experimental data showing the geographic distance (kilometer) from the cattle farm and the number of antibiotic resistant bacteria grown on Streptomycin plates.

Distance from Cattle (km) (Greeley, CO)	No. of AbR + Colonies (resistant bacteria)
5	3.7
10	4.1
23.6	3.1
28.2	3.8

(Continued)

Table 11.1 (Continued)

Distance from Cattle (km) (Greeley, CO)	No. of AbR + Colonies (resistant bacteria)
37	2.8
44.5	2.1
51.3	3.0
60	3.1
45	3.2
62	2.4
15	2.5
75	2.8

Experimental Results

According to the ANOVA table output in Figure 11.3, the p-value resulting from the linear regression was not significant ($p = 0.081$)

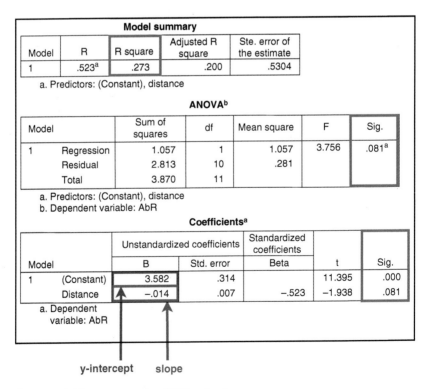

Figure 11.3 Linear regression SPSS output.

for the model. Therefore, we fail to reject the null hypothesis that there is no association between distance from Greeley, CO and the number of resistant bacteria. The R^2 value (0.273) indicates that 27.3% of the variation in the number of resistant bacteria is explained by the distance from Greeley, CO. Further, in the coefficients table, the p-value for the slope (0.081) suggests we should fail to reject the null hypothesis and it must be assumed that the slope of the line explaining the relationship between distance and number of resistant bacteria is equal to zero. Finally, the p-value for the y-intercept is significant ($p < 0.0001$); therefore, the regression line crosses the y-axis at a value other than zero. For a graphic illustration of the experimental results, see Figure 11.4. Taken in sum, the lack of significance in the ANOVA table and the slope indicate distance from Greeley, CO should not be used to predict the number of antibiotic resistant bacteria.

If the model were significant, scientists would use the coefficients in the linear regression equation to predict the number of resistant bacteria from distance to the cattle farming site. To construct this equation, the slope (b_1) and intercept (b_0) are input into the equation $Y = b_0 + b_1 X_1$. In this example, the equation would be as follows: $Y = -0.014X + 3.582$. A scientist wanting to estimate the number of

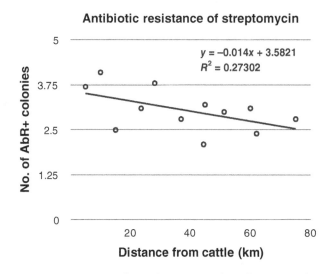

Figure 11.4 Scatter plot with regression line illustrating the relationship between distance from the cattle farm (kilometer) and the number of antibiotic resistant colonies.

resistant bacteria 61 km from the cattle farming site would plug 61 into the equation as X and solve for Y:

$$Y = -0.014(61) + 3.5821$$

In doing so, the scientist yields an answer of 2.73, thus predicting that 2.73 resistant bacteria exist 61 km from the cattle farming site.

Concluding Statement

The model generated from the simple linear regression cannot be used to predict the number of antibiotic resistant bacteria from the distance of the cattle farming site (F-value = 3.756, $p > 0.05$; R^2 = 0.273; slope t-value = -0.014, $p > 0.05$; y-intercept t-value = 3.582, $p \leq 0.05$).

Note: The tutorials following this chapter will illustrate cases in which the independent variable is an excellent predictor of the dependent variable, as well as additional examples that lack significance, and thus the tutorials will provide an opportunity to see different outcomes from regression analysis.

Tutorials

11.3 Linear Regression Excel Tutorial

During the Pleistocene era, a vast majority of the United States was covered in thick, cool forests. This quickly changed with the Holocene, which caused an environmental shift to warmer temperatures and deserts. The dramatic climate change sparked the establishment of what has become known as the Arizona Sky Islands and its rich biodiversity. Specifically, the Talus Snail *Sonorella*, which is endemic to these forested sky islands, has been used as a reliable indicator of biogeographical history. This study aimed to determine the biogeographical history of *Sonorella* on several mountain ranges throughout Arizona. The genetic distances between *Sonorella* populations were calculated using a phylogenetic analysis of 12S ribosomal DNA and were compared to geographical distances between populations.

*This example was taken from the research conducted by Nick D. Waters, Vanessa Morales, Dr. Lance Gilbertson, and Dr. Kathleen Weaver.

> **(R)** Formulate a question about the data that can be addressed by performing a linear regression analysis.
>
> ***
>
> Question: What is the relationship between geographic distance and genetic difference?

> **(H)** Based on the question, formulate the null and alternative hypotheses that address the question proposed.
>
> ***
>
> Association Hypotheses (ANOVA table)
>
> Null Hypothesis (H_0): There is no relationship between geographic distance and genetic difference.
> Alternative Hypothesis (H_1): There is relationship between geographic distance and genetic difference.
>
> Slope Hypotheses (Coefficients table)
>
> Null Hypothesis (H_0): The slope of distance is zero.
> Alternative Hypothesis (H_1): The slope of distance is not zero.
>
> y-intercept Hypotheses (Coefficients table)
>
> Null Hypothesis (H_0): The y-intercept is zero.
> Alternative Hypothesis (H_1): The y-intercept is not zero.

Now that an appropriate question has been developed, along with a set of testable hypotheses, you can run the statistical analysis.

- This tutorial focuses on running a simple linear regression in Excel.
- Refer to Chapter 12 for tips and tools when using Excel.
- Check all assumptions prior to running the test.

Linear Regression Excel Tutorial

1. Arrange your data similar to the following.

	A	B	C	D
1	A	B	Geographic Distance (km)	Genetic Difference
2	ChP1	PMCP1	111.13	8.94
3	ChP1	PM3	45.18	6.65
4	ChP1	PM4	101.23	8.62
5	ChP1	SC1	87.23	8.10
6	PMCP1	PM3	18.50	5.60
7	PMCP1	PM4	15.00	5.77
8	PMCP1	SC1	84.40	8.23
9	PM3	PM4	3.40	5.08
10	PM3	SC1	68.46	7.68
11	PM4	PTC1	2.80	5.21
12	PM4	SC1	51.47	6.78

2. In order to produce the regression equation, we must first graph the raw data and determine the equation of the best-fit line. Generate a scatter plot of **Geographic Distance** (independent variable) and **Genetic Difference** (dependent variable). The graph should look similar to the one below.

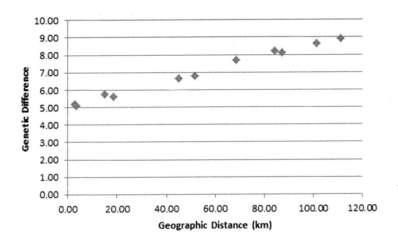

3. To add the best-fit line, select **Add Chart Element** in the toolbar, scroll down, and select the **More Trendline Options** selection.

4. Select **Linear**. Mark the boxes to **Display Equation on chart** and **Display R-squared value on chart**. Click **Close**.

5. The best-fit line and regression equation will appear on the graph.

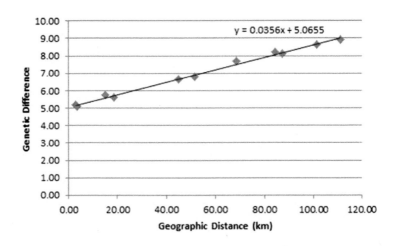

6. Label **E1** as **fitted values**. The fitted values are the predicted values derived from the regression equation. Use the regression equation to calculate the **fitted values**. In other words, plug in x (**Geographic Distance**) to calculate y (**fitted value**) in the equation $y = 0.0356x + 5.0655$ (regression equation). To start, select **E2** and insert the following equation making sure to select **C2** in replacement of x. Press enter/return.

	A	B	C	D	E	F
1	A	B	Geographic Distance (km)	Genetic Difference	Fitted Values	
2	ChP1	PMCP1	111.13	8.9	=(0.0356*C2)+5.0655	
3	ChP1	PM3	45.18	6.65		
4	ChP1	PM4	101.23	8.62		
5	ChP1	SC1	87.23	8.10		
6	PMCP1	PM3	18.50	5.60		
7	PMCP1	PM4	15.00	5.77		
8	PMCP1	SC1	84.40	8.23		
9	PM3	PM4	3.40	5.08		
10	PM3	SC1	68.46	7.68		
11	PM4	PTC1	2.80	5.21		
12	PM4	SC1	51.47	6.78		

7. The calculated **fitted value** will result as seen below.

	A	B	C	D	E	F
1	A	B	Geographic Distance (km)	Genetic Difference	Fitted Values	
2	ChP1	PMCP1	111.13	8.94	9.021728	
3	ChP1	PM3	45.18	6.65		
4	ChP1	PM4	101.23	8.62		
5	ChP1	SC1	87.23	8.10		
6	PMCP1	PM3	18.50	5.60		
7	PMCP1	PM4	15.00	5.77		
8	PMCP1	SC1	84.40	8.23		
9	PM3	PM4	3.40	5.08		
10	PM3	SC1	68.46	7.68		
11	PM4	PTC1	2.80	5.21		
12	PM4	SC1	51.47	6.78		

8. Now that the general equation has been inserted into **E2**, you can apply the same equation to the following cells. This will allow you to determine the remaining fitted values based on each respective x value. To apply the equation to **E3:E12**, use the fill handle and drag it down to **E12**.

	A	B	C	D	E	F
1	A	B	Geographic Distance (km)	Genetic Difference	Fitted Values	
2	ChP1	PMCP1	111.13	8.94	9.021728	
3	ChP1	PM3	45.18	6.65		
4	ChP1	PM4	101.23	8.62		
5	ChP1	SC1	87.23	8.10		
6	PMCP1	PM3	18.50	5.60		
7	PMCP1	PM4	15.00	5.77		
8	PMCP1	SC1	84.40	8.23		
9	PM3	PM4	3.40	5.08		
10	PM3	SC1	68.46	7.68		
11	PM4	PTC1	2.80	5.21		
12	PM4	SC1	51.47	6.78		

9. The calculated **fitted values** will result as seen below.

	A	B	C	D	E	F
1	A	B	Geographic Distance (km)	Genetic Difference	Fitted Values	
2	ChP1	PMCP1	111.13	8.94	9.02173	
3	ChP1	PM3	45.18	6.65	6.67391	
4	ChP1	PM4	101.23	8.62	8.66929	
5	ChP1	SC1	87.23	8.10	8.17089	
6	PMCP1	PM3	18.50	5.60	5.72410	
7	PMCP1	PM4	15.00	5.77	5.59950	
8	PMCP1	SC1	84.40	8.23	8.07014	
9	PM3	PM4	3.40	5.08	5.18654	
10	PM3	SC1	68.46	7.68	7.50268	
11	PM4	PTC1	2.80	5.21	5.16518	
12	PM4	SC1	51.47	6.78	6.89783	

10. Label **F1** as **residuals**. The residuals are the difference between the observed values for genetic difference and the predicted values (**fitted values**) using the regression equation. In **F2** subtract the **fitted values** (**E2**) from **Genetic Difference** (**D2**) for **ChP1** and **PMCP1**. Press enter/return.

	A	B	C	D	E	F
1	A	B	Geographic Distance (km)	Genetic Difference	Fitted Values	Residuals
2	ChP1	PMCP1	111.13	8.94	9.02173	=D2-E2
3	ChP1	PM3	45.18	6.65	6.67391	
4	ChP1	PM4	101.23	8.62	8.66929	
5	ChP1	SC1	87.23	8.10	8.17089	
6	PMCP1	PM3	18.50	5.60	5.72410	
7	PMCP1	PM4	15.00	5.77	5.59950	
8	PMCP1	SC1	84.40	8.23	8.07014	
9	PM3	PM4	3.40	5.08	5.18654	
10	PM3	SC1	68.46	7.68	7.50268	
11	PM4	PTC1	2.80	5.21	5.16518	
12	PM4	SC1	51.47	6.78	6.89783	

11. The calculated **residual** will result as seen below.

	A	B	C	D	E	F
1	A	B	Geographic Distance (km)	Genetic Difference	Fitted Values	Residuals
2	ChP1	PMCP1	111.13	8.94	9.02173	-0.081728
3	ChP1	PM3	45.18	6.65	6.67391	
4	ChP1	PM4	101.23	8.62	8.66929	
5	ChP1	SC1	87.23	8.10	8.17089	
6	PMCP1	PM3	18.50	5.60	5.72410	
7	PMCP1	PM4	15.00	5.77	5.59950	
8	PMCP1	SC1	84.40	8.23	8.07014	
9	PM3	PM4	3.40	5.08	5.18654	
10	PM3	SC1	68.46	7.68	7.50268	
11	PM4	PTC1	2.80	5.21	5.16518	
12	PM4	SC1	51.47	6.78	6.89783	

12. Apply the equation to **F3:F12**.

	A	B	C	D	E	F
1	A	B	Geographic Distance (km)	Genetic Difference	Fitted Values	Residuals
2	ChP1	PMCP1	111.13	8.94	9.02173	-0.081728
3	ChP1	PM3	45.18	6.65	6.67391	
4	ChP1	PM4	101.23	8.62	8.66929	
5	ChP1	SC1	87.23	8.10	8.17089	
6	PMCP1	PM3	18.50	5.60	5.72410	
7	PMCP1	PM4	15.00	5.77	5.59950	
8	PMCP1	SC1	84.40	8.23	8.07014	
9	PM3	PM4	3.40	5.08	5.18654	
10	PM3	SC1	68.46	7.68	7.50268	
11	PM4	PTC1	2.80	5.21	5.16518	
12	PM4	SC1	51.47	6.78	6.89783	

13. The calculated **residuals** will result as seen below.

	A	B	C	D	E	F
1	A	B	Geographic Distance (km)	Genetic Difference	Fitted Values	Residuals
2	ChP1	PMCP1	111.13	8.94	9.02173	-0.08173
3	ChP1	PM3	45.18	6.65	6.67391	-0.02391
4	ChP1	PM4	101.23	8.62	8.66929	-0.04929
5	ChP1	SC1	87.23	8.10	8.17089	-0.07089
6	PMCP1	PM3	18.50	5.60	5.72410	-0.12410
7	PMCP1	PM4	15.00	5.77	5.59950	0.17050
8	PMCP1	SC1	84.40	8.23	8.07014	0.15986
9	PM3	PM4	3.40	5.08	5.18654	-0.10654
10	PM3	SC1	68.46	7.68	7.50268	0.17732
11	PM4	PTC1	2.80	5.21	5.16518	0.04482
12	PM4	SC1	51.47	6.78	6.89783	-0.11783

14. Now, make a scatter plot to look for homogeneity of variance. Use the calculated **fitted values** (x-axis) and **residuals** (y-axis) to generate a scatter plot similar to the one below.

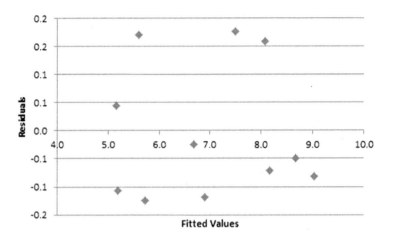

15. Once we have clearly met all of the assumptions, let us continue with running a linear regression analysis. Open **Data Analysis** under the **Data** tab in the toolbar.

16. The following window will appear. Scroll down until you highlight **Regression** and click **OK**.

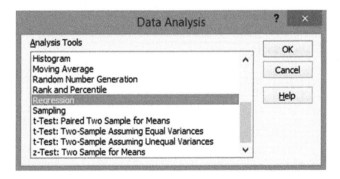

17. The following will appear. Make sure the boxes are empty. If they are not, simply delete the content in them. Input the *y* values by clicking on the icon to the right of **Input Y Range**.

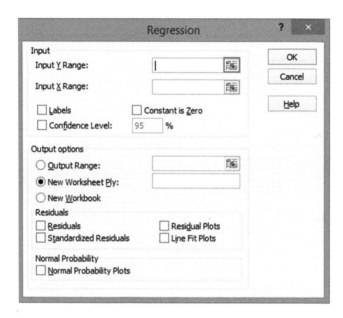

18. Select the *y* values, including the label. Click the small icon on the right when finished.

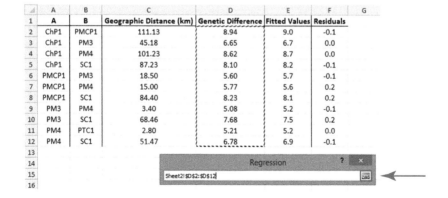

19. The following window will be returned. Input the *x* values by clicking on the icon to the right of **Input X Range**.

20. Select the *x* values, including the label. Click the small icon on the right when finished.

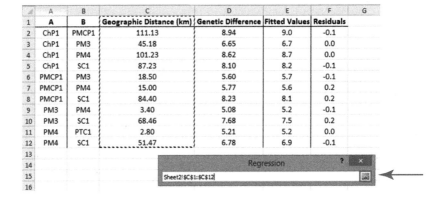

21. Make sure the box entitled **Labels** is selected so Excel can recognize the first cells selected as labels and not actual data points. Then under **Output** options, select **Output Range**.

Linear Regression | 495

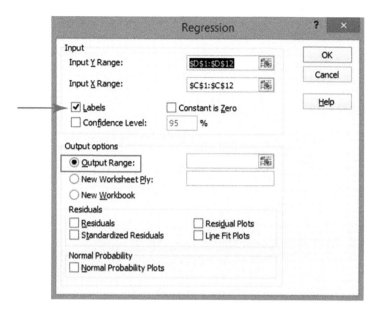

22. Click on the icon to the right of **Output Range**.

23. Select an empty cell in the worksheet where you would like the summary output to be placed. Then click on the icon to the right.

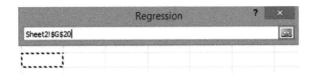

24. Click **OK**.

25. The following output will appear.

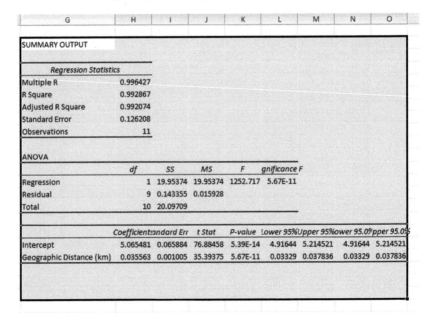

26. Adjust the width of the cells in order to view the entire contents of the output.

SUMMARY OUTPUT

Regression Statistics	
Multiple R	0.99642705
R Square	0.992866866
Adjusted R Square	0.992074295
Standard Error	0.12620761
Observations	11

ANOVA

	df	SS	MS	F	Significance F
Regression	1	19.95373566	19.95374	1252.717	5.673E-11
Residual	9	0.143355248	0.015928		
Total	10	20.09709091			

	Coefficients	Standard Error	t Stat	P-value	Lower 95%	Upper 95%	Lower 95.0%	Upper 95.0%
Intercept	5.065480606	0.065884216	76.88458	5.39E-14	4.916440155	5.214521058	4.916440155	5.214521058
Geographic Distance (km)	0.035563372	0.001004792	35.39375	5.67E-11	0.033290373	0.03783637	0.033290373	0.03783637

To formulate the results in a concluding statement for linear regression, you will simply state whether or not there is a predictive relationship between the two variables. At the end of your statement make sure to include the R^2. If your results are significant you would also provide the equation.

Concluding Statement

When looking at the genetic differences between species with respect to geographic distances (kilometer), distance can predict how closely related one species is to another ($F = 1252.72$, $p < 0.05$, $R^2 = 0.99$; slope: $t = 35.394$, $p < 0.05$; y-intercept: $t = 76.885$, $p < 0.05$).

11.4 Linear Regression SPSS Tutorial

As has been mentioned in other chapters, methods to derive body mass from the human skeleton do not produce highly accurate estimates needed for forensic contexts. As this would be a valuable tool to aid in positive identification from the skeleton, many approaches must be explored in an attempt to find one that performs with an acceptable accuracy rating. Bone weight (in kilograms) is a new avenue that might shed light on the relationship of the skeleton to soft tissue. In this study, a researcher weighed the long bones of the arms and legs, the pelvis, and the scapula of 30 individuals with documented sex, age, height (stature), and weight during life to examine their relationship to BMI. She read that muscle markings of the upper limb and pectoral girdle were better indicators of body mass, and subsequently BMI, than other areas of the skeleton. Therefore, she concentrated her initial analysis on a bone that is included in equations for estimating stature, as well as displays changes due to increased body mass: the humerus.

*This example was taken from the research conducted by Dr. Kanya Godde.

(R) Formulate a question about the data that can be addressed by performing a linear regression analysis.

****

Question: What is the relationship between BMI and humeral weight?

(H) Based on the question, formulate the null and alternative hypotheses that address the question proposed.

****

Association Hypotheses (ANOVA table)

Null Hypothesis (H_0): There is no relationship between BMI and humeral weight.
Alternative Hypothesis (H_1): There is a relationship between BMI and humeral weight.

Slope Hypotheses (Coefficients table)

Null Hypothesis (H_0): The slope of humeral weight is zero.
Alternative Hypothesis (H_1): The slope of humeral weight is not zero.

y-intercept Hypotheses (Coefficients table)

Null Hypothesis (H_0): The *y*-intercept is zero.
Alternative Hypothesis (H_1): The *y*-intercept is not zero.

Now that an appropriate testable question has been developed along with a set of testable hypotheses, you can run the statistical analysis.

- Utilize the following tutorial to run a simple linear regression in SPSS.
- Refer to Chapter 13 for getting started and understanding SPSS.
- Check all assumptions prior to running the test.

Linear Regression SPSS Tutorial

1. Type the data into the first two columns in SPSS.

	HumWeight	BMI
1	.19	21.69
2	.13	30.13
3	.16	25.29
4	.08	43.59
5	.10	25.68
6	.14	24.22
7	.17	21.89
8	.14	17.56
9	.08	29.40
10	.12	40.54
11	.14	26.51
12	.26	25.76
13	.15	32.00
14	.18	30.00
15	.12	31.00
16	.14	37.00
17	.14	41.00
18	.13	29.83
19	.11	18.55
20	.18	27.60
21	.22	15.30
22	.09	31.01
23	.16	38.46
24	.16	60.07
25	.18	23.95
26	.10	21.10
27	.09	39.98
28	.13	23.80
29	.12	23.20
30	.15	27.99

2. Click on the **Analyze** menu and select **Regression** and then click on **Linear**.

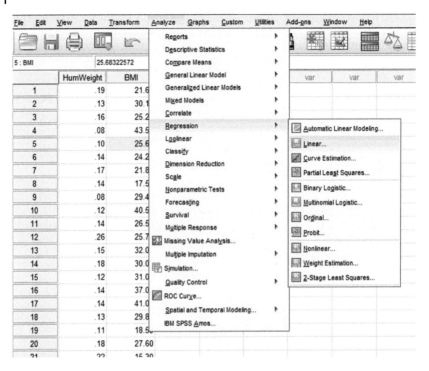

3. The Linear Regression box will appear similar to the picture below. Click on **BMI** to highlight it and click on the arrow next to the **Dependent** box.

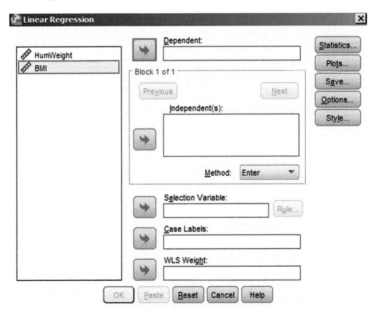

4. Click on the variable for humeral weight and click on the arrow next to the **Independent(s)** box.

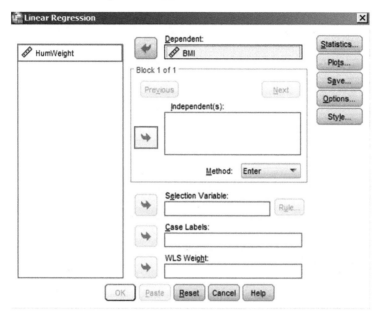

5. Once the two variables are in their respective places as dependent and independent variables, click **Save**.

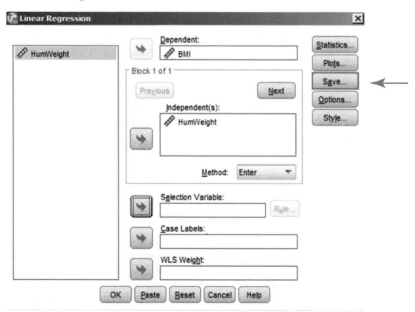

6. In the **Save** box, click the boxes next to **Unstandardized** in both the **Predicted Values** and **residuals** sections. Click **Continue** then click **OK**.

7. A separate document will appear. This is referred to as the output.

Model Summary

Model	R	R Square	Adjusted R Square	Std. Error of the Estimate
1	.225[a]	.051	.017	9.18921

a. Predictors: (Constant), HumWeight

ANOVA[a]

Model		Sum of Squares	df	Mean Square	F	Sig.
1	Regression	126.098	1	126.098	1.493	.232[b]
	Residual	2364.366	28	84.442		
	Total	2490.464	29			

a. Dependent Variable: BMI
b. Predictors: (Constant), HumWeight

Coefficients[a]

Model		Unstandardized Coefficients		Standardized Coefficients	t	Sig.
		B	Std. Error	Beta		
1	(Constant)	36.743	6.183		5.943	.000
	HumWeight	-51.212	41.908	-.225	-1.222	.232

a. Dependent Variable: BMI

Only one of the regression coefficients (y-intercept) is significant ($t = 5.943$, p-value < 0.0001), while the other is not (slope: $t = -1.222$, p-value $= 0.232$). In a good regression model, both the slope and intercept are significant at the alpha (significance level) the statistician chooses. The null hypothesis for each of these coefficients is that they equal zero, which according to our output, can be rejected for one and supported for the other. This means that humeral weight does not have a relationship with BMI.

To look at the strength of the model, or the amount of the variation in Y explained by X, we will look at the R^2. The value is 0.051, which means 5.1% of the variation in BMI is explainable by humeral weight (kilogram). Overall, this is a low R^2, indicating other factors are contributing to the variance in BMI that were not included here.

8. Before making our final conclusions, we want to test for homogeneity of variance of the residuals. *Note*: You should have already tested for the other assumptions before starting this tutorial. Step 7 saved the residuals and fitted values (predicted values from the regression equation) to your SPSS worksheet. Verify this is the case by saving your output and returning to the data in SPSS.

	HumWeight	BMI	PRE_1	RES_2
1	.19	21.69	27.01261	-5.31982
2	.13	30.13	30.08532	.04325
3	.16	25.29	28.54896	-3.25445
4	.08	43.59	32.64592	10.94889
5	.10	25.68	31.62168	-5.93846
6	.14	24.22	29.57320	-5.35366
7	.17	21.89	28.03685	-6.14411
8	.14	17.56	29.57320	-12.01649
9	.08	29.40	32.64592	-3.24933
10	.12	40.54	30.59744	9.93794
11	.14	26.51	29.57320	-3.05842
12	.26	25.76	23.42777	2.33256
13	.15	32.00	29.06108	2.93892
14	.18	30.00	27.52473	2.47527
15	.12	31.00	30.59744	.40256
16	.14	37.00	29.57320	7.42680
17	.14	41.00	29.57320	11.42680
18	.13	29.83	30.08532	-.25122
19	.11	18.55	31.10956	-12.56025
20	.18	27.60	27.52473	.07078
21	.22	15.30	25.47625	-10.17166
22	.09	31.01	32.13380	-1.12207
23	.16	38.46	28.54896	9.91535

9. To examine the pattern of the residuals, we will plot them against the fitted values. Click on **Graphs** and select **Chart Builder**.

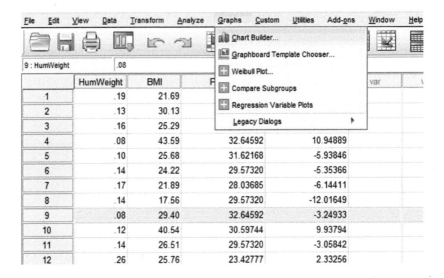

10. The following box may pop up. If it does not, skip to step 11. If it does, ensure you have complied with the instructions and press **OK**. If you have not, then click on **Define Variable Properties** and assign the proper variable type using the information in Chapter 13.

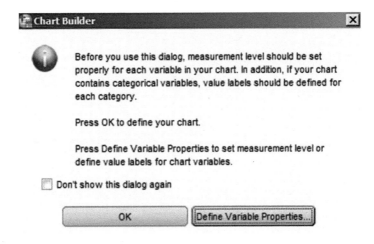

11. In the **Chart Builder** box, select the **Scatter/Dot** option.

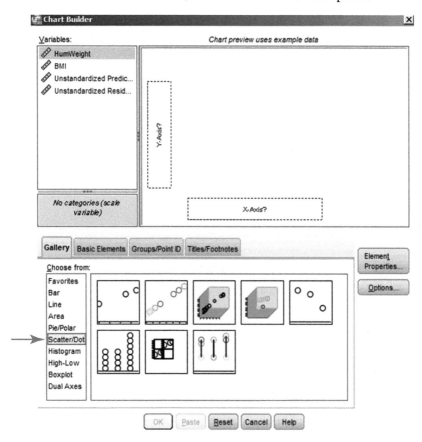

12. Drag the scatter plot option to the center of the **Chart Preview** area.

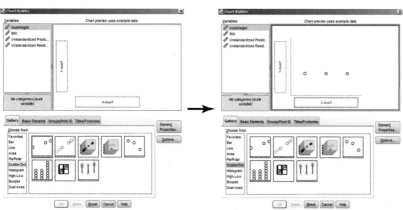

13. Drag the **Unstandardized residuals** to the **Y-axis?** area of the **Chart Preview** box.

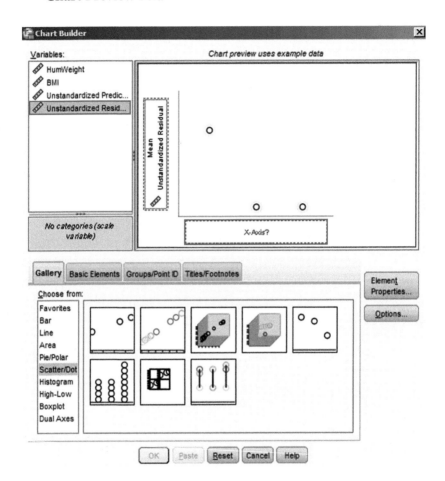

14. Drag the **Unstandardized Predicted** values to the **X-axis?** area of the **Chart Preview** box.

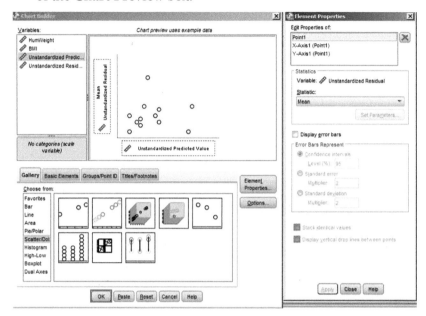

Press **OK.** (Please note there are two screens that pop up when working with the **Chart Builder** and we are only focusing on the one on the left.)

15. The following plot will be generated.

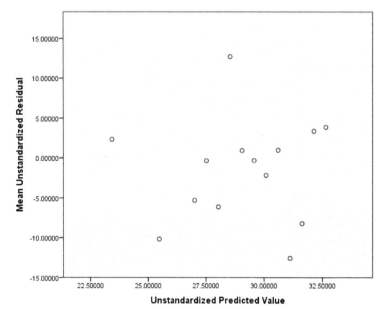

The residuals in the plot look relatively homogeneously distributed, indicating our residuals are mostly homogeneous, or homoscedastic. However, there are 2 outliers that should be investigated prior making definitive interpretations. We will now turn to making our final conclusions.

Concluding Statement

Humeral weight (kilograms) is not significant in predicting BMI (F = 1.493, p-value = 0.232, R^2 = 0.051; slope: $t = -1.222$, p-value 0.232; y-intercept: $t = 5.943$, p-value < 0.0001).

Note: If you want to save the SPSS file with the inserted data as well as the SPSS output with the results of the statistical analysis performed, then you must save each document separately (see Chapter 13).

11.5 Linear Regression Numbers Tutorial

Natural selection is a powerful mechanism of evolution that leads to adaptive changes in traits over time. One of the key concepts of natural selection is that it can only act on traits that are heritable. Thus, evaluating the heritability of a trait has become key to understanding the evolution of a particular trait in a population. A classic study by Grant and Grant (2002) evaluated the heritability of beak size in Darwin's finches on the Galapagos Islands. The goal of the research was to observe the process of natural selection in a natural setting. The Grants measured beak size (in millimeters) of both parents and offspring over a 2-year period.

> 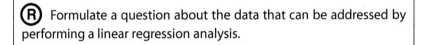 Formulate a question about the data that can be addressed by performing a linear regression analysis.
>
> **
>
> Question: Is beak size heritable? What is the relationship between parent beak size and offspring beak size in Darwin's finches?

> **(H)** Based on the question, formulate the null and alternative hypotheses that address the question proposed.
>
> **
>
> Association Hypotheses (ANOVA table)
>
> Null Hypothesis (H_0): Parent's beak size is not associated with offspring beak size.
> Alternative Hypothesis (H_1): Parent's beak size is associated with offspring beak size.
>
> Slope Hypotheses (Coefficients table)
>
> Null Hypothesis (H_0): The slope is equal to zero.
> Alternative Hypothesis (H_1): The slope is not equal to zero.
>
> y-intercept Hypotheses (Coefficients table)
>
> Null Hypothesis (H_0): The y-intercept is equal to zero.
> Alternative Hypothesis (H_1): The y-intercept is not equal to zero.

Now that an appropriate testable question has been developed, along with a set of testable hypotheses, you can run the statistical analysis.

- To run a simple linear regression in Numbers, utilize the following tutorial.
- More information on programming in Numbers is found in Chapter 14.
- Check all assumptions prior to running the test.

Linear Regression Numbers Tutorial

1. Arrange your data into two columns, **Midparent** beak size (the average trait value of the father and mother) and **Offspring** beak size.

An Introduction to Statistical Analysis in Research

Finch Beak Size

Midparent	Offspring
10.905	8.74
10.29	10.04
10.2	9.75
10.4	9.76
9.8	9.55
9.81	8.7
9.595	9.62
9.75	9.22
9.2	8.79
9.45	9.61
9.45	9.02
9.21	7.85
9.105	9.01
9.025	8.26
9.01	8.25
9.25	9.25
9.2	9.45
9	8.52
8.82	8.61
8.21	8.12
8.255	7.81
10.805	10.81

2. In order to make a regression graph, we must first graph the raw data and determine the equation of the best-fit line. Generate a scatter plot of the midparent beak size (independent variable) and offspring beak size (dependent variable). The graph should be similar to the one below.

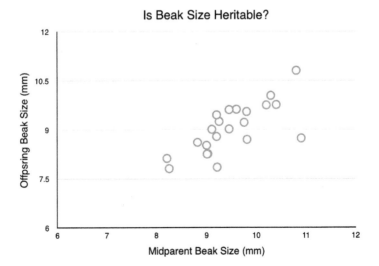

3. To add the best-fit line, select **Format → Series** in the toolbar. Under **Trendlines**, select **Linear** and click the boxes for **Show Equation** and **Show R² Value**.

4. The best-fit line and regression equation will appear on the graph.

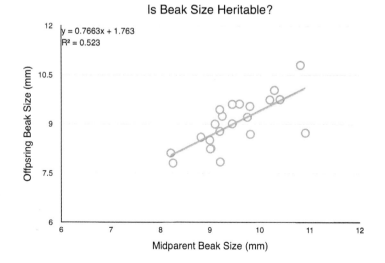

5. Label **D1** as **fitted values**. The fitted values are the predicted values derived from the regression equation. Use the regression equation to calculate the **fitted values**. In other words, plug in *x* (Midparent) to calculate *y* (fitted value) in the equation $y = 0.7663x + 1.763$ (regression equation). To start, select **D2** and insert the following equation making sure to select **B2** in replacement of *x*. Press return/enter.

B	C	D	E	F
Finch Beak Size				
Midparent	Offspring	Fitted Values		
10.905	8.74	• *fx* ˅ (.7663× B2 ▾)+1.763		
10.29	10.04			
10.2	9.75			
10.4	9.76			

6. Now that the general equation has been inserted into **D2**, you can apply the same equation to the following cells. This will allow you to determine the remaining fitted values based on each respective *x* value. To apply the equation to **D2:D23**, drag the yellow circle down to **D23**.

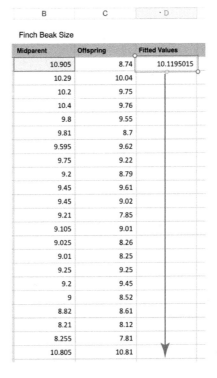

7. The calculated **fitted values** will result as seen below.

Finch Beak Size

Midparent	Offspring	Fitted Values
10.905	8.74	10.1195015
10.29	10.04	9.648227
10.2	9.75	9.57926
10.4	9.76	9.73252
9.8	9.55	9.27274
9.81	8.7	9.280403
9.595	9.62	9.1156485
9.75	9.22	9.234425
9.2	8.79	8.81296
9.45	9.61	9.004535
9.45	9.02	9.004535
9.21	7.85	8.820623
9.105	9.01	8.7401615
9.025	8.26	8.6788575
9.01	8.25	8.667363
9.25	9.25	8.851275
9.2	9.45	8.81296
9	8.52	8.6597
8.82	8.61	8.521766
8.21	8.12	8.054323
8.255	7.81	8.0888065
10.805	10.81	10.0428715

8. Label **F1** as residuals. The residuals are the difference between the observed values for beak size and the predicted values (fitted values) using the regression equation. In **E2** subtract the **fitted values** (**D2**) from offspring beak size (**C2**). Press enter/return or click the green check.

Finch Beak Size

Midparent	Offspring	Fitted Values	Residuals
10.905	8.74	10.1195015	• fx ⌄ (D2 ▼) - (C2 ▼)
10.29	10.04	9.648227	
10.2	9.75	9.57926	
10.4	9.76	9.73252	

9. Apply the equation to **E3:E23**.

Offspring	Fitted Values	Residuals
8.74	10.1195015	1.3795015
10.04	9.648227	
9.75	9.57926	
9.76	9.73252	
9.55	9.27274	
8.7	9.280403	
9.62	9.1156485	
9.22	9.234425	
8.79	8.81296	
9.61	9.004535	
9.02	9.004535	
7.85	8.820623	
9.01	8.7401615	
8.26	8.6788575	
8.25	8.667363	
9.25	8.851275	
9.45	8.81296	
8.52	8.6597	
8.61	8.521766	
8.12	8.054323	
7.81	8.0888065	
10.81	10.0428715	

10. The calculated residuals will result as seen below.

Finch Beak Size

Midparent	Offspring	Fitted Values	Residuals
10.905	8.74	10.1195015	1.3795015
10.29	10.04	9.648227	-0.391773000000001
10.2	9.75	9.57926	-0.170740000000002
10.4	9.76	9.73252	-0.0274799999999988
9.8	9.55	9.27274	-0.27726
9.81	8.7	9.280403	0.580403
9.595	9.62	9.1156485	-0.5043515
9.75	9.22	9.234425	0.0144249999999992
9.2	8.79	8.81296	0.0229600000000012
9.45	9.61	9.004535	-0.605465000000001
9.45	9.02	9.004535	-0.0154650000000007
9.21	7.85	8.820623	0.970623000000002
9.105	9.01	8.7401615	-0.269838500000001
9.025	8.26	8.6788575	0.4188575
9.01	8.25	8.667363	0.417363
9.25	9.25	8.851275	-0.398725000000001
9.2	9.45	8.81296	-0.637039999999999
9	8.52	8.6597	0.139700000000001
8.82	8.61	8.521766	-0.0882339999999999
8.21	8.12	8.054323	-0.0656769999999991
8.255	7.81	8.0888065	0.278806500000001
10.805	10.81	10.0428715	-0.767128500000002

11. Now, make a scatter plot to look for homogeneity of variance. Use the calculated fitted values (x-axis) and residuals (y-axis) to generate a scatter plot similar to the one below.

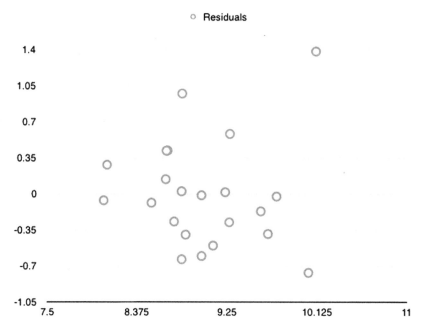

12. The residuals in the plot look relatively homogeneously distributed with the exception of one outlier. Thus, our residuals are not homogeneous, or homoscedastic. Check with your instructor for how they would like you to handle the outlier. For the sake of demonstration, we have left the outlier in. We can now turn to making our calculations.

13. To determine if there is a predictive relationship among your data, you will calculate the R^2 value. To calculate R^2, you must start by calculating r. The function for correlation or r is = **correl**. This function will return the correlation between the midparent and offspring beak size.

Finch Beak Size

Midparent	Offspring						
10.905	8.74						
10.29	10.04	Correlation	· fx ˅	CORREL ▼	y-values	x-values	✕ ✓
10.2	9.75	R^2					
10.4	9.76						

14. Highlight the y- and x-values and click the green check or press enter/return.

Finch Beak Size

Midparent	Offspring
10.905	8.74
10.29	10.04
10.2	9.75
10.4	9.76
9.8	9.55
9.81	8.7
9.595	9.62
9.75	9.22
9.2	8.79
9.45	9.61
9.45	9.02
9.21	7.85
9.105	9.01
9.025	8.26
9.01	8.25
9.25	9.25
9.2	9.45
9	8.52
8.82	8.61
8.21	8.12
8.255	7.81
10.805	10.81

Correlation · fx · CORREL ▼ B2:B23 ▼ . C2:C23 ▼
R^2

15. Finally, calculate the R^2 value. To calculate R^2, you will square the r-value that you calculated in steps 13 and 14. Type **=cell value ^2** and click the green check or press enter/return.

Finch Beak Size

Midparent	Offspring
10.905	8.74
10.29	10.04
10.2	9.75
10.4	9.76
9.8	9.55

Correlation 0.72
R^2 · fx · E3 ▼ ^2

16. The resulting correlation and R^2 value is listed below.

Correlation	0.72
R^2	0.52

The R^2 value indicates the quality of fit or the proportion of the variation in y-values that can be explained by the variation in the x-values. The closer the R^2 value is to 1, the better the fit. To formulate the results in a concluding statement for linear regression, you will simply state whether or not there is a predictive relationship between the two variables. At the end of your statement make sure to include the R^2.

Concluding Statement

In our finch data, 52% of the variation in offspring beak size is explainable by midparent beak size ($R^2 = 0.52$).

Note: Numbers does not offer the capability to determine the F-value and t-values, thus you cannot report the significance. You can manually calculate the F-value (ANOVA table) and t-values (t table) to determine significance.

11.6   Linear Regression R Tutorial

One of the important chemical features of freshwater ecosystems is salinity, or the amount of salt in the water. In freshwater rivers and streams, salinity accumulates through the weathering (or eroding away) of rocks and is one of the many abiotic conditions that may influence the distribution of species and dictate the composition of entire aquatic communities. An aquatic ecologist is interested in being able to predict the occurrence of salinity sensitive freshwater communities at different elevations on the island of Hispaniola. In particular, the researcher is interested in developing a model of how salinity varies by elevation. In other words, the researcher wants to know whether there is a predictable relationship between the elevation of different freshwater habitats and the salinity measured at those sites.

*This example was taken from the research conducted by Dr. Pablo Weaver.

> **(R)** Formulate a question about the data that can be addressed by performing a linear regression analysis.
>
> **
>
> Question: What is the relationship between the elevation of aquatic habitats and salinity?

> **(H)** Based on the question, formulate the null and alternative hypotheses that address the question proposed.
>
> **
>
> Association Hypotheses (ANOVA table)
>
> Null Hypothesis (H_0): There is no relationship between salinity and elevation.
> Alternative Hypothesis (H_1): There is a relationship between salinity and elevation.
>
> Slope Hypotheses (Coefficients table)
>
> Null Hypothesis (H_0): The slope of elevation is zero.
> Alternative Hypothesis (H_1): The slope of elevation is not zero.
>
> y-intercept Hypotheses (Coefficients table)
>
> Null Hypothesis (H_0): The y-intercept is zero.
> Alternative Hypothesis (H_1): The y-intercept is not zero.

Now that an appropriate testable question has been developed along with a set of testable hypotheses, you can run the statistical analysis.

- This tutorial focuses on running a linear regression analysis in R.
- Refer to Chapter 15 for R-specific terminology and instructions on how to invoke and construct code.
- Check all assumptions prior to running the test.

Linear Regression R Tutorial

1. Type the data into two vectors, one labeled **salinity** (reported in parts per thousand) and the other **elevation** (reported in meters). Press enter/return.

```
> salinity<-c(0.21, 0.12, 0.13, 0.11, 0.14, 0.22, 0.32, 0.21, 0.22, 0.41, 0.4,
0.12, 0.1, 0.11, 0.21, 0.2, 0.32, 0.1, 0.09, 0.08, 0.05, 0.01)

> elevation<-c(644, 681, 238, 289, 200, 712, 780, 158, 137, 650, 632, 659,
777, 124, 701, 52, 370, 550, 290, 1255, 1093, 1097)
```

2. Linear regression is part of the GLM (which includes an ANOVA and *t*-tests on the regression coefficients), and thus we will be using the linear model function, `lm()`, in R. As we will be pulling two different sets of output from R (an ANOVA table and regression coefficients table), we need to store the linear model we are creating to a name of our choice, in this case, **weathering**. The function reads the equation $Y \sim X$ or Dependent variable $\sim$ Independent variable. In our example, elevation will be used as a predictor (or independent variable) for salinity. Press enter/return.

```
> weathering<-lm(salinity~elevation)
```

3. Let us generate the two output tables that we need to check for homoscedasticity and which also provide us the output we will be interpreting. First, apply the `anova()` function to **weathering** to generate the ANOVA table.

```
> anova(weathering)
```

4. Second, we will pull the coefficients table. Use the `summary()` function to generate the table from **weathering**.

```
> summary(weathering)
```

5. Before interpreting the output, we want to test for homoscedasticity, or homogeneity of variance, of the residuals, an assumption for linear regression. You should have already tested for the other assumptions before starting this tutorial. To test for the homogeneity of variance, we need to plot the residuals against the fitted values using the `plot()` function. Type in the line of code below and press enter/return.

```
> plot(resid(weathering)~fitted.values(weathering))
```

The following plot will be generated.

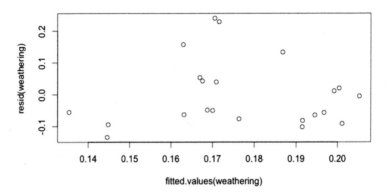

The residuals in the plot look relatively homogeneously distributed, indicating our residuals are mostly homogeneous, or homoscedastic. We can now turn to making our final conclusions.

6. The following output will appear.

```
> anova(weathering)
Analysis of Variance Table

Response: salinity
           Df  Sum Sq   Mean Sq  F value  Pr(>F)
elevation   1  0.008256 0.0082565 0.6995  0.4128
Residuals  20  0.236053 0.0118026
> summary(weathering)

Call:
lm(formula = salinity ~ elevation)

Residuals:
    Min      1Q   Median      3Q     Max
-0.13452 -0.07337 -0.04936 0.04162 0.23948

Coefficients:
              Estimate Std. Error t value Pr(>|t|)
(Intercept)  2.083e-01  4.468e-02   4.662 0.00015 ***
elevation   -5.816e-05  6.954e-05  -0.836 0.41282
---
Signif. codes:  0 '***' 0.001 '**' 0.01 '*' 0.05 '.' 0.1 ' ' 1

Residual standard error: 0.1086 on 20 degrees of freedom
Multiple R-squared:  0.0338,   Adjusted R-squared: -0.01452
F-statistic: 0.6995 on 1 and 20 DF,  p-value: 0.4128
```

In multiple regression, the ANOVA table and the coefficients table would have different p-values for the slope (elevation) and the p-value at the bottom right-hand corner of the output. However, as this is linear regression, the p-values are the same. The F-value in the ANOVA table is not significant ($F = 0.6995$, p-value $= 0.4128$), which means we fail to reject the null hypothesis that elevation is not associated with salinity.

Only one of the regression coefficients (y-intercept) is significant ($t = 4.662$, p-value $= 0.00015$), while the other is not (slope: $t = -0.836$, p-value $= 0.41282$). In a good regression model, both the slope and intercept are significant at the alpha (significance level) the statistician chooses. The null hypothesis for each of these coefficients is that they equal zero, which according to our output, can be rejected for one and supported for the other. This means that elevation does not have a relationship with salinity of the water.

To look at the strength of the model, or the amount of the variation in Y explained by X, we will look at the multiple R^2. The value is 0.0338, which means 3.38% of the variation in salinity is explainable by elevation. Overall, this is a low R^2, indicating other factors are contributing to the variance in salinity that were not included here.

Concluding Statement

Elevation is not significant in predicting salinity of water ($F = 0.6995$, p-value $= 0.4128$, $R^2 = 0.0338$; slope: $t = -0.836$, p-value $= 0.41282$; y-intercept: $t = 4.662$, p-value $= 0.00015$).

12

Basics in Excel

There are dozens of shortcuts and tools to know within Excel. The following chapter showcases a few of the most common ones that students have used or requested in our research methods courses (installing the Data Analysis ToolPak, expanding cell width, commonly used equations like average and standard deviation, etc.).

There are many different versions of Excel and the location of menus, titles of menus, etc. can differ. Thus, if you cannot locate a particular Excel feature with the tutorials provided, try clicking on a similarly worded item to locate the version of the menu, submenu, etc. for the tutorial you are working through. A quick Internet search can also help to locate an item.

This book was built with a few versions of Excel. There are free options available on the Internet (e.g., OpenOffice) for students/instructors who do not have access to Excel. The free platforms will provide a spreadsheet-based software similar to Excel and allow students to complete the tutorials.

The tutorials in this book are built to show a variety of approaches to using Excel, so students can find their own unique style in working with statistical software, as well as to enrich the student learning experience through exposure to more and varied examples.

12.1 Opening Excel

1. Open **Excel** by double clicking on the Excel icon.

2. The following window will appear. Double click the **Blank workbook**.

3. A blank spreadsheet will appear with the menu toolbar directly above.

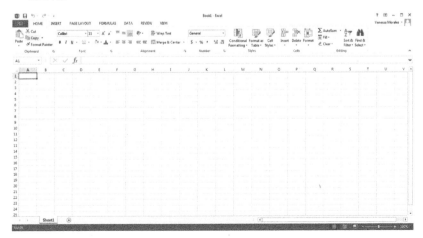

12.2 Installing the Data Analysis ToolPak

1. The Data Analysis ToolPak should be located along the toolbar under the **DATA** tab. If it does not appear, then you must manually install it.

2. To begin, select **FILE**.

3. Select the **Options** button located near the bottom of the window.

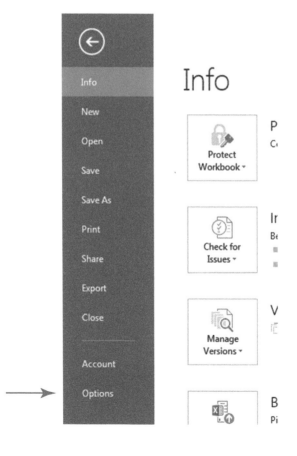

4. The following window will appear.

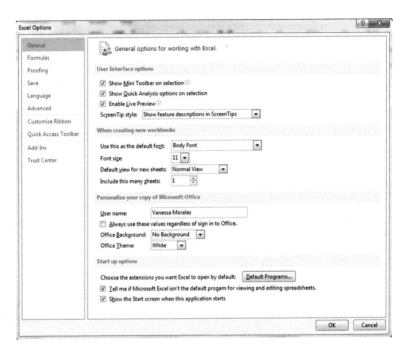

5. Select the **Add-Ins** option that appears along the left side.

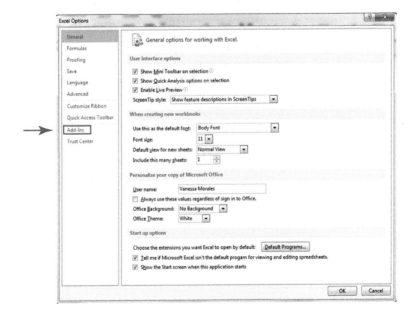

6. Towards the bottom of the window, under **Manage**, make sure the drop-down list has **Excel Add-Ins** selected. Then press **Go**.

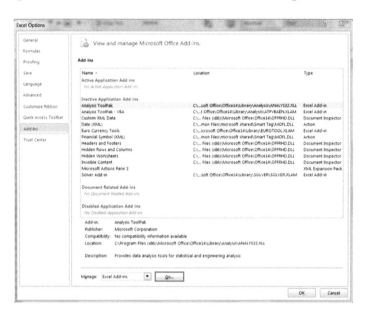

7. The following window will appear.

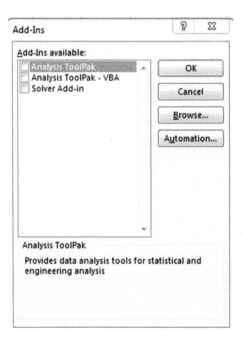

8. Check the box corresponding to the **Analysis ToolPak** and click **OK**.

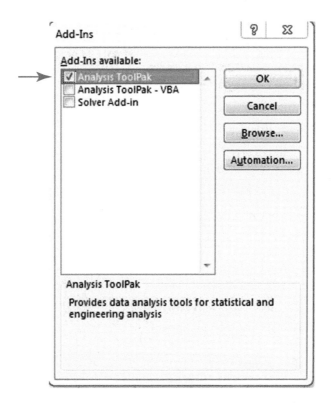

9. The application should now be installed in the toolbar, under the **DATA** tab.

Note: If you do not have the Data Analysis ToolPak, you can still complete the tutorials by inputting the equations into Excel for the specific statistic needed.

12.3 Cells and Referencing

Cells are the boxes in Excel where you type your data and formulas.

Cells are named by the column and row in which they appear. In the following example, the number **7** is in column **B** and row **2**, which means the cell is referred to as **B2** (column precedes the row).

How you refer to a cell within a formula is called referencing. There are two types: (1) **absolute** and (2) **relative**. An absolute reference is the one that does not change when it is copied and pasted into adjacent cells. It is denoted with a "$" immediately before the column or row reference. Either columns, rows, or both can be made absolute.

Absolute Column: $B2
Absolute Row: B$2
Absolute Column and Row: B2

Relative references change based on where they are copied and pasted, and are not denoted by any symbols.

Relative Column: B2
Relative Row: B2
Relative Columns and Row: B2

When copying and pasting a formula from one cell to another, Excel automatically adjusts the cell number based on the location where the formula is being copied and pasted from and to if relative referencing is used. So, for example, if you copy and paste a formula over three columns, Excel will add three columns to the original reference.

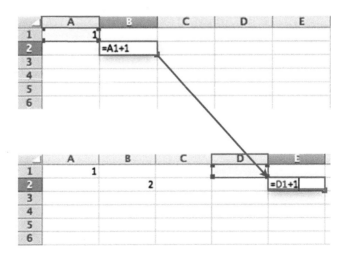

So, if our column reference in cell **B2** was **A** and we pasted it over three columns to **E2**, then the column reference would be **D**. Row references are handled similarly. If the reference in **B2** is copied and pasted to **B5**, the row reference changes from **1** in **B2** to **4** in **B5**.

	A	B	C	D	E
1		1			
2		=A1+1			
3					
4					
5					
6					

	A	B	C	D	E
1		1			
2			2		
3					
4					
5		=A4+1			
6					

If a cell reference is copied to a different column and a different row, then the column and row would change in the cell reference. The beauty of absolute referencing is that if you need to copy and paste a formula to another area of the spreadsheet but need the cell reference within the formula to stay the same, an absolute reference can be placed in the formula to prevent Excel from automatically changing the cell reference.

An **array** is a range of cells you want to analyze, separated by a ":". So, for example, if you wanted to analyze the contents of **C2** through **C7**, you would type in **C2:C7**. Then, **C2:C7** is the array.

	A	B	C
1			Data
2			89
3			243
4			234
5			5.70E+05
6			90
7			344
8			=C2:C7

12.4 Common Commands and Formulas

Subtraction

1. To find the difference between values, in an empty cell type in an equals sign (=) followed by the first selected value, a subtraction sign (−), and the second selected value.
   A generic version of this equation should look similar to the following.

   =(cell reference 1)−(cell reference 2)

2. For example, to find the difference between the two values given, type in an equals sign (=), then select the first value.

	A	B	C	D
1	Patient	pre LDL (mg/dL)	post LDL (mg/dL)	Difference
2	7	132	130	=B2
3	2	147	151	

3. Next, type in a subtraction sign (−), then select the **value** that you want to be subtracted from it.

	A	B	C	D
1	Patient	pre LDL (mg/dL)	post LDL (mg/dL)	Difference
2	7	132	130	=B2-C2
3	2	147	151	

4. Press enter/return and the value will appear.

	A	B	C	D
1	Patient	pre LDL (mg/dL)	post LDL (mg/dL)	Difference
2	7	132	130	2
3	2	147	151	

Addition

1. To find the sum of values, in an empty cell, type in an equals sign (=) followed by the word "**sum**." Then, double click on the **SUM** command that appears.

Girl Scouts	Candy Bars Sold
1	25
2	39
3	79
4	56
5	87
6	46
=su	

- *fx* SUBSTITUTE
- *fx* SUBTOTAL
- *fx* **SUM**
- *fx* SUMIF
- *fx* SUMIFS
- *fx* SUMPRODUCT
- *fx* SUMSQ
- *fx* SUMX2MY2
- *fx* SUMX2PY2
- *fx* SUMXMY2

→

Girl Scouts	Candy Bars Sold
1	25
2	39
3	79
4	56
5	87
6	46
=SUM(	
SUM(number1, [number2], ...)	

2. Highlight all the values you want to include in the sum.

Girl Scouts	Candy Bars Sold
1	25
2	39
3	79
4	56
5	87
6	46
	=SUM(C3:C8

3. Press enter/return. The resulting sum will appear.

Girl Scouts	Candy Bars Sold
1	25
2	39
3	79
4	56
5	87
6	46
Sum =	332

12.5 Applying Commands to Entire Columns

1. Take for instance the following dataset.

Sampled People	January(Before)	April (After)	Difference
1	4	15.5	
2	9	147	
3	6	20	
4	5	18	
5	3	15	
6	2	21	
7	8	16	
8	5.2	15	
9	4.9	18	
10	6.5	22	

2. Type in the formula for the difference. *Note:* Depending on which number is first in the subtraction equation, a positive or negative difference will appear.

Sampled People	January(Before)	April (After)	Difference
1	4	15.5	=D3-C3
2	9	147	

3. Press enter/return and the computation will appear.

Sampled People	January(Before)	April (After)	Difference
1	4	15.5	11.5
2	9	147	

4. Instead of typing in the same formula for the remaining nine samples, Excel allows you to drag the formula down to apply to all samples. To do so, select the cell that contains the original formula. Then click on the bottom right hand corner, a cross will appear. This is called the fill handle.

Sampled People	January(Before)	April (After)	Difference
1	4	15.5	11.5
2	9	147	

5. Click and drag down to the last row.

Sampled People	January(Before)	April (After)	Difference
1	4	15.5	11.5
2	9	147	
3	6	20	
4	5	18	
5	3	15	
6	2	21	
7	8	16	
8	5.2	15	
9	4.9	18	
10	6.5	22	

6. The relative references will be copied into each cell and resulting values will be the computed difference.

Sampled People	January(Before)	April (After)	Difference
1	4	15.5	11.5
2	9	147	138
3	6	20	14
4	5	18	13
5	3	15	12
6	2	21	19
7	8	16	8
8	5.2	15	9.8
9	4.9	18	13.1
10	6.5	22	15.5

7. This can be done for all formulas and calculation functions.
8. If you right click on the fill handle before dragging, after you are finished dragging the fill handle to the desired location, a menu will pop up with more options allowing you to copy and paste more than just a formula. For instance, you can repeat the same number many times by right clicking on the fill handle and dragging it the

number of rows that is equal to the number of times you want to repeat a value and then selecting **Copy Cells** from the menu that pops up.

9. The fill handle can also be dragged to the right or left following the same principles.

12.6 Inserting a Function

If you know the formula for the function you want to perform, in an empty cell, just type in an equals sign (=) followed by the function command as described above. However, if you do not know the function command, Excel provides a list of functions as a reference.

1. To access the function list, first click on the empty cell that you want the value to appear in.
2. In the toolbar along the top of the page, under the **FORMULAS** tab, is an **Insert Function** option. Double click on it.

3. A second way to access the function list is by clicking on the *fx* icon located just below the toolbar.

4. Finally, if you know the function (or what the function starts with), type in = and begin typing the function name. Options will appear as you type. For example, =**average** is used to calculate the mean of a group of numbers.

12.7 Formatting Cells

Sometimes, when inputting decimals, Excel will automatically round the numbers. If this occurs, select the cells that have been rounded and under the **HOME** tab, in the toolbar located along the top of the page, there are two options within the **Number** heading. The option to the left will increase the amount of decimal places that appear within the cell and the option to the right will decrease the amount of decimal places that appear.

13

Basics in SPSS

There are several shortcuts and tools to know in SPSS. The following chapter displays a few of the commonly used features within the program (how to type in your data, assign a label to the different variables, how to save your data and output, etc.).

The tutorials in this book are built to show a variety of approaches to using SPSS, so students can find their own unique style in working with statistical software, as well as to enrich the student learning experience through exposure to more and varied examples.

13.1 Opening SPSS

1. Open SPSS by selecting the SPSS icon located on the desktop. If not found on the desktop, click on the SPSS icon located either in your program menu in Windows or in your Applications folder in Mac.

2. Select the **New Dataset** option and click **OK**.

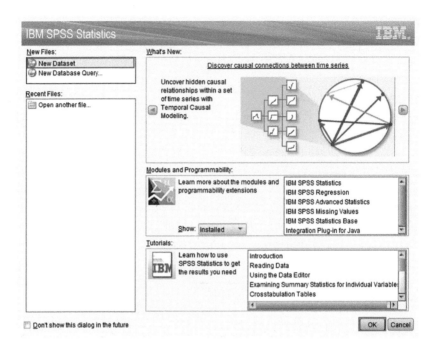

3. The screen that will appear is the spreadsheet that is associated with the **Data View** tab. Use this spreadsheet to enter your data.

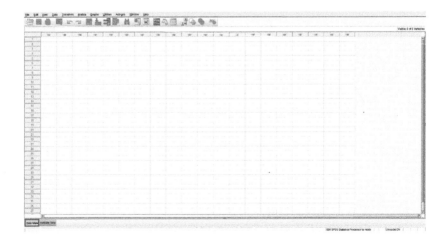

4. In order to switch from **Data View** to **Variable View** screen, use the tabs located at the bottom, left-hand corner of the SPSS screen.

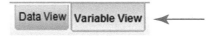

13.2 Labeling Variables

1. Select the **Data View** tab to begin entering your data.
2. In SPSS, all grouping variables, or independent variables, are designated by a numerical representation (i.e., 1, 2, 3, etc.). To organize the dataset for an independent test in the SPSS spreadsheet, there will be two separate columns, one for the independent variable and one for the dependent variable. Consider the following example.

Group	Change in HR (bpm)
Young Trained	10
Young Trained	28
Young Trained	7
Young Trained	16
Young Trained	5
Young Trained	10
Young Trained	3
Young Trained	11
Young Trained	13
Young Trained	12
Young Trained	2
Young Trained	17
Young Untrained	4
Young Untrained	3
Young Untrained	5
Young Untrained	8
Young Untrained	6
Young Untrained	6
Young Untrained	7
Young Untrained	3
Young Untrained	3
Young Untrained	2
Young Untrained	0
Young Untrained	3

In Excel

In SPSS

3. After all values have been inserted onto the spreadsheet, begin organizing your data by starting with the labels.

4. Along the bottom of the screen are two tabs labeled as **Variable View** and **Data View**. Select the **Variable View** tab.

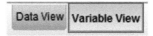

5. The new screen should look similar to the one below.

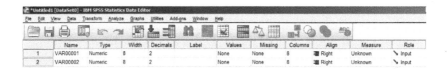

6. Start by typing in a label for Variable 1. For this example, the first column was the numerical representation of the group of individuals participating in the study. So a simple label such as "**Group**" would be suitable.

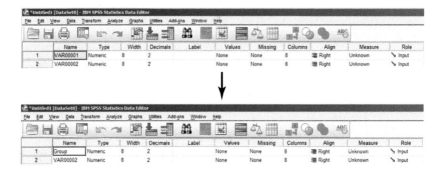

7. For the sake of this example, the second column represented Variable 2 or the change in heart rate after the relaxation demo (dependent variable). A label such as **HRChange** would be suitable. Keep

in mind that SPSS will not accept spacing or certain illegal characters such as a forward slash "/" and hyphen "-".

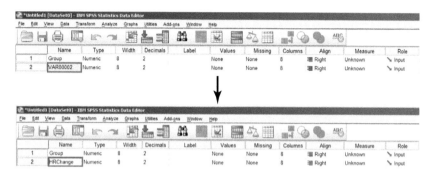

8. You may now return to **Data View** to start processing your data or you can continue on **Variable View** to change decimal placement and/or the measure for each variable. Review the following sections for assistance on how to do this.

13.3 Setting Decimal Placement

1. In order to set or make changes to the placement of the decimal, begin by selecting the **Variable View** tab.

2. If the independent variable is represented as a whole number, then the decimal placement should be "0." Considering the example used in the previous section, the independent variable is the **Group** and is represented by a whole number. Therefore, the decimal place can be set to "0."

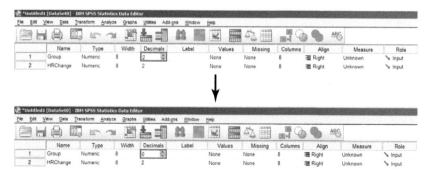

3. If the dependent variable is also represented as a whole number, then set the decimal place to "**0**" for the dependent variable. For the sake of this example, the change in heart rate was reported as a whole number, the decimal place can be set to "**0**" for **HRChange** as well.

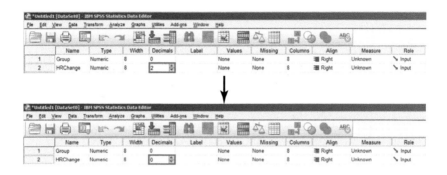

13.4 Determining the Measure of a Variable

Before running a statistical test, you must first determine the measure for the independent and dependent variables. This is a way of identifying your variables so that SPSS is able to recognize the independent and dependent variables.

1. Reflect on the **Group** variable and how it was reported. In this example, the measure of **Group** should be set as **Nominal** because the values correspond to the treatment groups.

2. For **HRChange**, the values were measured or observed on a numerical scale. Referring to the same example, the change in heart rate is a measurable value; therefore, the measure should be

set to **Scale**. For more information on the types of variables, see Chapter 1 under the Variables section.

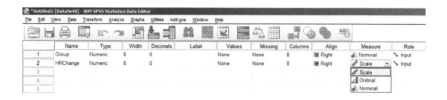

In other situations, the dependent variable(s) may be reported as ranked or ordinal data. If this is the case, then select **Ordinal** as the measure for the dependent variable(s).

13.5 Saving SPSS Data Files

1. Make sure the spreadsheet is selected.

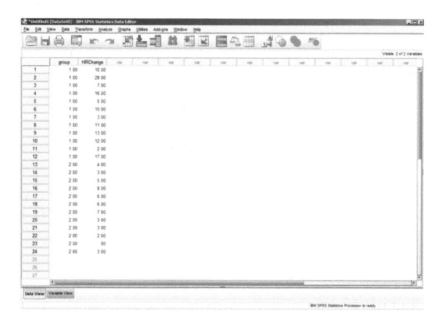

2. Go to **File** and select the **Save As** option.

3. Choose the **location** you would like to save the file in and then create a **file name**. Try to make the file name as descriptive as possible.

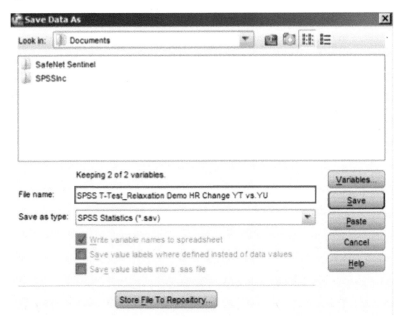

Basics in SPSS | 547

4. Then press **Save**.

13.6 Saving SPSS Output

1. Make sure the output is selected. If selected, it will say **Output [Document]** along the top.

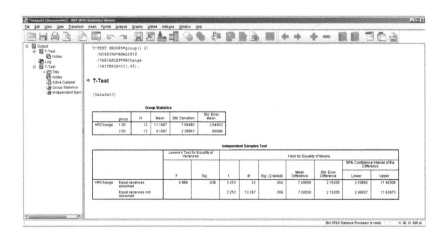

2. Go to **File** and select the **Save As** option.

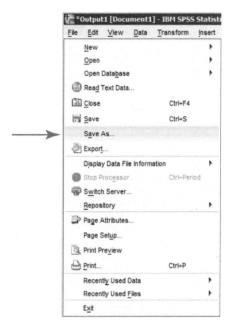

3. Choose the **location** you would like to save the file in and then create a **file name**. Make the file name of the output the same as the filename of the corresponding spreadsheet. To distinguish the output from the spreadsheet, follow the title with the word **Output**.

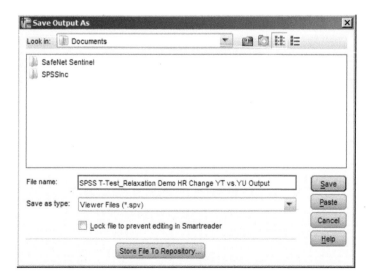

4. Then press **Save**.

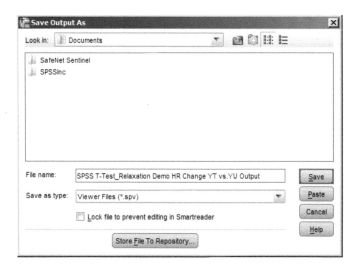

5. If you want to save a figure as a JPEG file, select the figure from the SPSS Statistics Viewer output, choose to **export** the selected figure from within the **File** menu. Within the **Export Output** box, select **None (Graphics only)** and save it as a JPEG (or other preferred file type). If you select the box for **Open the containing folder**, a screen showing you the location of the JPEG will appear.

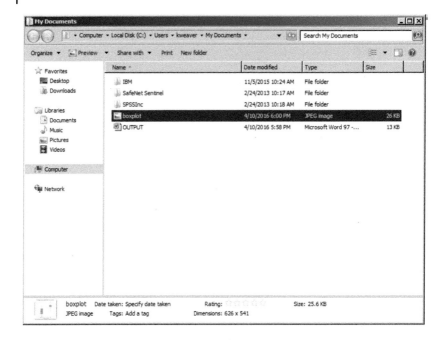

14

Basics in Numbers

Numbers is an ideal program for Mac users as it comes preinstalled on most Apple computers and laptops. Having familiarity with some short cuts will make navigating through the program much easier. Refer to this chapter for an introductory lesson on Numbers or if a refresher is needed.

The tutorials in this book are built to show a variety of approaches to using Numbers, so students can find their own unique style in working with statistical software, as well as to enrich the student learning experience through exposure to more and varied examples.

14.1 Opening Numbers

1. Open **Numbers** by double clicking on the Numbers icon.

An Introduction to Statistical Analysis in Research: With Applications in the Biological and Life Sciences, First Edition. Kathleen F. Weaver, Vanessa C. Morales, Sarah L. Dunn, Kanya Godde and Pablo F. Weaver.
© 2018 John Wiley & Sons, Inc. Published 2018 by John Wiley & Sons, Inc.
Companion Website: www.wiley/com/go/weaver/statistical_analysis_in_research

2. Double click the **Blank** spreadsheet on the **Template Chooser** window.

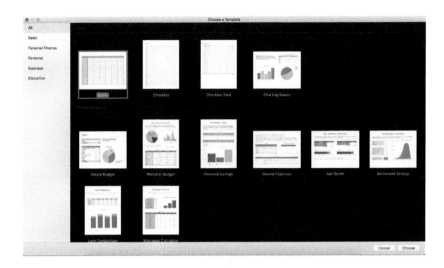

3. A blank spreadsheet will appear with the menu toolbar to the right.

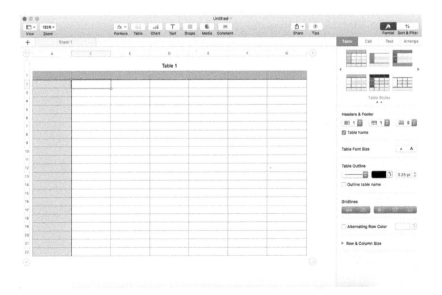

14.2 Common Commands

Addition

To calculate the summation between two or more values, refer to the following steps.

1. In an empty cell, type in an equals sign (=) followed by the word "**sum.**" Then, click on the SUM command that appears.

County Fair Field Trip	
School	Students Attending
1	254
2	391
3	145
4	174
5	286
6	103
7	243
fx ˇ SUM ▾ value	

2. Highlight all values you want to be calculated in the summation.

County Fair Field Trip	
School	Students Attending
1	254
2	391
3	145
4	174
5	286
6	103
7	243
fx ˇ SUM ▾ C5:C11 ▾	

An Introduction to Statistical Analysis in Research

3. Click the green checkmark or press enter/return.

County Fair Field Trip	
School	Students Attending
1	254
2	391
3	145
4	174
5	286
6	103
7	243

• fx ∨   SUM ▼   C5:C11 ▼

4. The calculated summation value will appear.

County Fair Field Trip	
School	Students Attending
1	254
2	391
3	145
4	174
5	286
6	103
7	243
Sum=	1596

Subtraction

To calculate the difference between two or more values, refer to the following steps.

1. In an empty cell, type in an equals sign (=).

Patient	pre LDL (mg/dL)	post LDL (mg/dL)	Difference
1	132	125 • fx ∨	
2	146	141	

2. Following the =, select the cell containing the first value.

3. Type in the subtraction sign (−), then select the cell containing the second value.

4. Click the green checkmark or press enter/return.

5. The calculated difference value will appear.

14.3 Applying Commands

Before you can apply a function or command to a series of cells, you must first manually insert the command/function you wish to use. Rather than manually inputting a command for all remaining cells, Numbers allows you to drag the formula down to apply to all cells.

To apply a command to a series of cells, refer to the following steps.

1. Select the cell that contains the original command. For this example, we computed the difference between the **post** and **pre LDL (mg/dL)**.

Patient	pre LDL (mg/dL)	post LDL (mg/dL)	Difference
1	132	125	-7
2	146	141	

2. Hover over the bottom center of the cell until a yellow circle appears. Click on the yellow circle and drag it down to the last row.

Patient	pre LDL (mg/dL)	post LDL (mg/dL)	Difference
1	132	125	-7
2	146	141	
3	159	150	
4	147	135	
5	163	149	
6	152	147	
7	155	144	
8	134	130	
9	150	146	
10	163	149	

3. The resulting values are the calculated differences for the remaining cells.

Patient	pre LDL (mg/dL)	post LDL (mg/dL)	Difference
1	132	125	-7
2	146	141	-5
3	159	150	-9
4	147	135	-12
5	163	149	-14
6	152	147	-5
7	155	144	-11
8	134	130	-4
9	150	146	-4
10	163	149	-14

14.4 Adding Functions

In each of the Numbers tutorials, we have provided the equations needed to run the most common functions you will use.

1. In the top bar, select the function box and select the drop down option for **Create Formula**.

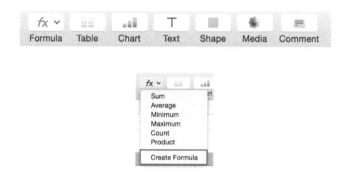

2. The selection will bring up the function bar in text.

3. If you select the drop down from *fx*, you are given the option of **Show Formula as Text** or **Convert Formula to Text in Cell**. Select **Show Formula as Text**. Converting the formula to text will allow you to see the equation, but you will not be able to use it within a calculation.

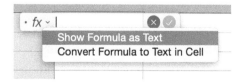

4. In addition to the function bar in text, a side bar of all functions will appear. You can search for a particular function in the search box at the top or you can limit the functions within a certain area, for instance **Statistical**.

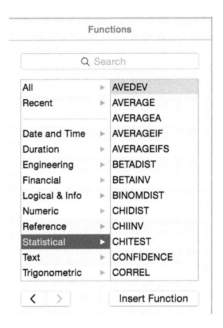

5. Select the function you want to use. This will bring up the definition of the function and directions within the spreadsheet.

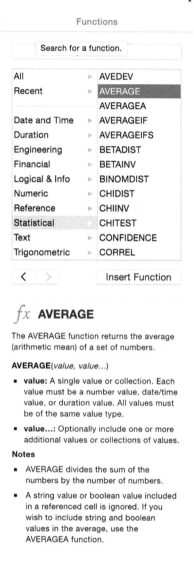

6. Follow the directions. In this case, select the values you want to average by clicking on the **value** in the function box and selecting the range of cells you want to evaluate.

15

Basics in R

R provides many benefits to students, including a cost benefit: it is free! The best way to understand R is to learn its programming as a language. R can seem intimidating at first when you do not understand the patterns behind how it runs tests, stores data, and recalls information. Learning the language of R will allow students to effectively work through problems and not worry about the programming aspect of the statistical platform. This chapter will introduce the basic linguistic features of R (e.g., storing variables, creating datasets, executing functions, customizing through arguments).

While this book focuses on the Mac version of R, there are R platforms designed for Windows and there is a product called RStudio. There are slight differences in scripting between each platform in more advanced programming, but for the scope of this book, the code should be universal. Slight differences may arise in menu names. If you notice this is the case, check several menus for the option listed in the tutorials.

The tutorials in this book are built to show a variety of approaches to using R, so the student can find their own unique style in working with statistical software, as well as to enrich the student learning experience through exposure to more and varied examples.

15.1 Opening R

1. To access R, click on the R icon located either in your program menu in Windows or in your Applications folder in Mac.

562 | An Introduction to Statistical Analysis in Research

2. When the program launches, the following window will appear.

15.2 Getting Acquainted with the Console

The R Console has several shortcuts located at the top. The icons you may use are as follows:

a. Runs an R executable file as compiled through the R Editor.

b. Toggles between the console and the graph output screen.

c. Opens R Editor files for editing.

d. Opens a new R Editor file.

e. Closes out of R.

Principles Behind R Programming

There are seven principles that must be followed for using R programming in this book, as well as composing your own scripts.

1. **R is case sensitive** – make sure to use the appropriate case.
2. **R is exact** – there is no "close enough" with R programming. Thus, you have to focus on making sure your code is 100% accurate.
3. **R is specific** – the order in which you place your coding matters.
4. **R is picky about spaces** – there are no spaces in variable names or names you wish to store.
5. **R does not recognize copy and paste** – never copy and paste into the R console or the R Editor. While the screen may show the exact information that you copied was successfully pasted into R, R does not necessarily recognize that information correctly internally. Thus, copying and pasting into R is NEVER recommended.
6. **R has different expectations for missing data** – it recognizes "." (when importing data from a file with continuous variables) and "NA" (when manually typing in the data and for factor variables).
7. **R loves the up and down arrows**– the up and down arrows on the keyboard can help you navigate through prior programming that you have already ran and want to repeat. At the command prompt, press the up arrow until you find the script you want to run and then press enter/return.

The Command Prompt and Bottom Bar

The command prompt "<" is where you place the code you wish to run. The bottom bar gives a preview of the arguments used for the function you are currently typing.

Vectors, Strings, Factors, and Data Frames

There are four key terms R users should be familiar with that refer to the data and data types: (1) vector, (2) string, (3), factor, and (4) data frame.

Vector – a series of values stored within a single variable of data. The length refers to the number of values stored within a vector. Variable and vector are used interchangeably in the R tutorials.

String – a value that consists of text, rather than numbers. Strings are contained within quotation marks when manually input into R.

Factor – a categorical variable. To test if a vector is recognized as a factor variable, use the function `is.factor()`.

Data frame – a series of vectors of equal length contained within a single dataset. Data frame and dataset are used interchangeably in the R tutorials.

Functions and Arguments

Functions

A function is the process that R executes on the data. For example, if you want R to run a *t*-test, the R function is "**t.test.**" In this book, functions are designated with parentheses "**()**" immediately succeeding the function name to designate the function. For example, the function **t.test** would be **t.test()**. An open parenthesis "(" succeeds function names, programming can be placed after this symbol, and a closed parenthesis ")" must end the function.

Arguments

An argument is a code that customizes the function being run in R. Arguments are found within the parentheses of a function and a comma separates it from other elements of the script within the function. Arguments usually have the equal sign to the right of the text:

```
col=
```

For example, when plotting, the "**col**" argument in the **plot()** function will instruct R that the value in quotation marks to the right of the "=" is the color of what is to be plotted:

```
plot(x,y, col="red")
```

Test a Particular Vector Within a Data Frame

To call up a particular vector within a data frame to be used in a function, input the following values: data frame name, dollar sign ($), and vector name. For example, for the data frame **animaldata** and the vector **catdata**:

```
animaldata$catdata
```

Within a function:

```
functionname(animaldata$catdata)
```

15.3 Loading Data

Data can be loaded into R in a few different ways. Data can be input into the R Editor and loaded from there, data can be manually input into the R console, and data can be loaded from .csv and .txt files. This tutorial will demonstrate how to manually input data into the R console and how to load data from a .csv file.

Manually Inputting Data

1. At the command prompt "<" type in the name for the vector you wish to assign, such as "**catdata**."

   > catdata

2. Next, we will use the symbol "<-" which tells R that everything to the right of the symbol should be stored to the name on the left of the symbol.

   > catdata<-

3. For a vector of numbers, the **c()** function must be invoked to tell R that a series of numbers or strings will be stored.

   > catdata<-c()

4. Within the parentheses, list your data with a comma "," separating each value.

   > catdata<-c(3,6,9,10,15,12)

5. Press enter/return. If the data were loaded successfully, R will return the command prompt. If the data were not loaded successfully, R will provide an error message in red.

   > catdata<-c(3,6,9,10,15,12)
   >

If the data that needed to be entered were strings, the strings would be contained within quotation marks and separated by a comma. Otherwise, the remaining programming applies.

```
> catdata<-c("cute", "cuddly", "purr", "silky")
>
```

6. Your vector is now loaded. Every time data are loaded into any statistical program, you should check whether the data are loaded properly (e.g., variable/vector names are in the correct position, the data appear correct). To do this, simply type in the vector or data frame name at the command prompt and press "return."

```
> catdata
[1]  3  6  9 10 15 12
```

The number in brackets designates the observation number, or the position of the value within the vector. This becomes handy when you have large numbers of text and are looking for a particular value. If you want to add additional vectors and make a dataset, continue with step 7. If you only want to work with separate vectors, repeat steps 1–5 to store additional vectors.

7. To create a dataset (a series of vectors where each row corresponds to an observation and vectors are of equal length), load all desired vectors of equal length to named vectors following steps 1–5.
8. Select a new name for the dataset (different from the vectors) and use the **data.frame()** function. Input the names of the vectors separated by a comma within the parentheses. In this case, the **catdata** vector will be joined to the **dogdata** vector to create the **animaldata** data frame.

```
> catdata<-c(3,6,9,10,15,12)
> dogdata<-c(71,612,903,44,12,7)
> animaldata<-data.frame(catdata, dogdata)
```

9. Press enter/return. Check that the data frame has been appropriately loaded by typing in the data frame name and pressing enter/return.

```
> animaldata
  catdata dogdata
1       3      71
2       6     612
3       9     903
4      10      44
5      15      12
6      12       7
>
```

Variable/vector names should be at the top and not have an observation number assigned to them (e.g., the column of numbers 1–6 on the left).

Loading a .csv File

1. Direct R to the area where your file is saved. To do this, go to the "**Misc**" menu and select **Change Working Directory** (menus may vary by R platform, so be sure to check several menus if you do not see **Change Working Directory** in the location referred to in this step).

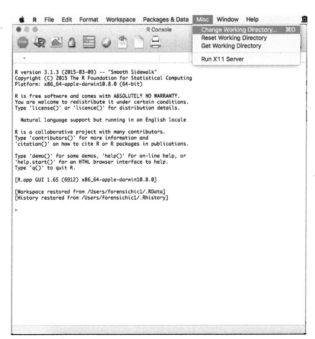

2. Select the location in which your file is saved, for example, the Desktop. Click "**Open**."

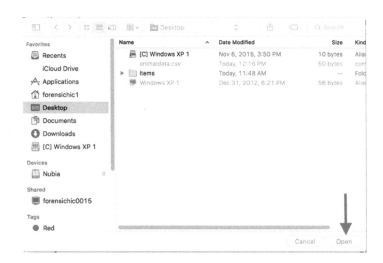

3. A screenshot of data saved as a .csv file through Excel that can be loaded by R is shown below. Please note the variable names are in the first row of each vector column.

	A	B
1	catadata	dogdata
2	3	71
3	6	612
4	9	903
5	10	44
6	15	12
7	12	7
8		

4. To load a .csv file, you will use a file name of your choice, the "<-", and the function **read.csv()**. Within the parentheses, you will place your file name, including the extension, within quotation marks. The argument **header=** will instruct R whether your data have variable names (**T** designates there are variable names and **F** indicates there are not).

> animaldata<-read.csv("animaldata.csv", header=T)

5. Press enter/return. Check if the data are loaded correctly by typing in the name at the command prompt and pressing return.

```
> animaldata<-read.csv("animaldata.csv", header=T)
> animaldata
  catadata dogdata
1        3      71
2        6     612
3        9     903
4       10      44
5       15      12
6       12       7
>
```

15.4 Installing and Loading Packages

Packages contain various functions for analyzing data. Packages are built by different people and need to be downloaded. Once downloaded, they can be loaded for execution. This tutorial will show you how to install a package and how to load it for using its functions.

Installing R Packages

1. Click on the **"Packages & Data"** menu and select **Package Installer** (menus may vary by R platform, so be sure to check several menus if you do not see **Packages & Data** or **Package Installer** in the location referred to in this step).

2. The Packages Repository will open and allow you to search for packages. You might be referred to a particular package in this book, or one might be suggested by your advisor. To find a specific package, click on the menu in the top left hand corner and select **Other Repository**.

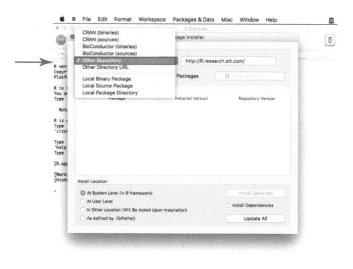

3. Next to the magnifying glass, type in the name of the package that you wish to search, in this case "`Hmisc,`" and click **Get List**.

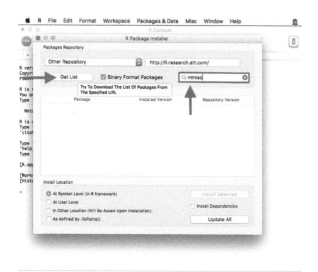

The first time you download a package, you will need to select an appropriate mirror to support the download (R will display a message asking you to select a mirror). Skip to step 4 if you have identified a mirror in the past.

Select **"Yes"** when prompted to choose a mirror. The screen shown below will pop up.

Select a mirror that is the closest to you geographically. For example, if you live in California, you will click on the **USA (CA 1)** mirror to select it and then click **"Ok"**.

After clicking **Ok**, the following screen will appear. Press **Yes** to permanently set your mirror.

4. R will return packages that contain the search terms. Click on the package you wish to download so that the **Install Selected** button in the bottom right hand portion of the window becomes available and is no longer grayed out. Click the box next to **Install Dependencies** (below the **Install Selected** button) and ensure the "**Install Location**" (bottom left hand side of window) is set to **At System Level** (**in R framework**). Click **Install Selected**.

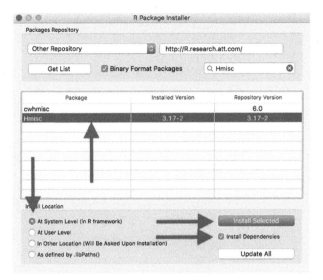

5. Because of the settings we selected in step 4, R will install the package you selected as well as any packages that your package relies upon to run. When R is done, there will be messages on screen indicating the package has been downloaded.

Basics in R | 575

If R shows an error in download (a message will be displayed indicating the package was not downloaded in its entirety), try restarting R and redo steps 1–5.

Loading a Package

1. Just having a package installed in R is not enough; you must also load the package when you want to use it. Simply type in the name of your package within the parentheses of the `library()` function. Press enter/return.

   ```
   > library(Hmisc)
   ```

2. If you are successful in loading the package, you may see messages indicating this is the case.

   ```
   > library(Hmisc)
   Loading required package: lattice
   Loading required package: survival
   Loading required package: Formula
   Loading required package: ggplot2
   ```

   You may also see nothing but a new command prompt, indicating R is ready for the next line of script.

   ```
   > library(ecodist)
   >
   ```

3. If you are unsuccessful, you will see a message along these lines.

   ```
   > library(cwhmisc)
   Error in library(cwhmisc) : there is no package called 'cwhmisc'
   ```

   Simply reinstall or install for the first time the package you wish to use by following steps 1–5 under **Installing R Packages**.

4. On occasion, you might want to use two different packages and find that they conflict, as in the packages **ecodist** and **vegan**.

```
> library(ecodist)
> library(vegan)
This is vegan 2.3-2

Attaching package: 'vegan'

The following object is masked from 'package:ecodist':

    mantel
```

The error means that both packages have a function called **mantel()**. If the functions are different, then either: (1) do not load both packages at once or (2) detach the second package when you get the error. To detach a package, use the **detach()** function. Type the argument **package:** in the parentheses and type the name of the package to detach (e.g., **vegan**).

```
> detach(package:vegan)
```

15.5 Troubleshooting

Below is an example of an R error message.

```
> x<-c(1,2,3,4,5,6,7)
> maen(x)
Error: could not find function "maen"
```

If you are receiving errors in R, the troubleshooting tips should help identify the source of the problem. Do not trust an error message. While many times it will help you isolate the problem by reading it and following the instructions, oftentimes it will not provide the information you need. Instead, follow these steps:

1. **Did you copy and paste into R?**

Remember, R will not necessarily recognize copied and pasted items correctly. Thus, if you have copied and pasted something into R, abandon it and manually type out the programming.

2. **Is your programming 100% accurate?**

Double check all of the code. Further, check for commas between all values and arguments. Make sure that every time there is an open parenthesis, there is also a closed parenthesis. Verify that any time there is a left quotation mark, there is also a right quotation mark. Also, check spelling.

3. **Are your data loaded correctly?**

If you skipped the step of verifying your data was loaded correctly, this might be the issue.

   a. Make sure your variable names are on top of the vector values and do not have an observation number on the left corresponding to them.

4. **Conduct an internet search of your error message.** You might find solutions to the problem.

16

Appendix

Flow Chart

Flow chart to determine the most appropriate statistical test:

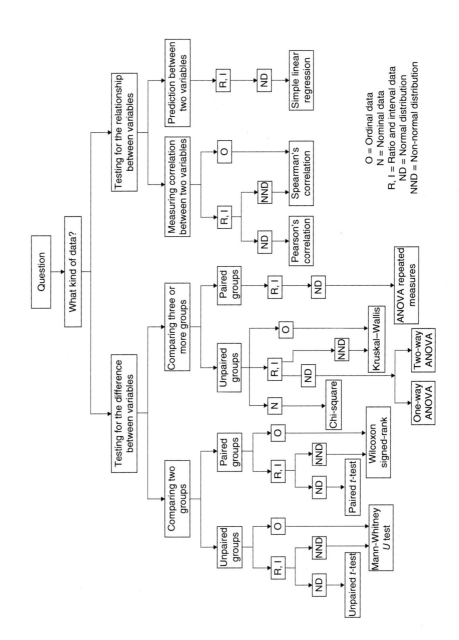

Literature Cited

Antón, S.C. (2012). Early homo: who, when, and where. *Current Anthropology* 53(S6): S278–S298.

Arnold, K. and Zuberbühler, K. (2006). The alarm-calling system of adult male putty-nosed monkeys, Cercopithecus nictitans martini. *Animal Behaviour* 72(3): 643–653.

Boeuf, G. and Payan, P. (2001). How should salinity influence fish growth? *Comparative Biochemistry and Physiology Part C: Toxicology and Pharmacology* 130(4): 411–423.

Bohannon, N. and Weaver, P. (2016). Behavioral Reproductive isolation in the Tiger Limia (*Limia isla*) and Humback Limia (*Limia nigrofasciata*). Poster Presentation West Coast Biological Sciences Undergraduate Research Conference, San Diego, CA.

Bruner, A.G., Gullison, R.E., Rice, R.E., and Da Fonseca, G.A. (2001). Effectiveness of parks in protecting tropical biodiversity. *Science* 291(5501): 125–128.

Candelaria, R., Morales, V., and Weaver, K. (2016). *Documenting the diversity of Oreohelix in central Washington*. Oral presentation at West Coast Biological Sciences Undergraduate Research Conference, San Diego, CA.

Castellanos, K., Cavazos, C., and Weaver, P. (2015). *Evolution in extreme environments: the effects of hypersalinity on the physiology and morphology of livebearing fish*. Unpublished dataset, Department of Biology, University of La Verne, La Verne, USA.

Castorani, M.C. and Hovel, K.A. (2016). Native predator chemical cues induce anti-predation behaviors in an invasive marine bivalve. *Biological Invasions* 18(1): 169–181.

Chesebro, J.W., McCroskey, J.C., Atwater, D.F., Bahrenfuss, R.M., Cawelti, G., Gaudino, J.L., and Hodges, H. (1992). Communication apprehension and self-perceived communication competence of at-risk students. *Communication Education* 41(4): 345–360.

Debinski, D.M. and Kelly, L. (1998). Decline of Iowa populations of the regal fritillary (*Speyeria idalia*) Drury. *Journal of the Iowa Academy of Science* 105(1): 16–22.

Dunn, S.L., Boutcher, Y., Trapp, G., and Boutcher, S.H. Recovery metabolism following high intensity intermittent and steady state exercise. *Med Sci Sports Exerc*. 2006; 38(5):S2806.

Fabião, J., Teixeira, A.T., and Araújo, M.A.V.C. (2013). Hydraulic stability of tetrapod armour layers – physical model study. In: *Proceedings of 6th International Short Course/Conference on Applied Coastal Research*.

Godde, K. (2011). *Bone weight's relationship to BMI*. Unpublished dataset, Sociology/Anthropology, University of La Verne, La Verne, USA.

Godde, K. (2015). *An Osteological Investigation of Naga-ed-Der; Evaluation of Sexual Dimorphism, Anemia, and their Relationship with Other Predynastic and Early Dynastic Egyptians*. Unpublished dataset, Sociology/Anthropology, University of La Verne, La Verne, USA.

Godde, K. (2014). Secular trends in cranial morphological traits: a socioeconomic perspective of change and sexual dimorphism in North Americans 1849–1960. *Annals of Human Biology* 42(3): 253–259.

Godde, K. and Wilson Taylor, R. (2011). Obesity in the skeleton: correlating long bone external dimensions to body mass index (BMI). *American Journal of Physical Anthropology* 144: 291–291.

Gopen, G.D. and Swan, J.A. (1990). The science of scientific writing. *American Scientist* 78: 550–558.

Grant, P.R. and Grant, B.R. (2002). Unpredictable evolution in a 30-year study of Darwin's finches. *Science* 296(5568): 707–711.

Hilu, R.A., Dudeen, O., and Barghouthi, S.A. (2015). Correlation between serum levels of vitamin B12 and anti-Helicobacter pylori IgA antibodies in vitamin B12 deficient Palestinian patients. *Turkish Journal of Medical Sciences* 45(3): 627–633.

Jolly, C.J. and Phillips-Conroy, J.E. (2003). Testicular size, mating system, and maturation schedules in wild anubis and hamadryas baboons. *International Journal of Primatology* 24(1): 125–142.

Kmiecik, K., O'Leary, K., and Yiing, L.C. (2008). Comparison of Tiger Leech Attraction to Heat and Movement Stimuli in Maliau Basin, Sabah. Retrieved from http://phylodiversity.net/bb08

Lindquist, E.J., Annunzio, R.D., Gerrand, A., MacDicken, K., Achard, F., Beuchle, R., Brink, A., Eva, D.E., Mayaux, P., San-Miguel-Ayanz, J., and Stibig, H.J. (2012). Global forest land-use change 1990–2005. Food and Agriculture Organization of the United Nations and European Commission Joint Research Centre, FAO Forestry Paper 169, pp. 1–53.

Meeks, L., Vera, D.L., and Dunn, S.L. (2014). Body composition changes within the first four weeks of the freshmen year at a private university. *FASEB J.* 28, 1031.6.

Moczygemba, B., Bohannon, N., and Weaver, P. (2015). *Limia*: Separation of two species. Oral Presentation at Southern California Conference on Undergraduate Research, Los Angeles, CA.

Morales, V., Candelaria, R., and Weaver, K. (2016). *Documenting the diversity of Oreohelix in central Washington.* Unpublished dataset, Department of Biology, University of La Verne, La Verne, USA.

NASA's Goddard Institute for Space Studies (GISS), http://climate.nasa.gov/

National Weather Service for National Oceanic and Atmospheric Administration, http://weather.gov/

Nunez, J., Vera, D.L., and Dunn, S.L. (2016). The Gender Differences in the Change of Dietary Intake and Body Composition Following the First Semester for Undergraduate First Year Students at a Small Liberal Arts University. *FASEB J.* 30, 686.3.

Ries, L. and Debinski, D.M. (2001). Butterfly responses to habitat edges in the highly fragmented prairies of central Iowa. *Journal of Animal Ecology* 70(5): 840–852.

Rios, M., Morales, V., Schutz, H., and Weaver, K. (2010). *Utilizing Shell Morphometrics and Radula Count for Species Identification in the Land Snail Oreohelix.* Poster session presented at the Evolution conference, Portland, OR.

Rosenfeld, L.B., Grant III, C.H., and McCroskey, J.C. (1995). Communication apprehension and self-perceived communication competence of academically gifted students. *Communication Education* 44: 79–86.

Ruiz, R., Garcia, J., and Weaver, K.F. (2015). STEAM: Putting the Arts and Advocacy into STEM; Improving Written Communication in the Freshman Year. Oral Presentation at Association of American Colleges and Universities, Crossing Boundaries: Transforming STEM Education Conference, Seattle, WA.

Tutin, C.E.G., Ham R.M., White, L.J.T., and Harrison, M.S. (1999). The primate community of the Lopé reserve, Gabon: diets, responses to fruit scarcity, and effects on biomass. *American Journal of Primatology* 42: 1–24.

Vera, D.L., Castillo, M., Oliver, K., Rosario, E., Granquist, M., and Dunn, S.L. (2014). Metabolic profile and inflammatory markers in an endurance trained and untrained population. *FASEB J.* 28, 884.20.

Villalobos, D. and Contreras, H. *The effects of meal energy content on the specific dynamic sction and respiratory pattern of the hissing cockroach (Gromphadorhina portentosa)*. Unpublished dataset, Department of Biology, University of La Verne, La Verne, USA.

Waters, N.D., Morales, V., Gilbertson, L., and Weaver, K. (2015). *Biogeography of Sonorella in the Arizonia sky islands.* Unpublished dataset, Department of Biology, University of La Verne, La Verne, USA.

Weaver, P. Unpublished dataset, Department of Biology, University of La Verne, La Verne, CA.

Weaver, P. and Weaver, K. Unpublished dataset, Department of Biology, University of La Verne, La Verne, CA.

Williams, H.L. and Salamy, A. (1972). Alcohol and sleep. In: *The Biology of Alcoholism*, pp. 435–483. Springer.

Wilson Taylor, R. and Godde, K. (2011). Obesity in the skeleton: correlating long bone external dimensions to Body Mass Index (BMI). *Program of the 80th Annual Meeting of the American Association of Physical Anthropologists* S52: 291.

Yund, P.O., Tilburg, C.E., and McCartney, M.A. (2015). Across-shelf distribution of the blue mussel larvae in the northern Gulf of Maine: consequence for population connectivity and a species range boundary. *Royal Society Open Science* 2(12): 1–16.

Glossary

absolute reference Cell reference in Excel that when copied and pasted the contents do not change. The column and row references are designated with a "$".

alternative hypotheses (H_1, H_2, H_3, etc.) Possible explanations for the significant differences observed between groups.

arcsine transformations The arcsine of the square root of each observed value.

argument The coding that customizes the function being run in R (e.g., col=).

array Numbers organized in a row or column in Excel.

bar chart (or bar graph) Consist of bars (vertical or horizontal) that represent the values of some type of categorical variables. Values can represent the means or medians for the groups or the actual value for a single observation.

bias Occurs when certain individuals are more likely to be selected than others in a sample.

bimodal (double-peaked) distributions Two or more distributions; split nonhomogeneous groups within one data set.

box plot Shows the median, as well as the distribution of the data through the use of quartiles; divides ranked data into four equal groups, each consisting of a quarter of the data.

categorical variables Variables that fall into two or more categories (e.g., nominal variables and ordinal variables).

central tendency A typical value that characterizes a population or data set.

clumped bar charts Demonstrate trends within each category of data.

clustered bar graphs Provides a clear depiction of the values associated; error bars (standard deviations) help to illustrate the variance in each one of the groups of data.

command prompt Denoted by ">" and where you can input programming into R.

confidence interval Indicates the reliability of an estimate.
continuous variables Quantitative variable(s) measured on a continuous scale.
controlled variables Factors that cause direct changes to the dependent variable(s) unrelated to the changes caused by the independent variable.
correlation The relationship or association between two variables.
counterbalancing Means to offset the slight differences that may be present in our data due bias.
data frame A collection of vectors of equal length that comprise a data set.
data transformation Method applied on a data set that strongly violates statistical assumptions in an attempt to "normalize" the data.
dependent variables The response variable; the variable that is influenced by the independent variable.
descriptive statistics Characterize the population, often represented with mean and variance of the sample, among other measures.
discrete variables Variables that are counted.
estimate a mathematical approximation of the value.
frequency distribution curve The frequency of each measured variable.
function The process that R executes on the data; designated with parentheses (e.g., plot()).
heteroscedasticity The assumption that the variance of the populations is uneven between residual data points; the residuals have an unequal spread.
histogram Another form of bar charts used to display continuous categories along the bottom of the chart with a set range of values; Both axes will be represented on a numerical scale.
homoscedasticity The assumption that the variances of the populations have relatively the same variance; the populations have an equal spread around the mean.
hypothesis A testable statement that provides a possible explanation to an observable event or phenomenon; an educated guess.
independent samples One individual contributed to only one observation.
independent variable The treatment variable; part of the experiment established by or directly manipulated by the research that causes a potential change in another variable, typically the dependent variable(s).
inferential statistics Used to determine the quality of our estimate in describing the sample and determine our ability to make predictions about the larger population.
interval variable Have an arbitrarily assigned zero point; unlike ratio data, comparisons of magnitude among different values on an interval scale are not possible.
kurtosis Symmetrical curves that are not shaped like the normal bell-shaped curve; Tails are either heavier or lighter than expected.

leptokurtic Tails are heavier; more outliers than normal distribution.
line graph Utilizes a line to connect the data points when both the independent and dependent variables are continuous.
linear regression Models the relationship between an explanatory and response variable. Statistical measures used are the R^2, which reflects the fit of the data to the trend-line.
log transformation The log of each observation, whether in the form of the natural log or base-10 log.
mean (μ) Average value, calculated by adding all the reported numerical values together and dividing that total by the number of observations; describes the location of the distribution heap or average of the samples.
median Middle value after arranging all the values in the data set in numerical order. If there is an odd number of observations, then the middle value will serve as the median. If there is an even number of observations, then the average of the two middle values must be calculated.
mode Most frequent value.
negatively skewed The values of the mean and median are less than the mode. The lingering tail region is on the left.
nominal variable The variable that is counted not measured, and they have no numerical value or rank; classify information into two or more categories.
nonparametric statistics Statistical tests that compare medians, do not calculate parameters, lack any assumptions about the data set or population, and data do not need to follow a normal distribution (where the distribution is unknown or not made clear).
normal distribution Bell-shaped curve (mesokurtic shape = normal heap); symmetric, convex shape; no lingering tail region on either side; homogenous group with the local maximum (mean) in the middle.
null hypothesis (H_0) Assumes that there is no difference between groups.
one-way analysis of variance (ANOVA) Parametric statistical test that compares the means of three or more sampling groups.
ordinal variables Variables are ranked and have two or more categories; however, the order of the categories is significant.
outlier Numerical value extremely distant from the rest of the data.
parametric statistics Compares means and associated values (e.g., standard deviations) under the assumption that data sets (1) are on the interval or ratio scale, (2) are normally distributed in some sense (e.g., errors), (3) are randomly sampled from the data set or population, (4) have equality of variances in select circumstances, and (5) are typically large data sets.
pie chart (or pie graph) Shows the different categories as pieces in relation to the total as slices (or sectors) in a pie.

plateau distributions Extreme versions of multimodal distributions; curve lacks a convex shape.

platykurtic The distribution tails are lighter; fewer outliers from the mean.

population All possible test subjects within a sampling group of research interest.

positively skewed The values of the mean and median are greater than the mode. The lingering tail region is on the right.

probability value (p) Indicates the significance among the variables being analyzed and the likelihood of making a type I error.

quantitative variables Variables that are counted or measured on a numerical scale.

random sample All individuals within a population have an equal chance of being selected, and the choice of one individual does not influence the choice of any other individual.

ratio variable Have a true zero point and comparisons of magnitude can be made.

repeated measures ANOVA (rANOVA) Analysis of samples applied to the framework of the one-way ANOVA but the observations lack independence due to the same individual(s) being sampled multiple times (e.g., across multiple conditions or time).

replication Involves repeating the same experiment in order to improve the chances of obtaining an accurate result.

R^2 value The coefficient of determination. The coefficient is used in linear regression analysis and indicates how well the data fits the regression line. It is the square of the correlation coefficient.

sample A subset of individuals from a larger population that will serve as a representative group.

sample of convenience Samples are often chosen based on the availability of a sample to the researcher.

scatterplot Looks at overall trends in a scattering of unrelated data points across a continuous scale.

skewed distributions Asymmetrical distribution curves; distribution heap either towards the left or right with a lingering tail region.

sphericity The variance of the differences between the levels of a categorical variable is equal.

square-root transformations The square root of each observation and are commonly used in counts.

stacked bar charts Illustrate the relative contributions of parts to the whole.

standard deviation (σ) The square root of variance.

statistical power The probability of correctly rejecting a false null hypothesis.

stratified sample The population is organized first by category (i.e., strata) and then random individuals are selected from each category.

string A vector that contains text, rather than numbers.

Student's *t*-test Statistical test that compares the means of one group to the expected mean or two groups to one another.

systematic sample Participants are ordered (e.g., alphabetically), a random 1st individual is identified, and every *k*th individual afterwards is selected for inclusion in the sample.

two-way analysis of variance (ANOVA) Statistical analysis comparing the means of two or more subgroups within multiple comparison groups.

two-way repeated measures ANOVA (two-way rANOVA) Statistical analysis comparing the differences between multiple groups and repeated measures.

type I error Null hypothesis is rejected incorrectly (false positive).

type II error Null hypothesis fails to be rejected when it should have been (false negative).

variance (σ^2) Takes into account the spread of the distribution curve.

vector A series of values stored in R that represent a single variable.

volunteer sample Used when participants volunteer for a particular study.

Index

Note: Page numbers in bold denote description of the terms and those followed by "f" denote figures.

a

absolute reference   530
alligator   119
analysis of variance (ANOVA)   227–236
   assumptions   233
   one-way   228
   repeated measures   230–231
   two-way   229–230
   two-way repeated measures   231–232
aquatic salinity   517
arcsine transformation   33–34
argument   565
array   531
asymmetrical distribution
   left skew   22, 23, 23f
   negatively skewed   23, 23f
   right skew   22, 22f
   positively skewed   22, 22f
athletic training   228

b

bar chart or bar graph   63–66
   clumped   64–65
   clustered   63–64
   error bars   63–64
   stacked   65

bell curve   13, 19, 21, 23
bias   **2**, 11
bimodal distribution   **25**
biodiversity   106, 174, 299
blood lactate levels   237
blue mussels   138
Body Mass Index (BMI)   2, 291, 288, 498
body weight   2, 122
Bonferroni correction   236
box plots   67–69
   quartiles   67
brain size   163
butterfly habitat patch   418

c

caffeine   7
Canada geese   196–199
candy wrapper   241
Cartesian coordinates   138
central tendency   13
chimpanzee testicular volume   467
chi-square   393–405
   assumptions   397
   chi-square value   354, 367, 397
   formula   397
command prompt (R)   564
concussions   228, 229

An Introduction to Statistical Analysis in Research: With Applications in the Biological and Life Sciences, First Edition. Kathleen F. Weaver, Vanessa C. Morales, Sarah L. Dunn, Kanya Godde and Pablo F. Weaver.
© 2018 John Wiley & Sons, Inc. Published 2018 by John Wiley & Sons, Inc.
Companion Website: www.wiley/com/go/weaver/statistical_analysis_in_research

confidence interval 8
correlation 435–448
　correlation coefficients **438**, 476, 479
　Pearson's, *see* Pearson's correlation
　Spearman's, *see* Spearman's rank order correlation
counterbalancing **6–7**

d

data distribution 193
data frame 565
data transformation 32
degrees of freedom 354, 398
descriptive statistics 7–8
discrete variables **12**
distribution 13, 18, 20,
Dunnett's C post-hoc test 354, 357

e

equality of variances 193
equation of a line 138, 475
error 11, 27, 30, 31
　false negative 31
　false positive 31
　type I and type II error **31**
estimate 2, 7–8
experimental design 1, 9–10

f

factor (R) 564
finch 436, 508
fish 135
　Cyprinodon gene expression 305
　Limia osmoregulation 328
　Limia schooling 218
　Poecillidae temperature preference 350
food consumption 188
forest cover 168
frequency distribution curve **18**
function (R) 565
F-value 476

g

gender identity 181
General Linear Model (GLM) 473
germination 376
global temperature 158
glycerol levels 247, 261

h

heart rate 205, 209, 213
height 127, 223
Helicobacter pylori 462
heteroscedasticity **440**
histogram **18**, 21, 65
homogeneity 200
homogeneous variance 479
homoscedasticity **440**, 480
hydrangea 272
hypothesis **9**, 29,
　alternative hypothesis **9**, 29
　null hypothesis **9**, 29

i

independent samples **233**
inferential statistics 8, 28
Institutional Animal Care and Use Committee (IACUC) 6
Institutional Review Board (IRB) 6

k

Klamath pH concentration 91, 201
Kruskal-Wallis 353–361
　assumptions 357
kurtosis **23**, 44
　leptokurtic **24**
　mesokurtic 23
　platykurtic **24**

l

least significant differences (LSD) 235
level of significance 200
Levene's test 200
line graphs 136–137

linear equation 138, 474–475
linear regression 442, 473–484
   assumptions 479
linear relationship 435, 473
log transformation 33
low-density lipoprotein (LDL) levels 34, 313
low variability 32

m

Madagascar hissing cockroach 302
Mann–Whitney 297–302
   assumptions 298
marine iguana 19
mean 7, **14**, 20
mean rank 302, 354, 358
median 14, **16–17**
mode 14, **17–18**
multiple regression 474
mussels 109

n

non-normal distribution 21
nonparametric statistical analyses 192
normal distribution 20, **21**

o

observation 2
one tail test 201
outliers 16, **26**, 68, 478

p

paired groups **194**
parametric statistical analyses 191
   assumptions 191–192
Pearson's correlation 439–445
   assumptions 440
pie charts or pie graph 165–167
plateau distribution **25**
polar bears 406

population 2
post hoc analyses 228, 234
probability value **438**
putty-nosed monkeys 386
p-value 8, **9**, 23

r

regression analysis 476
related samples 194
relative reference 530
replication **6**
R^2 value 476, 478

s

sample 2
sample size 2, **31**
sampling design 3–6
   random sample **3**
   sample of convenience **6**
   stratified sample **5**
   systematic sample **4**
   volunteer sample **5**
scatter plot 136–139
science communication 337
significance 23, 29, 31
significance criterion 32
skeletal
   cranial capacity 116, 319, 324
   femur length 449
   humerus weight 497
   lung volume 28
   skull 429
skewed distribution **22**
sleep 358
snails
   Oreohelix tooth count 362
   parasite 354
   radula 70
   Sonorella genetic differences 484
snowfall patterns 63, 140
Spearman's rank order correlation 439–442, 445–448
   assumptions 440

sphericity   233
square-root transformation   33
standard deviation   **21**, 41
statistical power   **31**
Streptomycin resistance   480
string   564
studying   442

t

tiger leech   394, 426
trend line   138
t-test   195–205
  assumptions   200
  one-sample   196
  two-sample independent   196–198
  two-sample paired   198–199
Tukey Honestly Significant Difference (HSD)   235
two-sample tests   194

u

unpaired groups   **194**
unrelated samples   194

v

variables   9–12, 297
  categorical   **11**
  continuous   **12**
  controlled   **11**
  dependent   9, **10**
  explanatory   474–475
  independent   9, **10**, 297
  interval   **12**
  nominal   **11**
  ordinal   **11**, **12**
  predictor   484, 519
  quantitative   **12**
  ratio   **12**
  response   9, **10**
variance   7, 20, 39
vector   135, 564, 566
video games   368

w

water temperature   151
whisker plot, *see* box plot
Wilcoxon   297–299, 302–305
  assumptions   298